Teacher Edition

Eureka Math Algebra II Module 4

Special thanks go to the Gordan A. Cain Center and to the Department of Mathematics at Louisiana State University for their support in the development of *Eureka Math*.

Published by the non-profit Great Minds

Printed in the U.S.A.
This book may be purchased from the publisher at eureka-math.org
10 9 8 7 6 5 4 3 2 1

ISBN 978-1-63255-642-4

Table of Contents[1]

Inferences and Conclusions from Data

[1]Each lesson is ONE day, and ONE day is considered a 45-minute period.

Algebra II • Module 4

Inferences and Conclusions from Data

OVERVIEW

The concepts of probability and statistics covered in Algebra II build on students' previous work in Grade 7 and Algebra I. Topics A and B address standards **S-CP.A.1–5** and **S-CP.B.6–7**, which deal primarily with probability. In Topic A, fundamental ideas from Grade 7 are revisited and extended to allow students to build a more formal understanding of probability. More complex events are considered (unions, intersections, complements) (**S-CP.A.1**). Students calculate probabilities based on two-way data tables and interpret them in context (**S-CP.A.4**). They also see how to create "hypothetical 1000" two-way tables as a way of calculating probabilities. Students are introduced to conditional probability (**S-CP.A.3**, **S-CP.A.5**), and the important concept of independence is developed (**S-CP.A.2**, **S-CP.A.5**). The final lessons in this topic introduce probability rules (**S-CP.B.6**, **S-CP.B.7**).

Topic B is a short topic consisting of four lessons. This topic introduces the idea of using a smooth curve to model a data distribution, describes properties of the normal distribution, and asks students to distinguish between data distributions for which it would be reasonable to use a normal distribution as a model and those for which a normal distribution would not be a reasonable model. In the final two lessons of this topic, students use tables and technology to find areas under a normal curve and interpret these areas in the context of modeling a data distribution (**S-ID.A.4**).

Topics C and D develop students' understanding of statistical inference and introduce different types of statistical studies (observational studies, surveys, and experiments) (**S-IC.B.3**). In Topic C, students explore using data from a random sample to estimate a population mean or a population proportion. Building on what they learned about sampling variability in Grade 7, students use simulation to create an understanding of margin of error. Students calculate the margin of error and interpret it in context (**S-IC.B.4**). Students also evaluate reports from the media using sample data to estimate a population mean or proportion (**S-IC.B.6**).

Topic D focuses on drawing conclusions based on data from a statistical experiment. Given data from a statistical experiment, students use simulation to create a randomization distribution and use it to determine if there is a significant difference between two treatments (**S-IC.B.5**). Students also critique and evaluate published reports based on statistical experiments that compare two treatments (**S-IC.B.6**).

Focus Standards

Summarize, represent, and interpret data on a single count or measurement variable.

S-ID.A.4 Use the mean and standard deviation of a data set to fit it to a normal distribution and to estimate population percentages. Recognize that there are data sets for which such a procedure is not appropriate. Use calculators, spreadsheets, and tables to estimate areas under the normal curve.

Understand and evaluate random processes underlying statistical experiments.

S-IC.A.1 Understand statistics as a process for making inferences about population parameters based on a random sample from that population.

S-IC.A.2 Decide if a specified model is consistent with results from a given data-generating process, e.g., using simulation. *For example, a model says a spinning coin falls heads up with probability* 0.5. *Would a result of* 5 *tails in a row cause you to question the model?*

Make inferences and justify conclusions from sample surveys, experiments, and observational studies.

S-IC.B.3 Recognize the purposes of and differences among sample surveys, experiments, and observational studies; explain how randomization relates to each.

S-IC.B.4 Use data from a sample survey to estimate a population mean or proportion; develop a margin of error through the use of simulation models for random sampling.

S-IC.B.5 Use data from a randomized experiment to compare two treatments; use simulations to decide if differences between parameters are significant.

S-IC.B.6 Evaluate reports based on data.

Understand independence and conditional probability and use them to interpret data.

S-CP.A.1 Describe events as subsets of a sample space (the set of outcomes) using characteristics (or categories) of the outcomes, or as unions, intersections, or complements of other events ("or," "and," "not").

S-CP.A.2 Understand that two events A and B are independent if the probability of A and B occurring together is the product of their probabilities, and use this characterization to determine if they are independent.

S-CP.A.3 Understand the conditional probability of A given B as $P(A \text{ and } B)/P(B)$, and interpret independence of A and B as saying that the conditional probability of A given B is the same as the probability of A, and the conditional probability of B given A is the same as the probability of B.

S-CP.A.4 Construct and interpret two-way frequency tables of data when two categories are associated with each object being classified. Use the two-way table as a sample space to decide if events are independent and to approximate conditional probabilities. *For example, collect data from a random sample of students in your school on their favorite subject among math, science, and English. Estimate the probability that a randomly selected student from your school will favor science given that the student is in tenth grade. Do the same for other subjects and compare the results.*

S-CP.A.5 Recognize and explain the concepts of conditional probability and independence in everyday language and everyday situations. *For example, compare the chance of having lung cancer if you are a smoker with the chance of being a smoker if you have lung cancer.*

Use the rules of probability to compute probabilities of compound events in a uniform probability model.

S-CP.B.6 Find the conditional probability of A given B as the fraction of B's outcomes that also belong to A, and interpret the answer in terms of the model.

S-CP.B.7 Apply the Addition Rule, $P(A \text{ or } B) = P(A) + P(B) - P(A \text{ and } B)$, and interpret the answer in terms of the model.

Foundational Standards

Use random sampling to draw inferences about a population.

7.SP.A.1 Understand that statistics can be used to gain information about a population by examining a sample of the population; generalizations about a population from a sample are valid only if the sample is representative of that population. Understand that random sampling tends to produce representative samples and support valid inferences.

7.SP.A.2 Use data from a random sample to draw inferences about a population with an unknown characteristic of interest. Generate multiple samples (or simulated samples) of the same size to gauge the variation in estimates or predictions. *For example, estimate the mean word length in a book by randomly sampling words from the book; predict the winner of a school election based on randomly sampled survey data. Gauge how far off the estimate or prediction might be.*

Draw informal comparative inferences about two populations.

7.SP.B.3 Informally assess the degree of visual overlap of two numerical data distributions with similar variabilities, measuring the difference between the centers by expressing it as a multiple of a measure of variability. *For example, the mean height of players on the basketball team is* 10 cm *greater than the mean height of players on the soccer team, about twice the variability (mean absolute deviation) on either team; on a dot plot, the separation between the two distributions of heights is noticeable.*

7.SP.B.4 Use measures of center and measures of variability for numerical data from random samples to draw informal comparative inferences about two populations. *For example, decide whether the words in a chapter of a seventh-grade science book are generally longer than the words in a chapter of a fourth-grade science book.*

Investigate chance processes and develop, use, and evaluate probability models.

7.SP.C.8 Find probabilities of compound events using organized lists, tables, tree diagrams, and simulation.

a. Understand that, just as with simple events, the probability of a compound event is the fraction of outcomes in the sample space for which the compound event occurs.

b. Represent sample spaces for compound events using methods such as organized lists, tables and tree diagrams. For an event described in everyday language (e.g., "rolling double sixes"), identify the outcomes in the sample space which compose the event.

c. Design and use a simulation to generate frequencies for compound events. *For example, use random digits as a simulation tool to approximate the answer to the question: If 40% of donors have type A blood, what is the probability that it will take at least 4 donors to find one with type A blood?*

Summarize, represent, and interpret data on a single count or measurement variable.

S-ID.A.2 Use statistics appropriate to the shape of the data distribution to compare center (median, mean) and spread (interquartile range, standard deviation) of two or more different data sets.

S-ID.A.3 Interpret differences in shape, center, and spread in the context of the data sets, accounting for possible effects of extreme data points (outliers).

Summarize, represent, and interpret data on two categorical and quantitative variables.

S-ID.B.5 Summarize categorical data for two categories in two-way frequency tables. Interpret relative frequencies in the context of the data (including joint, marginal, and conditional relative frequencies). Recognize possible associations and trends in the data.

Focus Standards for Mathematical Practice

MP.2 **Reason abstractly and quantitatively.** Students use data from a sample to estimate a population mean or proportion and generalize from a sample to the population. They associate a margin of error with estimates based on a sample and interpret them in the context of generalizing from a sample to the population. Students also make conjectures or claims about independence and use arguments based on probabilities to support them.

MP.3 **Construct viable arguments and critique the reasoning of others.** Students test conjectures about treatment differences in the context of a statistical experiment. Students critique and evaluate reports based on data from random samples and reports based on data from experiments. Students frequently develop conjectures and use statistical reasoning to evaluate them.

MP.4 **Model with mathematics.** Students use smooth curves to model data distributions. Students use the normal distribution as a model in order to answer questions about a data distribution. Students use probability models to describe real-world contexts.

MP.5 **Use appropriate tools strategically.** Students use technology to carry out simulations in order to study sampling variability. Students also use technology to compute estimates of population characteristics (such as the mean and standard deviation) and to calculate margin of error. Students use simulation to investigate statistical significance in the context of comparing treatments in a statistical experiment.

Terminology

New or Recently Introduced Terms

- **Complement of an Event** (The *complement of an event*, A, denoted by A^C, is the event that A does not occur.)
- **Conditional Probability** (The probability of an event given that some other event occurs. The *conditional probability* of A given B is denoted by $P(A|B)$.)
- **Experiment** (An *experiment* is a study in which subjects are assigned to treatments for the purpose of seeing what effect the treatments have on some response.)
- **Hypothetical 1000 Table** (A *hypothetical 1000 table* is a two-way table that is constructed using given probability information. It represents a hypothetical population of 1000 individuals that is consistent with the given probability distribution and also allows calculation of other probabilities of interest.)
- **Independent Events** (Two events A and B are independent if $P(A|B) = P(A)$. This implies that knowing that B has occurred does not change the probability that A has occurred.)
- **Intersection of Two Events** (The *intersection of two events*, A and B, denoted by $A \cap B$, is the event that A and B both occur.)
- **Lurking Variable** (A *lurking variable* is one that causes two variables to have a high relationship even though there is no real direct relationship between the two variables.)
- **Margin of Error** (The *margin of error* is the maximum likely error when data from a sample is used to estimate a population characteristic, such as a population proportion or a population mean.)
- **Normal Distribution** (A *normal distribution* is a distribution that is bell shaped and symmetric.)
- **Observational Study** (An *observational study* is one in which the values of one or more variables are observed with no attempt to affect the outcomes.)
- **Random Assignment** (*Random assignment* is the process of using a chance mechanism to assign individuals to treatments in an experiment.)
- **Random Selection** (*Random selection* is the process of selecting individuals for a sample using a chance mechanism that ensures that every individual in the population has the same chance of being selected.)
- **Sample Survey** (A *sample survey* is an observational study in which people respond to one or more questions.)
- **Treatment** (A *treatment* is something administered in an experimental study.)
- **Union of Two Events** (The *union of two events*, A and B, denoted by $A \cup B$, is the event that either A or B or both occur.)

Familiar Terms and Symbols[2]

- Association
- Chance experiment
- Conditional relative frequency
- Distribution shape (skewed, symmetric)
- Event
- Mean
- Sample space
- Sampling variability
- Standard deviation

Suggested Tools and Representations

- Graphing calculator or graphing software
- Random number tables
- Random number software
- Normal distribution
- Two-way frequency tables
- Spreadsheets

Assessment Summary

Assessment Type	Administered	Format	Standards Addressed
Mid-Module Assessment Task	After Topic B	Constructed response with rubric.	S-ID.A.4, S-IC.A.2, S-CP.A.1, S-CP.A.2, S-CP.A.3, S-CP.A.4, S-CP.A.5, S-CP.B.6, S-CP.B.7
End-of-Module Assessment Task	After Topic D	Constructed response with rubric.	S-ID.A.4, S-IC.A.1, S-IC.A.2, S-IC.B.3, S-IC.B.4, S-IC.B.5, S-IC.B.6

[2]These are terms and symbols students have seen previously.

Topic A

Probability

S-IC.A.2, S-CP.A.1, S-CP.A.2, S-CP.A.3, S-CP.A.4, S-CP.A.5, S-CP.B.6, S-CP.B.7

Focus Standards:	S-IC.A.2	Decide if a specified model is consistent with results from a given data-generating process, e.g., using simulation. *For example, a model says a spinning coin falls heads up with probability* 0.5. *Would a result of* 5 *tails in a row cause you to question the model?*
	S-CP.A.1	Describe events as subsets of a sample space (the set of outcomes) using characteristics (or categories) of the outcomes, or as unions, intersections, or complements of other events ("or," "and," "not").
	S-CP.A.2	Understand that two events A and B are independent if the probability of A and B occurring together is the product of their probabilities, and use this characterization to determine if they are independent.
	S-CP.A.3	Understand the conditional probability of A given B as $P(A \text{ and } B)/P(B)$, and interpret independence of A and B as saying that the conditional probability of A given B is the same as the probability of A, and the conditional probability of B given A is the same as the probability of B.
	S-CP.A.4	Construct and interpret two-way frequency tables of data when two categories are associated with each object being classified. Use the two-way table as a sample space to decide if events are independent and to approximate conditional probabilities. *For example, collect data from a random sample of students in your school on their favorite subject among math, science, and English. Estimate the probability that a randomly selected student from your school will favor science given that the student is in tenth grade. Do the same for other subjects and compare the results.*
	S-CP.A.5	Recognize and explain the concepts of conditional probability and independence in everyday language and everyday situations. *For example, compare the chance of having lung cancer if you are a smoker with the chance of being a smoker if you have lung cancer.*
	S-CP.B.6	Find the conditional probability of A given B as the fraction of B's outcomes that also belong to A, and interpret the answer in terms of the model.
	S-CP.B.7	Apply the Addition Rule, $P(A \text{ or } B) = P(A) + P(B) - P(A \text{ and } B)$, and interpret the answer in terms of the model.

Instructional Days:	7
Lesson 1:	Chance Experiments, Sample Spaces, and Events (E)[1]
Lesson 2:	Calculating Probabilities of Events Using Two-Way Tables (P)
Lessons 3-4:	Calculating Conditional Probabilities and Evaluating Independence Using Two-Way Tables (P, P)
Lesson 5:	Events and Venn Diagrams (P)
Lessons 6-7:	Probability Rules (P, P)

Fundamental ideas from Grade 7 are revisited and extended to allow students to build a more formal understanding of probability. Students expand their understanding of chance experiments, sample space, and events to the more complex understanding of events defined as unions, intersections, and complements (**S-CP.A.1**). Students develop this understanding as they consider events that can be described as unions and intersections in the context of a game involving cards and spinners. One such game is introduced in Lesson 1, and then students explore further variations of the game in the lesson's Problem Set. Students also consider whether observations from a chance experiment are consistent with a given probability model (**S-IC.A.2**).

Students calculate probabilities of unions and intersections using data in two-way data tables and interpret them in context (**S-CP.A.4**). Students deepen their understanding by creating *hypothetical 1000* two-way tables (i.e., tables based on a hypothetical population of 1,000 observations) and then use these tables to calculate probabilities. Students use given probability information to determine the marginal totals and individual cell counts. This table then allows students to calculate conditional probabilities, as well as probabilities of unions, intersections, and complements, without the need for formal probability rules.

Students are introduced to conditional probability (**S-CP.A.3**, **S-CP.A.5**), which is used to illustrate the important concept of independence by describing two events, A and B, as independent if the conditional probability of A given B is not equal to the unconditional probability of A. In this case, knowing that event B has occurred does not change the assessment of the probability that event A has also occurred (**S-CP.A.2**, **S-CP.A.5**). Students use two-way tables to determine if two events are independent by calculating and interpreting conditional probabilities. In Lesson 3, students are presented with athletic participation data from Rufus King High School in two-way frequency tables, and conditional probabilities are calculated using column or row summaries. The conditional probabilities are then used to investigate whether or not there is a connection between two events.

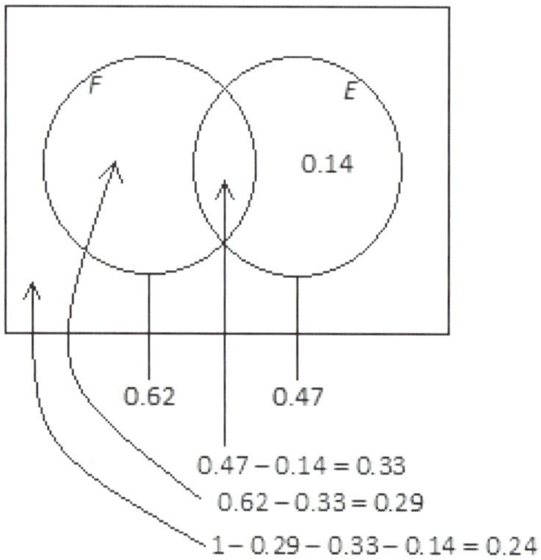

Students are also introduced to Venn diagrams to represent the sample space and various events. Students see how the regions of a Venn diagram connect to the cells of a two-way table. Venn diagrams also help students understand probability formulas involving the formal symbols of union, intersection, and complement.

[1]Lesson Structure Key: **P**-Problem Set Lesson, **M**-Modeling Cycle Lesson, **E**-Exploration Lesson, **S**-Socratic Lesson

Additionally, a Venn diagram can show how subtracting the probability of an event from 1 enables students to acquire the probability of the complement of the event and why the probability of the intersection of two events is subtracted from the sum of event probabilities when calculating the probability of the union of two events.

The final lessons in this topic introduce probability rules (the multiplication rule for independent events, the addition rule for the union of two events, and the complement rule for the complement of an event) (**S-CP.B.6**, **S-CP.B.7**). Students use the multiplication rule for independent events to calculate the probability of the intersection of two events. Students interpret independence based on the conditional probability and its connection to the multiplication rule.

Lesson 1: Chance Experiments, Sample Spaces, and Events

Student Outcomes

- Students determine the sample space for a chance experiment.
- Given a description of a chance experiment and an event, students identify the subset of outcomes from the sample space corresponding to the complement of an event.
- Given a description of a chance experiment and two events, students identify the subset of outcomes from the sample space corresponding to the union or intersection of two events.
- Students calculate the probability of events defined in terms of unions, intersections, and complements for a simple chance experiment with equally likely outcomes.

Materials

Each group needs the following:

- Copy of rules of game
- Fair coin
- Spinner with three equal area sectors
- Spinner with six equal area sectors
- Card bag with six cards (four blue with letters A, B, C, and D and two red with letters E and F)
- Scenario cards (several per person)
- Scoring cards (several per person)

Lesson Notes

This lesson provides a review of probability topics first encountered in Grade 7. In Grade 7, students were introduced to chance experiments, events, equally likely events, and sample spaces. This lesson reviews these topics to prepare students for the more advanced probability topics developed in this grade level, which provide the foundation for inferential thinking. This lesson asks students to think about events that are described with *and*, *or*, and *not* and to identify associated outcomes from the sample space. The more formal language of intersection, union, and complement is defined in Lessons 3 and 4. The structure of this lesson is exploratory, providing several opportunities to review chance experiments, events, equally likely events, and sample spaces. The vocabulary addressed in this lesson should be familiar to some students based on their previous work. For other students, this lesson may be their first exposure to these probability topics, and an explanation of the terms is necessary.

Be prepared to carefully explain the rules and procedures of the game that is the basis of this lesson. Conduct a few practice rounds with students before they complete the exercises. Consider using parts of this lesson as an opportunity to informally assess student understanding of probability. Provide a handout of the rules for each student. (A template is provided at the end of this teacher lesson.) This relatively simple game is designed to let students calculate and interpret simple probabilities based on coins, spinners, and picking names from a bag. Teachers are encouraged to either simplify or make the game more complex based on students' experience with games. This lesson also provides an opportunity to remind students that the origins of probability were to better understand games of chance.

Before presenting this lesson, prepare copies of the two spinners (Spinners 1 and 2) and several card bags. A description of the spinners and the card bags are provided for students. Preparing the spinners and the card bags before students start the game provides more time to focus on the game and the outcomes. Spinner 1 can be constructed by tracing a large circle on an $8.5'' \times 11''$ sheet of paper. Draw three approximately equal area sectors from the center of the circle (an estimate of the center is fine for this game). Students can use a paper clip positioned at the center of the circle by the tip of a pencil or pen as the pointer. Encourage students to attempt a few spins of the paper clip before continuing with the exercises. The card bag can be a bag or jar containing six equally sized small slips of paper with four slips designated as blue and the other two slips designated as red. The letters *A*, *B*, *C*, and *D* are written on the blue slips (one letter per card), and the letters *E* and *F* are written on the red slips. Make sure that students mix up the slips of paper with each card selected and that they cannot see the cards as they make their selections.

Spinner 2 can also be constructed on an $8.5'' \times 11''$ sheet of paper with approximately six equal sectors. Spinner 2 is used in the Problem Set. Prepare a spinner template so that each student has a copy of it when completing the problems for the Problem Set.

Classwork

Opening (5 minutes)

Spend five minutes discussing the following game with students. Read through the explanation as a class.

Alan is designing a probability game. He plans to present the game to people who will consider financing his idea. Here is a description of the game:

- **The game includes the following materials:**
  - **A fair coin with a head and a tail**
  - **Spinner 1 with three equal area sectors identified as 1, 2, and 3**
  - **Spinner 2 with six equal area sectors identified as 1, 2, 3, 4, 5, and 6**
  - **A card bag containing six cards. Four cards are blue with the letter *A* written on one card, *B* on another card, *C* on a third card, and *D* on the fourth card. Two cards are red with the letter *E* written on one card and the letter *F* written on the other. (Although actually using colored paper is preferable, slips of paper with the words *blue* or *red* written will also work.)**
  - **A set of scenario cards, each describing a chance experiment and a set of five possible events based on the chance experiment**

Spinner 1	**Spinner 2**
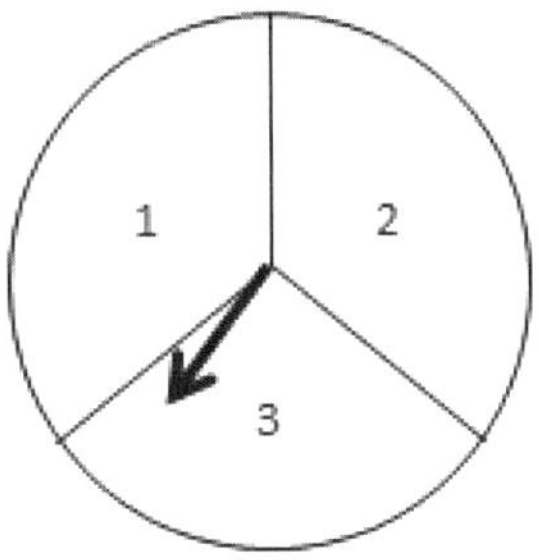	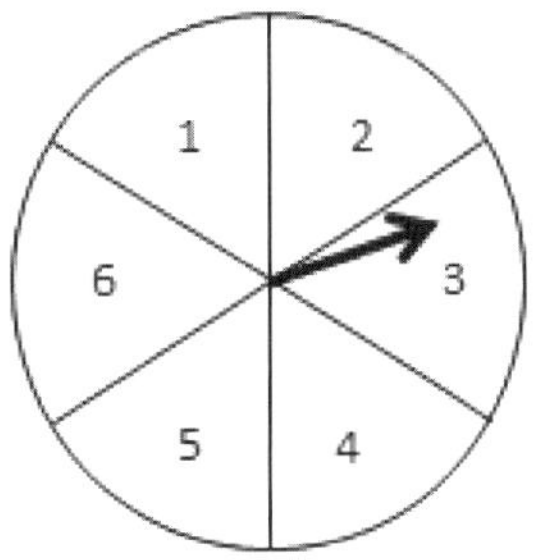

Card Bag:

Blue A	Blue B	Blue C	Blue D	Red E	Red F

- The game is played by two players (or two small groups of players) identified as Player 1 and Player 2.
- Rules of the game:
  - The scenario cards are shuffled, and one is selected.
  - Each player reads the description of the chance experiment and the description of the five possible outcomes.
  - Players independently assign the numbers 1–5 (no repeats) to the five events described on the scenario card based on how likely they think the event is to occur, with 5 being most likely and 1 being least likely.
  - Once players have made their assignments, the chance experiment described on the scenario card is performed. Points are then awarded based on the outcome of the chance experiment. If the event described on the scenario card has occurred, the player earns the number of points corresponding to the number that player assigned to that event (1–5 points). If an event occurs that is not described on the scenario card, then no points are awarded for that event.
  - If an outcome is described by two or more events on the scenario card, the player selects the higher point value.
  - The chance experiment is repeated four more times with points being awarded each time the chance experiment is performed.
  - The player with the largest number of points at the end of the game is the winner.

Alan developed two scenario cards for his demonstration to the finance people. A table in which the players can make their assignments and keep track of their scores accompanies each scenario card. Consider the first scenario card Alan developed.

Scenario Card 1

Game Tools: Spinner 1 (three equal sectors with the number 1 in one sector, the number 2 in the second sector, and the number 3 in the third sector)
Card bag (Blue-A, Blue-B, Blue-C, Blue-D, Red-E, Red-F)

Directions (chance experiment): Spin Spinner 1, and randomly select a card from the card bag (four blue cards and two red cards). Record the number from your spin and the color of the card selected.

Five Events of Interest:

Outcome is an odd number on Spinner 1 and a red card from the card bag.	Outcome is an odd number on Spinner 1.	Outcome is an odd number on Spinner 1 and a blue card from the card bag.	Outcome is an even number from Spinner 1 or a red card from the card bag.	Outcome is not a blue card from the card bag.

Player:

Scoring Card for Scenario 1:

Turn	Outcome from Spinner 1	Outcome from the Card Bag	Points
1			
2			
3			
4			
5			

Here is an example of Alan demonstrating the first scenario card. The chance experiment for Scenario Card 1 is "Spin Spinner 1, and record the number. Randomly select a card from the card bag (four blue cards and two red cards). Record the color of the card selected."

Alan assigned the numbers 1–5 to the descriptions, as shown below. Once a number is assigned, it cannot be used again.

Five Events of Interest:

Outcome is an odd number on Spinner 1 and a red card from the card bag.	Outcome is an odd number on Spinner 1.	Outcome is an odd number on Spinner 1 and a blue card from the card bag.	Outcome is an even number from Spinner 1 or a red card from the card bag.	Outcome is not a blue card from the card bag.
3	1	4	2	5

Alan is now ready to take his five turns. The results were recorded from the spinner and the card bag. Based on the results, Alan earned the points indicated for each turn.

Player: Player 1

Scoring Card for Scenario 1:

Turn	Outcome from Spinner 1	Outcome from the Card Bag	Points Based on Alan's Assignment of the Numbers to the Five Events
1	2	Blue	2
2	1	Red	5
3	1	Red	5
4	3	Blue	4
5	2	Blue	2

Alan earned a total of 18 points. The game now turns to Player 2. Player 2 assigns the numbers 1–5 to the same description of outcomes. Player 2 does not have to agree with the numbers Alan assigned. After five turns, the player with the most number of points is the winner.

Exploratory Challenge/Exercises 1–13 (30 minutes)

Let students work with a partner (Player 1 and Player 2) or in two small groups of players. As an Exploratory lesson, students should experiment with the game as they begin making sense of the rules and procedures. The exercises are designed to help students understand the strategy of winning the game based on analyzing the probabilities of the events.

- Let's look more closely at Scenario Card 1.

Exploratory Challenge/Exercises 1–13

1. **Would you change any of the assignments of 1–5 that Alan made? Explain your answer. Assign the numbers 1–5 to the event descriptions based on what you think is the best strategy to win the game.**

Outcome is an odd number on Spinner 1 and a red card from the card bag.	Outcome is an odd number on Spinner 1.	Outcome is an odd number on Spinner 1 and a blue card from the card bag.	Outcome is an even number from Spinner 1 or a red card from the card bag.	Outcome is not a blue card from the card bag.

Answers will vary.

Encourage students to explore different strategies for assigning the numbers, and have them share their thinking with the class. Some students may not have formed a strategy yet, and they may assign the numbers by simply guessing. Other students may have a sense of the probabilities of the various outcomes and assign numbers that reflect that thinking. The questions that follow help them organize their thinking to develop a strategy for playing the game.

2. **Carry out a turn by observing an outcome from spinning Spinner 1 and picking a card. How many points did you earn from this first turn?**

Answers will vary.

Allow students to explain to others the points they earned and how they calculated the points. This is an opportunity to review the meaning of the *or* and *and* language. The *or* (to be defined in several later lessons as a *union*) is indicating that either one or both of the descriptions is necessary for this event. The event "odd number or a red card" means that if the spinner outcome was a 1 or 3, this event has occurred regardless of the card selected. It also means that if one of the red cards was selected, and the result from the spinner is a 2, this event has occurred. The event "odd number or a red card" also occurs if the outcome is an odd number and a red card. An *and* event (for example, "odd number and a red card") means both that the number must be odd and that the card must be red.

3. **Complete four more turns (for a total of five), and determine your final score.**

Player: Your Turn

Scoring Card for Scenario 1:

Trial	Outcome from Spinner 1	Outcome from the Card Bag	Points Based on Your Assignment of Numbers to the Events
1			
2			
3			
4			
5			

Answers will vary depending on the outcomes when the student plays the game.

Use Exercise 3 to determine if students understand the directions of the game. Provide some time to discuss outcomes and to help students explain how points were obtained either with individual students or in small groups as students work through the questions.

MP.2 Encourage students to explain their reasoning in how they assigned numbers to the events and what the number of points obtained from spinning the spinner and selecting a card indicate about their decisions. Students could use the outcomes as information about which outcomes are more likely to occur.

4. **If you changed the numbers assigned to the descriptions, was your score better than Alan's score? Did you expect your score to be better? Explain. If you did not change the numbers from those that Alan assigned, explain why you did not change them.**

   Note: This question provides students an opportunity to begin explaining the strategy they are using in this game. Anticipate that in most cases students are assigning the larger numbers (the 5 or the 4) to the outcomes they think are most likely. At this point in the lesson, students have not been asked to calculate the actual probabilities of the events of interest. However, some students may have already calculated or estimated the probabilities. Anticipate that students change the assignment of numbers to try to improve their scores.

   MP.3 *As students discuss strategies for assigning numbers based on which events are most likely to occur, they are also evaluating the thinking of others.*

5. **Spinning Spinner 1 and drawing a card from the card bag is a *chance experiment*. One possible outcome of this experiment is $(1$, Blue-A$)$. Recall that the *sample space* for a chance experiment is the set of all possible outcomes. What is the sample space for the chance experiment of Scenario Card 1?**

   The sample space consists of 18 outcomes.

   {(1, Blue-A), (1, Blue-B), (1, Blue-C), (1, Blue-D), (1, Red-E), (1, Red-F), (2, Blue-A), (2, Blue-B), (2, Blue-C), (2, Blue-D), (2, Red-E), (2, Red-F), (3, Blue-A), (3, Blue-B), (3, Blue-C), (3, Blue-D), (3, Red-E), (3, Red-F)}

> *Scaffolding:*
>
> For English language learners, the term *space* may need clarification and rehearsal, as it is a noun with several meanings and also a verb.

6. **Are the outcomes in the sample space equally likely? Explain your answer.**

   Yes, each outcome is equally likely to occur because the spinner is equally likely to land in any of the three segments, and each of the six cards is equally likely to be selected. The selection of a card and the result of spinning the spinner do not depend on each other, so the 18 outcomes in the sample space should be equally likely.

7. **Recall that an *event* is a collection of outcomes from the sample space. One event of interest for someone with Scenario Card 1 is "odd number on Spinner 1 and a red card." What are the outcomes that make up this event? List the outcomes of this event in the first row of Table 1 (see Exercise 9).**

   See the completed chart in Exercise 9.

8. **What is the probability of getting an odd number on Spinner 1 and picking a red card from the card bag? Also enter this probability in Table 1 (see Exercise 9).**

   See the completed table in Exercise 9. Direct students to write their probabilities as fractions or as decimals. Fractions do not need to be reduced, as the unreduced fractions are meaningful based on the context.

9. **Complete Table 1 by listing the outcomes for the other events and their probabilities based on the chance experiment for this scenario card.**

The following table organizes the responses to Exercises 7–9:

Table 1

Event	Outcomes	Probability
Odd number on Spinner 1 and a red card from the card bag	*Note: If students struggle with the "and" language, discuss the outcomes with students.* *(1, Red-E), (1, Red-F), (3, Red-E), (3, Red-F)*	*The probability is* $\frac{4}{18}$, *which is approximately* 0.222. *Assign* 1 *point to this event.*
Odd number on Spinner 1	*Note: Students may indicate that the probability of this outcome is the same as the probability of just getting an odd number on the spinner without considering the color of the card selected.* *(1, Blue-A), (1, Blue-B), (1, Blue-C), (1, Blue-D), (1, Red-E), (1, Red-F), (3, Blue-A), (3, Blue-B), (3, Blue-C), (3, Blue-D), (3, Red-E), (3, Red-F)*	*The probability is* $\frac{12}{18}$ *or* $\frac{2}{3}$, *which is approximately* 0.667. *Assign* 5 *points to this event.*
Odd number on Spinner 1 and a blue card from the card bag	*(1, Blue-A), (1, Blue-B), (1, Blue-C), (1, Blue-D), (3, Blue-A), (3, Blue-B), (3, Blue-C), (3, Blue-D)*	*The probability is* $\frac{8}{18}$, *which is approximately* 0.444. *Assign* 3 *points to this event.*
Even number on Spinner 1 or a red card from the card bag	*Note: This event involves the "or." Make sure students identify all of the outcomes that apply. This is an excellent event to discuss with students.* *(1, Red-E), (1, Red-F), (2, Blue-A), (2, Blue-B), (2, Blue-C), (2, Blue-D), (2, Red-E), (2, Red-F), (3, Red-E), (3, Red-F)*	*The probability is* $\frac{10}{18}$, *which is approximately* 0.556. *Assign* 4 *points to this event.*
Not picking a blue card from the card bag	*Note: Although formally defined in Lesson 3, this is an example of the complement of the event "picking a blue card." An informal discussion of the word "complement" may be considered; however, it is not necessary at this time.* *(1, Red-E), (1, Red-F), (2, Red-E), (2, Red-F), (3, Red-E), (3, Red-F)*	*The probability is* $\frac{6}{18}$, *which is approximately* 0.333. *Assign* 2 *points to this event.*

10. **Based on the above probabilities, how would you assign the numbers 1–5 to each of the game descriptions? Explain.**

If players assigned the points based on a 5 *assigned to the event most likely to occur, followed by* 4*, etc., then there would be changes made to the assignment of points.*

11. **If you changed any of the points assigned to the game descriptions, play the game again at least three times and record your final scores for each game. Do you think you have the best possible assignment of numbers to the events for this scenario card? If you did not change the game descriptions, also play the game so that you have at least three final scores. Compare your scores with scores of other members of your class. Do you think you have the best assignment of numbers to the events for this scenario card?**

Answers will vary. A good way to discuss this question with students is to compare the score Alan received with his assignment of points and the scores students received based on their assignments of points. Based on the assignment of points described in Exercise 10, one possible student response is shown below.

MP.2 This is another opportunity for students to demonstrate their reasoning using the results of various turns, linking the assignment of numbers to events and the scores obtained.

Turn	Outcome from Spinner 1	Outcome from the Card Bag	Points Based on the Assignment of Points in Exercise 10
1	2	*Blue*	4
2	1	*Red*	5
3	1	*Red*	5
4	3	*Blue*	5
5	2	*Blue*	4

The final score for the above five turns is 23 points. This is a better score than Alan received from his assignment of points.

12. **Why might you not be able to answer the question of whether or not you have the best assignments of numbers to the game descriptions with at least three final scores?**

To decide the best assignment of points involves playing the game many times.

Note: Use this question to point out that probability is a long-run relative frequency. For this example, to get a good sense of the probabilities, students would need to play the game many times.

13. **Write your answers to the following questions independently, and then share your responses with a neighbor:**

    a. **How did you make decisions about what to bet on?**

    Answers will vary. I calculated the probability of each outcome and bet in the order of probability, 5 (greatest) to 1 (least). I used probability but changed a few bets based on the earlier outcomes.

    b. **How do the ideas of probability help you make decisions?**

    Answers will vary. Calculating probability gives you a good idea of the number of times an outcome will occur and helps you make better decisions.

Closing (5 minutes)

- How would you change the strategy of assigning the numbers 1–5 if the lowest score was the winner of the game?
    - *The event with the greatest probability will be assigned the number 1, the event with the second greatest probability would be assigned the number 2, etc. The event with the least probability will be assigned the number 5.*

Ask students to summarize the main ideas of the lesson in writing or with a neighbor. Use this opportunity to informally assess their comprehension. The Lesson Summary below offers some important ideas that should be included.

Lesson Summary

- **SAMPLE SPACE: The *sample space* of a chance experiment is the collection of all possible outcomes for the experiment.**
- **EVENT: An *event* is a collection of outcomes of a chance experiment.**
- **For a chance experiment in which outcomes of the sample space are equally likely, the probability of an event is the number of outcomes in the event divided by the number of outcomes in the sample space.**
- **Some events are described in terms of *or*, *and*, or *not*.**

Exit Ticket (5 minutes)

Name ______________________________ Date ______________

Lesson 1: Chance Experiments, Sample Spaces, and Events

Exit Ticket

1. For the chance experiment described in Scenario Card 1, why is the probability of the event "spinning an odd number and randomly selecting a blue card" not the same as the probability of the event "spinning an even number and randomly selecting a blue card"? Which event would have the greater probability of occurring, and why?

2. Why is the probability of the event "spinning an odd number from Spinner 1 *and* randomly selecting a blue card" not equal to the probability of "spinning an odd number from Spinner 1 *or* randomly selecting a blue card"?

3. If one of the red cards is changed to a blue card, what is the probability of the event "spinning an odd number from Spinner 1 and randomly selecting a red card from the card bag"?

Exit Ticket Sample Solutions

1. For the chance experiment described in Scenario Card 1, why is the probability of the event "spinning an odd number and randomly selecting a blue card" not the same as the probability of the event "spinning an even number and randomly selecting a blue card"? Which event would have the greater probability of occurring, and why?

   The first event includes the following outcomes from the sample space: (1, Blue-A), (1, Blue-B), (1, Blue-C), (1, Blue-D), (3, Blue-A), (3, Blue-B), (3, Blue-C), (3, Blue-D).

   The second event includes the following outcomes: (2, Blue-A), (2, Blue-B), (2, Blue-C), (2, Blue-D).

   Because the spinner has two sectors representing odd numbers and only one sector representing an even number, the numbers of outcomes in the two events are different. The probability of the first event is $\frac{8}{18}$ *compared to the probability of* $\frac{4}{18}$ *for the second event.*

2. Why is the probability of the event "spinning an odd number from Spinner 1 *and* randomly selecting a blue card" not equal to the probability of "spinning an odd number from Spinner 1 *or* randomly selecting a blue card"?

   The outcomes for the first event must include both the odd number and a blue card or (1, Blue-A), (1, Blue-B), (1, Blue-C), (1, Blue-D), (3, Blue-A), (3, Blue-B), (3, Blue-C), and (3, Blue-D). The probability of that event would be $\frac{8}{18}$*, which is approximately* 0.444*. The outcomes of the second event would include each of the above outcomes, but it would also include the outcomes of odd numbers from the spinner with a red card and even numbers with a blue card. The outcomes for this event include (1, Blue-A), (1, Blue-B), (1, Blue-C), (1, Blue-D), (1, Red-E), (1, Red-F), (2, Blue-A), (2, Blue-B), (2, Blue-C), (2, Blue-D), (3, Blue-A), (3, Blue-B), (3, Blue-C), (3, Blue-D), (3, Red-E), and (3, Red-F). The only outcomes not included would be red cards with an even number. The probability for this event is* $\frac{16}{18}$*, which is approximately* 0.889*.*

3. If one of the red cards is changed to a blue card, what is the probability of the event "spinning an odd number from Spinner 1 and randomly selecting a red card from the card bag"?

   Changing one red card to a blue card will result in five blue cards and one red card in the card bag. Each of the five blue cards could be paired with each odd number that is possible from the spinner (the number 1 or the number 3). There would be a total of ten outcomes for this event. The number of outcomes in the sample space, however, would not change. As a result, the probability of this event is $\frac{10}{18}$*, which is approximately* 0.556*.*

Problem Set Sample Solutions

Consider a second scenario card that Alan created for his game:

Scenario Card 2

Tools: Spinner 1
Spinner 2: a spinner with six equal sectors (Place the number 1 in a sector, the number 2 in a second sector, the number 3 in a third sector, the number 4 in a fourth sector, the number 5 in a fifth sector, and the number 6 in the last sector.)

Directions (chance experiment): Spin Spinner 1, and spin Spinner 2. Record the number from Spinner 1, and record the number from Spinner 2.

Five Events of Interest:

Outcome is an odd number on Spinner 2.	Outcome is an odd number on Spinner 1 and an even number on Spinner 2.	Outcome is the sum of 7 from the numbers received from Spinner 1 and Spinner 2.	Outcome is an even number on Spinner 2.	Outcome is the sum of 2 from the numbers received from Spinner 1 and Spinner 2.

Player:

Scoring Card for Scenario 2:

Turn	Outcome from Spinner 1	Outcome from Spinner 2	Points
1			
2			
3			
4			
5			

1. Prepare Spinner 1 and Spinner 2 for the chance experiment described on this second scenario card. (Recall that Spinner 2 has six equal sectors.)

   Prepare Spinner 2 as described. You can use the same Spinner 1 used for Scenario Card 1.

2. What is the sample space for the chance experiment described on this scenario card?

   There are 18 outcomes on this scenario card. Students can list all of the outcomes or describe them. Here is the list (the first number is the outcome from Spinner 1, and the second number is the outcome from Spinner 2);

   $(1,1)$, $(1,2)$, $(1,3)$, $(1,4)$, $(1,5)$, $(1,6)$, $(2,1)$, $(2,2)$, $(2,3)$, $(2,4)$, $(2,5)$, $(2,6)$, $(3,1)$, $(3,2)$, $(3,3)$, $(3,4)$, $(3,5)$, $(3,6)$

3. Based on the sample space, determine the outcomes and the probabilities for each of the events on this scenario card. Complete the table below.

Event	Outcomes	Probability
Outcome is an odd number on Spinner 2.	$(1,1)$, $(1,3)$, $(1,5)$, $(2,1)$, $(2,3)$, $(2,5)$, $(3,1)$, $(3,3)$, $(3,5)$	*The probability is* $\frac{9}{18}$, *which is* 0.5.
Outcome is an odd number on Spinner 1 and an even number on Spinner 2.	$(1,2)$, $(1,4)$, $(1,6)$, $(3,2)$, $(3,4)$, $(3,6)$	*The probability is* $\frac{6}{18}$, *which is approximately* 0.333.
Outcome is the sum of 7 from the numbers received from Spinner 1 and Spinner 2.	$(1,6)$, $(2,5)$, $(3,4)$	*The probability is* $\frac{3}{18}$, *which is approximately* 0.167.
Outcome is an even number on Spinner 2.	$(1,2)$, $(1,4)$, $(1,6)$, $(2,2)$, $(2,4)$, $(2,6)$, $(3,2)$, $(3,4)$, $(3,6)$	*The probability is* $\frac{9}{18}$, *which is* 0.5
Outcome is the sum of 2 from the numbers received from Spinner 1 and Spinner 2.	$(1,1)$	*The probability is* $\frac{1}{18}$, *which is approximately* 0.056.

Note that although each event is different, some events are subsets of another event. As a result, students want to assign a larger number to the event with more outcomes. Expect that students obtain scores of 4 or 5 for each turn. As an extension, students may be asked to revise the descriptions of the events on the strategy cards in order to make the game more challenging.

4. **Assign the numbers 1–5 to the events described on the scenario card.**

The following assignments would be based on the 5 assigned to the event with the greatest probability (the most likely outcome), 4 to the event with the next largest, etc.:

Five Events of Interest: Scenario 2

Outcome is an odd number on Spinner 2.	Outcome is an odd number on Spinner 1 and an even number on Spinner 2.	Outcome is the sum of 7 from the numbers received from Spinner 1 and Spinner 2.	Outcome is an even number on Spinner 2.	Outcome is the sum of 2 from the numbers received from Spinner 1 and Spinner 2.
4	3	2	5	1

5. **Determine at least three final scores based on the numbers you assigned to the events.**

Responses will vary. Provided are three final scores based on outcomes from carrying out the game.

Player: Scott

Trial	Outcome from Spinner 1	Outcome from Spinner 2	Points (see Problem 4)
1	2	6	5
2	1	5	4
3	2	6	5
4	3	3	4
5	2	2	5

Final Score: 23 points

Player: Scott

Trial	Outcome from Spinner 1	Outcome from Spinner 2	Points (see Problem 4)
1	3	3	4
2	3	6	5
3	1	5	4
4	3	1	4
5	2	4	5

Final Score: 22 points

Player: Scott

Trial	Outcome from Spinner 1	Outcome from Spinner 2	Points (see Problem 4)
1	2	2	5
2	1	1	4
3	1	4	5
4	3	3	4
5	2	2	5

Final Score: 23 points

6. **Alan also included a fair coin as one of the scenario tools. Develop a scenario card (Scenario Card 3) that uses the coin and one of the spinners. Include a description of the chance experiment and descriptions of five events relevant to the chance experiment.**

Answers will vary. Encourage students to be creative with this part of their assignment. Anticipate language similar to that used in the examples. A sample response card is included.

The following is an example of a completed Scenario Card 3:

Scenario Card 3

Tools: Fair coin (head or tail)
Spinner 1 *(three equal sectors with the number 1 in one sector, the number 2 in the second sector, and the number 3 in the third sector)*

Directions (chance experiment): *Toss fair coin, and spin Spinner 1. Record the head or tail from your toss and the number from your spin.*

Five Events of Interest:

Outcome is an odd number on Spinner 1.	*Outcome is a prime number on Spinner 1.*	*Outcome is a tail.*	*Outcome is a head and is not an even number on Spinner 1.*	*Outcome is a tail and a 1 on Spinner 1.*
5	3	4	2	1

7. **Determine the sample space for your chance experiment. Then, complete the table below for the five events on your scenario card. Assign the numbers 1–5 to the descriptions you created.**

Evaluate this chart based on the sample space and the descriptions developed by students. To evaluate, encourage several students to explain their game scenario cards to other students, or provide a sample of the scenario cards developed for students to try. The sample response is based on the scenario card presented in Problem 6.

Event	Outcomes	Probability
Outcome is an odd number on Spinner 1.	$(1,H)$, $(1,T)$, $(3,H)$, $(3,T)$	*The probability is* $\frac{4}{6}$, *which is approximately* 0.667.
Outcome is a prime number on Spinner 1.	$(2,H)$, $(2,T)$, $(3,H)$, $(3,T)$	*The probability is* $\frac{4}{6}$, *which is approximately* 0.667.
Outcome is a tail.	$(1,T)$, $(2,T)$, $(3,T)$	*The probability is* $\frac{3}{6}$, *which is* 0.5.
Outcome is a head and is not an even number on Spinner 1.	$(1,H)$, $(3,H)$	*The probability is* $\frac{2}{6}$, *which is approximately* 0.333.
Outcome is a tail and a 1 on Spinner 1.	$(1,T)$	*The probability is* $\frac{1}{6}$, *which is approximately* 0.167.

8. **Determine a final score for your game based on five turns.**

Turn			Points
1			
2			
3			
4			
5			

Answers vary based on the descriptions developed by students. Note: If time permits, encourage selected students to explain their games to other members of the class.

Rules of the game for Scenario Card 1 described in the lesson:

- The scenario cards are shuffled, and one is selected.
- Each player reads the description of the chance experiment and the description of the five events described on the scenario card.
- Players independently assign the numbers 1–5 (no repeats) to the five events described on the scenario card based on how likely they think the event is to occur, with 5 being most likely and 1 being least likely.
- Once players have made their assignments, the chance experiment described on the scenario card is performed. Points are then awarded based on the outcome of the chance experiment. If the event described on the scenario card has occurred, the player earns the number of points corresponding to the number that player assigned to that event (1–5 points). If an event occurs that is not described on the scenario card, then no points are awarded for that event.
- If an outcome is described by two or more events on the scenario card, the player selects the higher point value.
- The chance experiment is repeated four more times with points being awarded each time the chance experiment is performed.
- The player with the largest number of points at the end of the game is the winner.

Spinner 1

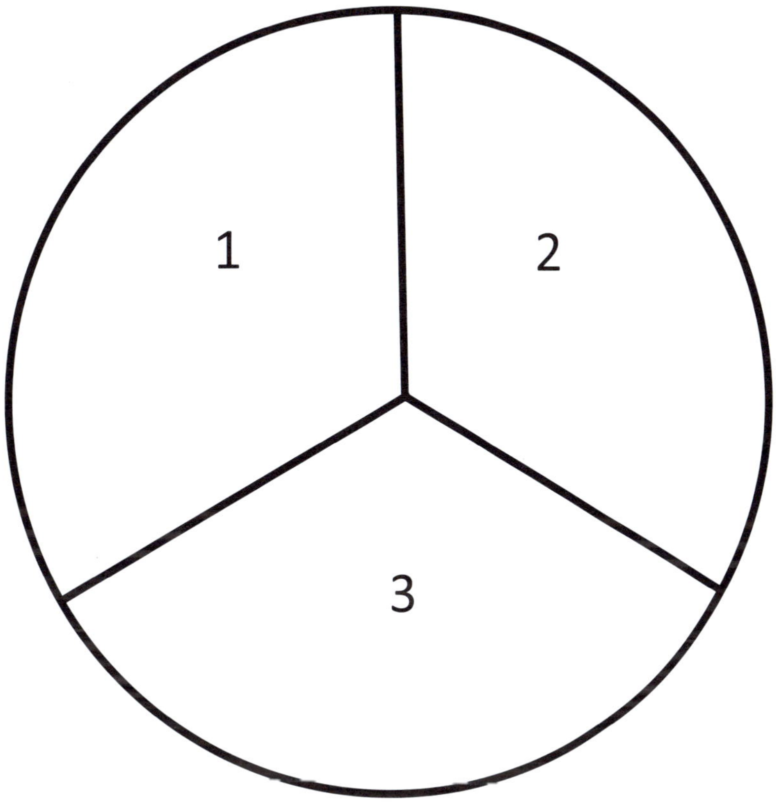

Spinner 2

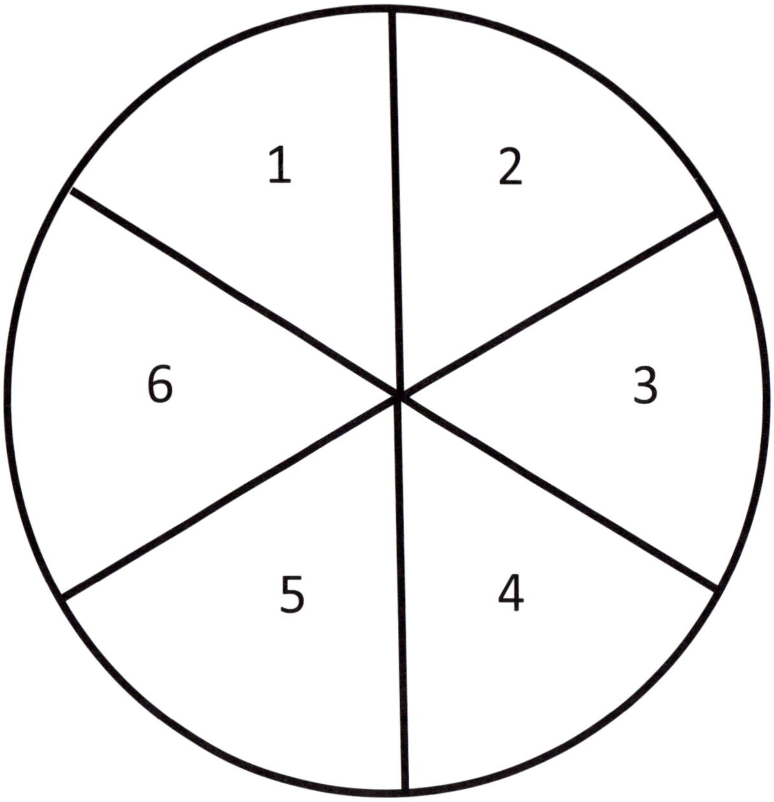

Lesson 2: Calculating Probabilities of Events Using Two-Way Tables

Student Outcomes

- Students calculate probabilities given a two-way table of data.
- Students construct a hypothetical 1000 two-way table given probability information.
- Students interpret probabilities in context.

Lesson Notes

In this lesson, students construct and interpret data in two-way tables. Questions are designed to help students understand what the tables are summarizing. Two-way frequency tables were introduced in Grade 8 (Module 6, Lesson 13) and revisited in Algebra I (Module 2, Lessons 9–11) as a way to organize and interpret bivariate categorical data. This lesson reviews and extends those concepts associated with two-way tables. Students also construct and interpret a hypothetical 1000 table. Research indicates that a hypothetical 1000 table makes understanding probabilities formed from a two-way table easier for students to understand. In Exercises 9–14, students create a hypothetical 1000 table (i.e., a two-way frequency table based on a population of $1{,}000$ students) to answer probability questions. The probabilities from this table are approximately equal to the probabilities from the actual frequencies, which students also interpret in the next lesson. The probability questions build on students' previous work with probabilities in Grade 7.

Lessons 2–4 work together to develop the standards for the cluster "Understand independence and conditional probability and use them to interpret data (**S-CP**)." Lesson 2 requires students to calculate several probabilities that are identified in Lesson 3 as conditional probabilities. However, for this lesson, students calculate the probabilities using the table, the question posed, and the context of the data. Lessons 3 and 4 offer a more formal description of a conditional probability and how it is used to determine if two events are independent.

Classwork

Example 1 (2 minutes): Building a New High School

Students encounter references to two-way tables often in this module. It may be important to discuss the word *table* with students, especially for students learning the language. Highlight that the word *table* is used in these lessons to describe a tool for organizing data. Data organized in rows and columns are referenced as two-way tables or two-way frequency tables.

A formal definition of tables is difficult for all students. Carefully point out the tables organized in these lessons, and point out the cells by describing what they represent.

> *Scaffolding:*
> For English language learners, provide visuals of the two different meanings for *table,* and practice saying the word chorally.

Example 1: Building a New High School

The school board of Waldo, a rural town in the Midwest, is considering building a new high school primarily funded by local taxes. They decided to interview eligible voters to determine if the school board should build a new high school facility to replace the current high school building. There is only one high school in the town. Every registered voter in Waldo was interviewed. In addition to asking about support for a new high school, data on gender and age group were also recorded. The data from these interviews are summarized below.

	Should Our Town Build a New High School?					
	Yes		No		No Answer	
Age (in years)	Male	Female	Male	Female	Male	Female
18–25	29	32	8	6	0	0
26–40	53	60	40	44	2	4
41–65	30	36	44	35	2	2
66 and Older	7	26	24	29	2	0

Scaffolding:

Consider using the following example to help students understand how to build and interpret a two-way table. If needed, Grade 8, Module 6, Lesson 13 can be used for remediation.

In a class of 20 students, 5 boys and 7 girls like chocolate ice cream. Six boys like vanilla ice cream.

- What are the variables?
- How could we organize the variables in the following table?

	Chocolate	*Vanilla*
Boys	5	6
Girls	7	

- What does the missing cell represent? How can you determine the missing value?

Exercises 1–8: Building a New High School (15–20 minutes)

MP.2

The following exercises ask students to interpret the data summarized in the table. The exercises allow students to reason abstractly by making summaries based on data. Discuss how the data summaries are being used to answer each question. Although the questions ask students to predict, the word *probability* is not always used in the questions. This is intentional, as it helps students to revisit their previous work with probability. While working with students, emphasize the goal of predicting outcomes based on the data.

Allow students to work in small groups as they answer the following questions. Select questions to discuss as a whole group based on student work. Allow students to use a calculator. For these questions, a scientific calculator is sufficient.

Exercises 1–8: Building a New High School

1. Based on this survey, do you think the school board should recommend building a new high school? Explain your answer.

   273 out of the 515 eligible voters (approximately 53%) indicated "yes." As a result, I think the voters will recommend building a new high school.

2. An eligible voter is picked at random. If this person is 21 years old, do you think he would indicate that the town should build a high school? Why or why not?

   Most of the eligible voters ages 18–25 indicated "yes." 61 of the 75 eligible voters in this age group indicated "yes." I would predict a 21-year-old in this age group to have answered "yes."

3. An eligible voter is picked at random. If this person is 55 years old, do you think she would indicate that the town should build a high school? Why or why not?

   Most of the eligible voters ages 41–65 indicated "no" to the question about building a high school. I would predict that a person in this age group would indicate that the town should not build a high school.

4. The school board wondered if the probability of recommending a new high school was different for different age categories. Why do you think the survey classified voters using the age categories 18–25 years old, 26–40 years old, 41–65 years old, and 66 years old and older?

The age groups used in the table represent people with different interests or opinions regarding the building of a new high school. For example, people ages 26–40 are more likely to have children in school than people in the other age groups. The probability that a person in each of these age categories would recommend building a new high school might vary.

5. It might be helpful to organize the data in a two-way frequency table. Use the given data to complete the following two-way frequency table. Note that the age categories are represented as rows, and the possible responses are represented as columns.

	Yes	No	No Answer	Total
18–25 Years Old	61	14	0	75
26–40 Years Old	113	84	6	203
41–65 Years Old	66	79	4	149
66 Years Old and Older	33	53	2	88
Total	273	230	12	515

Exercise 6 challenges students to think about a "headline summary" of data (something they may encounter in a newspaper, blog, or news report). Discuss how these types of short summaries of data need to be examined by readers. MP.3 The exercise provides an opportunity for students to form their own conclusions from the data and to reconcile their conclusions with summaries expressed in the form of headlines. In this way, students are also critiquing the reasoning of others.

6. A local news service plans to write an article summarizing the survey results. Three possible headlines for this article are provided below. Is each headline accurate or inaccurate? Support your answer using probabilities calculated using the table above.

Headline 1: *Waldo Voters Likely to Support Building a New High School*

Yes, this is accurate. We see that 273 out of 515, which is approximately 53.0%, support building a new school. The probability that an eligible voter would vote "yes" is greater than 0.50, so you would think it is likely that voters will support building a new high school.

Headline 2: *Older Voters Less Likely to Support Building a New High School*

Yes, this headline is accurate. If you define older voters as 41 or older, then 132 out of 237 voters 41 or older, which is approximately 55.7%, indicated they would not support building a new high school.

Headline 3: *Younger Voters Not Interested in Building a New High School*

This headline is not accurate. We see that 61 out of 75 eligible voters, which is approximately 81.3% representing the youngest eligible voters, indicated "yes" to building a new high school.

Exercise 7 is challenging. Students are expected to use both the probability of voting "yes" and the new information about the number of eligible voters expected to vote to answer this question. It may be necessary to help some students develop the final answer by working through a specific age group before they combine all age groups. Consider tackling this question as a whole group. This question also presents another opportunity to discuss the role of probability. Students previously indicated that most voters would vote "yes" to building a new high school. This question raises the possibility that voters vote "no" by adding information about past voter turnout.

7. **The school board decided to put the decision on whether or not to build the high school up for a referendum in the next election. At the last referendum regarding this issue, only 25 of the eligible voters ages 18–25 voted, 110 of the eligible voters ages 26–40 voted, 130 of the eligible voters ages 41–65 voted, and 80 of the eligible voters ages 66 and older voted. If the voters in the next election turn out in similar numbers, do you think this referendum will pass? Justify your answer.**

   Use the probabilities that a voter from an age group would vote "yes." Multiply the probabilities that a person would vote "yes" by the number of people estimated to vote in each age category.

   $$25(0.813) + 110(0.557) + 130(0.443) + 80(0.375) \approx 169$$

   169 out of the estimated 345 voters will vote "yes." Since this is less than half, we would predict that the vote will indicate the high school should not be built.

8. **Is it possible that your prediction of the election outcome might be incorrect? Explain.**

   Yes. The above is only a prediction. The actual results could be different. It depends on the actual voter turnout and whether people actually vote as they indicated in the survey.

Example 2 (2 minutes): Smoking and Asthma

Read through the example as a class.

Example 2: Smoking and Asthma

Health officials in Milwaukee, Wisconsin, were concerned about teenagers with asthma. People with asthma often have difficulty with normal breathing. In a local research study, researchers collected data on the incidence of asthma among students enrolled in a Milwaukee public high school.

Students in the high school completed a survey that was used to begin this research. Based on this survey, the probability of a randomly selected student at this high school having asthma was found to be 0.193. Students were also asked if they had at least one family member living in their house who smoked. The probability of a randomly selected student having at least one member in his (or her) household who smoked was reported to be 0.421.

Exercises 9–14 (10–15 minutes)

Make sure students understand how the information is organized in the given table. If necessary, point to a cell in the table, and ask students to describe what this cell represents in the context of the data.

This scenario is revisited in Lesson 4, where students explore the relationship between having asthma and having at least one family member who smokes.

Allow students to continue to work in small groups as they answer the questions that follow.

Scaffolding:

Consider challenging students working above grade level to build the hypothetical table independently or with a neighbor.

Exercises 9–14

It would be easy to calculate probabilities if the data for the students had been organized into a two-way table like the one used in Exercise 5. But there is no table here, only probability information. One way around this is to think about what the table might have been if there had been $1,000$ students at the school when the survey was given. This table is called a ***hypothetical 1000 two-way table***.

What if the population of students at this high school was $1,000$? The population was probably not exactly $1,000$ students, but using an estimate of $1,000$ students provides an easier way to understand the given probabilities. Connecting these estimates to the actual population is completed in a later exercise. Place the value of $1,000$ in the cell representing the total population. Based on a hypothetical 1000 population, consider the following table:

	No Household Member Smokes	At Least One Household Member Smokes	Total
Student Has Asthma	Cell 1	Cell 2	Cell 3
Student Does Not Have Asthma	Cell 4	Cell 5	Cell 6
Total	Cell 7	Cell 8	$1,000$

9. The probability that a randomly selected student at this high school has asthma is 0.193. This probability can be used to calculate the value of one of the cells in the table above. Which cell is connected to this probability? Use this probability to calculate the value of that cell.

   Cell 3

   The value for this cell would be 193 *students.*

10. The probability that a randomly selected student has at least 1 household member who smokes is 0.421. Which cell is connected to this probability? Use this probability to calculate the value of that cell.

    Cell 8

    The value for this cell would be 421 *students.*

11. In addition to the previously given probabilities, the probability that a randomly selected student has at least one household member who smokes and has asthma is 0.120. Which cell is connected to this probability? Use this probability to calculate the value of that cell.

    Cell 2

    The value for this cell would be 120 *students.*

12. Complete the two-way frequency table by calculating the values of the other cells in the table.

	No Household Member Smokes	At Least One Household Member Smokes	Total
Student Has Asthma	73	120	193
Student Does Not Have Asthma	506	301	807
Total	579	421	$1,000$

13. **Based on your completed two-way table, estimate the following probabilities as a fraction and also as a decimal (rounded to three decimal places):**

  a. **A randomly selected student has asthma. What is the probability this student has at least 1 household member who smokes?**

  The probability is $\frac{120}{193}$, *which approximately* 0.622.

  b. **A randomly selected student does not have asthma. What is the probability this student has at least one household member who smokes?**

  The probability is $\frac{301}{807}$, *which approximately* 0.373.

  c. **A randomly selected student has at least one household member who smokes. What is the probability this student has asthma?**

  The probability is $\frac{120}{421}$, *which approximately* 0.285.

14. **Do you think that whether or not a student has asthma is related to whether or not this student has at least one family member who smokes? Explain your answer.**

  Yes, the probability a student with asthma has a household member who smokes is noticeably greater than the probability a student who does not have asthma has a household member who smokes.

Closing (5 minutes)

- What was the role of probability in our two examples?
  - *To use data to predict outcomes of events, such as selecting a voter who will vote "yes" or selecting a student who has asthma and at least one household member who smokes*
- As you used probabilities to make predictions, did you learn anything about the context of the data?
  - *Answers will vary. Anticipate that students indicate that younger voters were less likely to vote or that older voters in Waldo were less likely to support the new high school or that students with asthma were more likely to have a household family member who smokes. Emphasize that probabilities must be interpreted in context.*

Ask students to summarize the main ideas of the lesson in writing or with a neighbor. Use this as an opportunity to informally assess comprehension of the lesson. The Lesson Summary below offers some important ideas that should be included.

> Lesson Summary
>
> **Data organized in a two-way frequency table can be used to calculate probabilities.**
>
> **In certain problems, probabilities that are known can be used to create a hypothetical 1000 two-way table. The hypothetical population of $1,000$ can then be used to calculate probabilities.**
>
> **Probabilities are always interpreted in context.**

Exit Ticket (5 minutes)

EUREKA MATH™

Name ______________________________ Date ______________

Lesson 2: Calculating Probabilities of Events Using Two-Way Tables

Exit Ticket

Did male and female voters respond similarly to the survey question about building a new high school? Recall the original summary of the data.

	Should Our Town Build a New High School?					
	Yes		No		No Answer	
Age (in years)	Male	Female	Male	Female	Male	Female
18–25	29	32	8	6	0	0
26–40	53	60	40	44	2	4
41–65	30	36	44	35	2	2
66 and Older	7	26	24	29	2	0

1. Complete the following two-way frequency table:

	Yes	**No**	**No Answer**	**Total**
Male	119		6	
Female				
Total		230	12	515

2. Use the above two-way frequency table to answer the following questions:
   a. If a randomly selected eligible voter is female, what is the probability she will vote to build a new high school?

   b. If a randomly selected eligible voter is male, what is the probability he will vote to build a new high school?

3. An automobile company has two factories assembling its luxury cars. The company is interested in whether consumers rate cars produced at one factory more highly than cars produced at the other factory. Factory A assembles 60% of the cars. A recent survey indicated that 70% of the cars made by this company (both factories combined) were highly rated. This same survey indicated that 10% of all cars made by this company were both made at Factory B *and* were *not* highly rated.

   a. Create a hypothetical 1000 two-way table based on the results of this survey by filling in the table below.

	Car Was Highly Rated by Consumers	**Car Was Not Highly Rated by Consumers**	**Total**
Factory A			
Factory B			
Total			

   b. A randomly selected car was assembled in Factory B. What is the probability this car is highly rated?

Exit Ticket Sample Solutions

Did male and female voters respond similarly to the survey question about building a new high school? Recall the original summary of the data.

	Should Our Town Build a New High School?					
	Yes		No		No Answer	
Age (in years)	Male	Female	Male	Female	Male	Female
18–25	29	32	8	6	0	0
26–40	53	60	40	44	2	4
41–65	30	36	44	35	2	2
66 and Older	7	26	24	29	2	0

1. Complete the following two-way frequency table:

	Yes	No	No Answer	Total
Male	119	116	6	241
Female	154	114	6	274
Total	273	230	12	515

2. Use the above two-way frequency table to answer the following questions:

   a. If a randomly selected eligible voter is female, what is the probability she will vote to build a new high school?

   $\frac{154}{274} \approx 0.562$

   b. If a randomly selected eligible voter is male, what is the probability he will vote to build a new high school?

   $\frac{119}{241} \approx 0.494$

3. An automobile company has two factories assembling its luxury cars. The company is interested in whether consumers rate cars produced at one factory more highly than cars produced at the other factory. Factory A assembles 60% of the cars. A recent survey indicated that 70% of the cars made by this company (both factories combined) were highly rated. This same survey indicated that 10% of all cars made by this company were both made at Factory B *and* were *not* highly rated.

   a. Create a hypothetical 1000 two-way table based on the results of this survey by filling in the table below.

	Car Was Highly Rated by Consumers	Car Was Not Highly Rated by Consumers	Total
Factory A	400	200	600
Factory B	300	100	400
Total	700	300	1,000

   b. A randomly selected car was assembled in Factory B. What is the probability this car is highly rated?

   The probability a car from Factory B is highly rated is $\frac{300}{400}$, *which is* 0.750.

Problem Set Sample Solutions

1. The Waldo School Board asked eligible voters to evaluate the town's library service. Data are summarized in the following table:

	How Would You Rate Our Town's Library Services?							
	Good		Average		Poor		Do Not Use Library	
Age (in years)	Male	Female	Male	Female	Male	Female	Male	Female
18–25	10	8	5	7	5	5	17	18
26–40	30	28	25	30	20	30	20	20
41–65	30	32	26	21	15	10	5	10
66 and Older	21	25	8	15	2	10	2	5

a. What is the probability that a randomly selected person who completed the survey rated the library as good?

$\frac{184}{515} \approx 0.357$

b. Imagine talking to a randomly selected male voter who had completed the survey. How do you think this person rated the library services? Explain your answer.

Answers will vary. A general look at the table indicates that most male voters rated the library as good. As a result, I would predict that this person would rate the library as good.

c. Use the given data to construct a two-way table that summarizes the responses on gender and rating of the library services. Use the following template as your guide:

	Good	Average	Poor	Do Not Use	Total
Male	91	64	42	44	241
Female	93	73	55	53	274
Total	184	137	97	97	515

d. Based on your table, answer the following:

i. A randomly selected person who completed the survey is male. What is the probability he rates the library services as good?

$\frac{91}{241} \approx 0.378$

ii. A randomly selected person who completed the survey is female. What is the probability she rates the library services as good?

$\frac{93}{274} \approx 0.339$

e. Based on your table, answer the following:

i. A randomly selected person who completed the survey rated the library services as good. What is the probability this person is male?

$\frac{91}{184} \approx 0.495$

ii. A randomly selected person who completed the survey rated the library services as good. What is the probability this person is female?

$\frac{93}{274} \approx 0.339$

f. Do you think there is a difference in how male and female voters rated library services? Explain your answer.

Answers will vary. Yes, I think there is a difference. For example, 37.9% *of the male voters rated the library services as good, but only* 33.9% *of the female voters rated the services this way. There are also differences between the ratings from male and female voters for the other categories.*

2. Obedience School for Dogs is a small franchise that offers obedience classes for dogs. Some people think that larger dogs are easier to train and, therefore, should not be charged as much for the classes. To investigate this claim, dogs enrolled in the classes were classified as large (30 pounds or more) or small (under 30 pounds). The dogs were also classified by whether or not they passed the obedience class offered by the franchise. 45% of the dogs involved in the classes were large. 60% of the dogs passed the class. Records indicate that 40% of the dogs in the classes were small and passed the course.

a. Complete the following hypothetical 1000 two-way table:

	Passed the Course	Did Not Pass the Course	Total
Large Dogs	200	250	450
Small Dogs	400	150	550
Total	600	400	$1,000$

b. Estimate the probability that a dog selected at random from those enrolled in the classes passed the course.

$\frac{600}{1000} = 0.600$, *meaning* 60% *passed the course.*

c. A dog was randomly selected from the dogs that completed the class. If the selected dog was a large dog, what is the probability this dog passed the course?

$\frac{200}{450} \approx 0.444$, *meaning that approximately* 44.4% *of large dogs passed the course.*

d. A dog was randomly selected from the dogs that completed the class. If the selected dog is a small dog, what is the probability this dog passed the course?

$\frac{400}{550} \approx 0.727$, *meaning that approximately* 72.7% *of small dogs passed the course.*

e. Do you think dog size and whether or not a dog passes the course are related?

Answers will vary. Yes, there is a noticeably greater probability that a dog passed the obedience class if a dog is small than if the dog is large.

f. Do you think large dogs should get a discount? Explain your answer.

Answers will vary. No, large dogs should not get a discount. Large dogs are not as likely to have passed the obedience class as small dogs.

Lesson 3: Calculating Conditional Probabilities and Evaluating Independence Using Two-Way Tables

Student Outcomes

- Students construct a hypothetical 1000 two-way table from given probability information and use the table to calculate the probabilities of events.
- Students calculate conditional probabilities given a two-way data table or using a hypothetical 1000 two-way table.
- Students interpret probabilities, including conditional probabilities, in context.

Lesson Notes

This lesson is a continuation of the work started in Lesson 2. In this lesson, students learn a more formal definition of conditional probability and are asked to interpret conditional probabilities. Data are presented in two-way frequency tables, and conditional probabilities are calculated using column or row summaries. The work in this lesson leads up to the definition of independent events (Lesson 4).

Classwork

Example 1 (2–3 minutes)

The initial activity asks students to investigate a real-world context using probability. In this hypothetical example, students determine if female students are more likely to be involved in an athletic program as compared to male students. While questions are provided, consider using this opening problem without scaffolding to allow students to apply their knowledge from the first two lessons.

Example 1

Students at Rufus King High School were discussing some of the challenges of finding space for athletic teams to practice after school. Part of the problem, according to Kristin, is that female students are more likely to be involved in an after-school athletics program than male students. However, the athletic director assigns the available facilities as if male students are more likely to be involved. Before suggesting changes to the assignments, the students decided to investigate.

Suppose the following information is known about Rufus King High School: 40% of students are involved in one or more of the after-school athletics programs offered at the school. It is also known that 58% of the school's students are female. The students decide to construct a hypothetical 1000 two-way table, like Table 1, to organize the data.

Scaffolding:

Discuss types of sports offered by the school and where the teams practice. Take a poll in the class to find out the number of male and female students who participate in sports, what type, and when and where they practice. Calculate the class statistics before doing this example.

Table 1: Participation in After-School Athletics Programs (Yes or No) by Gender

	Yes—Participate in After-School Athletics Programs	No—Do Not Participate in After-School Athletics Programs	Total
Female	Cell 1	Cell 2	Cell 3
Male	Cell 4	Cell 5	Cell 6
Total	Cell 7	Cell 8	Cell 9

Exercises 1–6 (10–15 minutes): Organizing the Data

MP.2 Let students work with a partner and then confirm answers as a class. Exercise 3 requires students to express their thinking, reasoning from the table abstractly and quantitatively. As students calculate the probabilities, they need to interpret them in context.

Exercises 1–6: Organizing the Data

1. **What cell in Table 1 represents a hypothetical group of $1,000$ students at Rufus King High School?**

   Cell 9

2. **What cells in Table 1 can be filled based on the information given about the student population? Place these values in the appropriate cells of the table based on this information.**

   Cells 3 and 7 can be completed from the given information. See the completed table below.

3. **Based only on the cells you completed in Exercise 2, which of the following probabilities can be calculated, and which cannot be calculated? Calculate the probability if it can be calculated. If it cannot be calculated, indicate why.**

   a. **The probability that a randomly selected student is female**

      Yes, this can be calculated. The probability is 0.58.

   b. **The probability that a randomly selected student participates in an after-school athletics program**

      Yes, this can be calculated. The probability is 0.40.

   c. **The probability that a randomly selected student who does not participate in an after-school athletics program is male**

      No, this probability cannot be calculated. We need to know the value of cell 5 to calculate this probability.

   d. **The probability that a randomly selected male student participates in an after-school athletics program**

      No, this probability cannot be calculated. We need to know the value of cell 4 to calculate this probability.

4. **The athletic director indicated that 23.2% of the students at Rufus King are female and participate in after-school athletics programs. Based on this information, complete Table 1.**

	Yes—Participate in After-School Athletic Program	*No—Do Not Participate in After-School Athletic Program*	*Total*
Female	232	348	580
Male	168	252	420
Total	400	600	$1,000$

5. **Consider the cells 1, 2, 4, and 5 of Table 1. Identify which of these cells represent students who are female or who participate in after-school athletics programs.**

Cells 1, 2, and 4

6. **What cells of the two-way table represent students who are male and do not participate in after-school athletics programs?**

Cell 5

Example 2 (2–3 minutes)

The following definitions were first introduced in Grade 7. It is important, however, that students revisit the definitions of complement, union, and intersection. The definitions in this lesson are connected to the context of the data and do not focus on a symbolic representation of these terms. The use of Venn diagrams and sets to represent these events is developed in Lessons 7 and 8.

Example 2

The completed hypothetical 1000 table organizes information in a way that makes it possible to answer various questions. For example, you can investigate whether female students are more likely to be involved in the after-school athletic programs.

Consider the following events:

- **Let A represent the event "a randomly selected student is female."**
- **Let "not A" represent "the *complement* of A." The complement of A represents the event "a randomly selected student is not female," which is equivalent to the event "a randomly selected student is male."**
- **Let B represent the event "a randomly selected student participates in an after-school athletics program."**
- **Let "not B" represent "the *complement* of B." The complement of B represents the event "a randomly selected student does not participate in an after-school athletics program."**
- **Let "A or B" (described as A *union* B) represent the event "a randomly selected student is female or participates in an after-school athletics program."**
- **Let "A and B" (described as A *intersect* B) represent the event "a randomly selected student is female and participates in an after-school athletics program."**

Exercises 7–9 (8–10 minutes)

Let students continue to work with their partners and confirm answers as a class.

Exercises 7–9

7. **Based on the descriptions above, describe the following events in words:**

   a. **Not A or not B**

   Male students or students not participating in an after-school athletics program

   b. **A and not B**

   Female students not participating in an after-school athletics program

EUREKA MATH™

8. **Based on the above descriptions and Table 1, determine the probability of each of the following events:**

   a. A

   $$\frac{580}{1000} = 0.58$$

   b. B

   $$\frac{400}{1000} = 0.40$$

   c. Not A

   $$\frac{420}{1000} = 0.42$$

   d. Not B

   $$\frac{600}{1000} = 0.60$$

   e. A or B

   $$\frac{(232 + 348 + 168)}{1000} = \frac{748}{1000} = 0.748$$

   f. A and B

   $$\frac{232}{1000} = 0.232$$

9. **Determine the following values:**

   a. **The probability of A plus the probability of not A**

   $0.580 + 0.420 = 1.000$

   The sum is 1.

   b. **The probability of B plus the probability of not B**

   $0.400 + 0.600 - 1.000$

   The sum is 1.

   c. **What do you notice about the results of parts (a) and (b)? Explain.**

   Both probabilities total 1. This makes sense since both parts are asking for the probability of A and not A or B and not B. The probabilities of an event and its complement always sum to 1.

Example 3 (2 minutes): Conditional Probability

Read through the example as a class. Help students identify how the conditional probabilities are not based on the whole population but rather on a specific subgroup within the whole population that is represented by a row total or a column total. The visual of pulling apart the two-way table by rows or columns is intended to help in developing this idea with students.

Example 3: Conditional Probability

Another type of probability is called a *conditional probability*. Pulling apart the two-way table helps to define a conditional probability.

	Yes—Participate in After-School Athletics Program	No—Do Not Participate in After-School Athletics Program	Total
Female	Cell 1	Cell 2	Cell 3

Suppose that a randomly selected student is female. What is the probability that the selected student participates in an after-school athletics program? This probability is an example of what is called a *conditional probability*. This probability is calculated as the number of students who are female and participate in an after-school athletics program (or the students in cell 1) divided by the total number of female students (or the students in cell 3).

Exercises 10–15 (8–10 minutes)

The following exercises are designed to have students calculate and interpret conditional probabilities based on data in two-way tables. The conditional probabilities are based on focusing on either a row or a column in the table. Let students work with a partner and confirm answers as a class. Students are calculating probabilities and connecting the probability to the context. In these questions, students must make the connection to a subgroup within the population identified by the desired conditional probability. Both the calculation of the conditional probability and its meaning in context are developed with these questions.

MP.2

Exercises 10–15

10. **The following are also examples of conditional probabilities. Answer each question.**

    a. **What is the probability that if a randomly selected student is female, she participates in the after-school athletic program?**

    The probability that if a randomly selected student is female then she participates in the after-school athletic program is the value in Cell 1 divided by the value in Cell 3.

    $$\frac{232}{580} = 0.40$$

    b. **What is the probability that if a randomly selected student is female, she does not participate in after-school athletics?**

    The probability that if a randomly selected student is female then she does not participate in the after-school athletic program is the value in Cell 2 divided by the value in Cell 3.

    $$\frac{348}{580} = 0.60$$

11. **Describe two conditional probabilities that can be determined from the following row in Table 1:**

	Yes—Participate in After-School Athletics Program	No—Do Not Participate in After-School Athletics Program	Total
Male	Cell 4	Cell 5	Cell 6

The probability that if a randomly selected student is male, he participates in an after-school athletics program

The probability that if a randomly selected student is male, he does not participate in an after-school athletics program

EUREKA MATH™

12. **Describe two conditional probabilities that can be determined from the following column in Table 1:**

	Yes—Participate in After-School Athletics Program
Female	Cell 1
Male	Cell 4
Total	Cell 7

The probability that if a randomly selected student participates in an after-school athletics program, the student is female

The probability that if a randomly selected student participates in an after-school athletics program, the student is male

13. **Determine the following conditional probabilities:**

   a. **A randomly selected student is female. What is the probability she participates in an after-school athletics program? Explain how you determined your answer.**

   $\frac{232}{580} = 0.40$

   Since it is known that the selected student is female, I looked at the row for female students and used that information.

   b. **A randomly selected student is male. What is the probability he participates in an after-school athletics program?**

   $\frac{168}{420} = 0.40$

   c. **A student is selected at random. What is the probability this student participates in an after-school athletics program?**

   $\frac{400}{1000} = 0.40$

14. **Based on the answers to Exercise 13, do you think that female students are more likely to be involved in after-school athletics programs? Explain your answer.**

No, the conditional probabilities indicate male and female students are equally likely to be involved in an after-school athletics program.

15. **What might explain the concern female students expressed in the beginning of this lesson about the problem of assigning practice space?**

It is interesting that at this school, the probability that a randomly selected student is female is 0.58. *There are more female students than male students at this school. As a result, if facilities are assigned equally (given that both female and male students were found to be equally likely to be involved), the number of female students involved in after-school athletics programs is greater than the number of male students and might explain the concern regarding facilities.*

Closing (5 minutes)

- What did the probabilities tell us about the students at Rufus King High School?
  - *Answers will vary. Anticipate that students might mention that there are more female than male students or that male and female students are equally likely to be involved in after-school athletics programs (based on the conditional probabilities). Use this question to help students understand the difference between probabilities based on the whole population and probabilities based on a row or column of the table (conditional probabilities).*

Ask students to summarize the main ideas of the lesson in writing or with a neighbor. Use this opportunity to informally assess student comprehension of the lesson. The Lesson Summary below offers some important ideas that should be included.

Lesson Summary

Data organized in a two-way frequency table can be used to calculate probabilities. Two-way frequency tables can also be used to calculate conditional probabilities.

In certain problems, probabilities that are known can be used to create a hypothetical 1000 two-way table. This hypothetical population of $1,000$ can be used to calculate conditional probabilities.

Probabilities are always interpreted by the context of the data.

Exit Ticket (5 minutes)

Name ______________________________ Date ______________

Lesson 3: Calculating Conditional Probabilities and Evaluating Independence Using Two-Way Tables

Exit Ticket

A state nonprofit organization wanted to encourage its members to consider the State of New York as a vacation destination. They are investigating whether their online ad campaign influenced its members to plan a vacation in New York within the next year. The organization surveyed its members and found that 75% of them have seen the online ad. 40% of its members indicated they are planning to vacation in New York within the next year, and 15% of its members did not see the ad and do not plan to vacation in New York within the next year.

1. Complete the following hypothetical 1000 two-way frequency table:

	Plan to Vacation in New York Within the Next Year	**Do Not Plan to Vacation In New York Within the Next Year**	**Total**
Watched the Online Ad			
Did Not Watch the Online Ad			
Total			

2. Based on the two-way table, describe two conditional probabilities you could calculate to help decide if members who saw the online ad are more likely to plan a vacation in New York within the next year than those who did not see the ad.

3. Calculate the probabilities you described in Problem 2.

4. Based on the probabilities calculated in Problem 3, do you think the ad campaign is effective in encouraging people to vacation in New York? Explain your answer.

Exit Ticket Sample Solutions

A state nonprofit organization wanted to encourage its members to consider the State of New York as a vacation destination. They are investigating whether their online ad campaign influenced its members to plan a vacation in New York within the next year. The organization surveyed its members and found 75% of them have seen the online ad. 40% of its members indicated they are planning to vacation in New York within the next year, and 15% of its members did not see the ad and do not plan to vacation in New York within the next year.

1. Complete the following hypothetical 1000 two-way frequency table:

	Plan to Vacation in New York Within the Next Year	Do Not Plan to Vacation in New York Within the Next Year	Total
Watched the Online Ad	300	450	750
Did Not Watch the Online Ad	100	150	250
Total	400	600	$1,000$

2. Based on the two-way table, describe two conditional probabilities you could calculate to help decide if members who saw the online ad are more likely to plan a vacation in New York within the next year than those who did not see the ad.

   The probability that a randomly selected member who watched the ad is planning to vacation in New York

   The probability that a randomly selected member who did not watch the ad is planning to vacation in New York

3. Calculate the probabilities you described in Problem 2.

   The probability that a member who watched the ad is planning a vacation in New York: $\frac{300}{750} = 0.400$

   The probability that a member who did not watch the ad is planning a vacation in New York: $\frac{100}{250} = 0.400$

4. Based on the probabilities calculated in Problem 3, do you think the ad campaign is effective in encouraging people to vacation in New York? Explain your answer.

   The conditional probabilities are the same. It does not appear that the ad campaign is encouraging people to vacation in New York.

Problem Set Sample Solutions

Oostburg College has a rather large marching band. Engineering majors were heard bragging that students majoring in engineering are more likely to be involved in the marching band than students from other majors.

1. If the above claim is accurate, does that mean that most of the band is engineering students? Explain your answer.

   No. It means that if a randomly selected student is an engineering major, the probability this person is in the marching band is greater than if this person was not an engineering major.

2. **The following graph was prepared to investigate the above claim:**

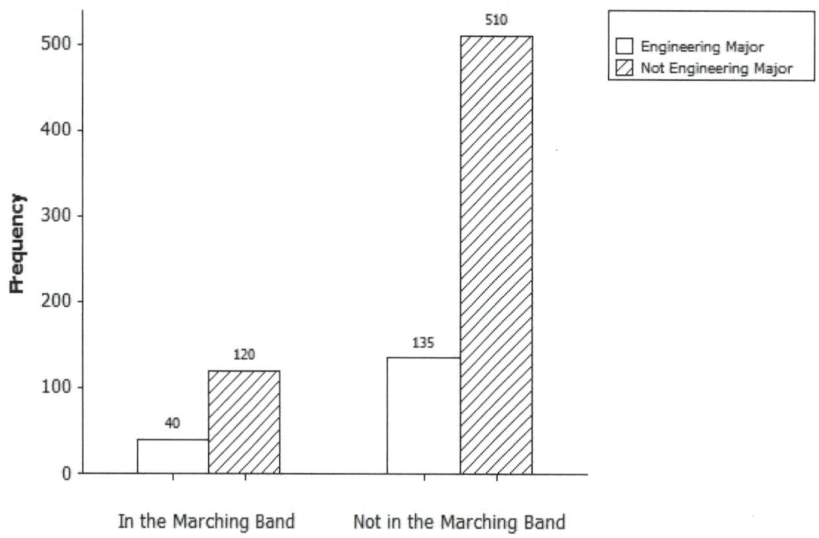

Based on the graph, complete the following two-way frequency table:

	In the Marching Band	Not in the Marching Band	Total
Engineering Major	40	135	175
Not Engineering Major	120	510	630
Total	160	645	805

3. **Let M represent the event that a randomly selected student is in the marching band. Let E represent the event that a randomly selected student is an engineering major.**

   a. **Describe the event represented by the complement of M.**

   A randomly selected student is not in the marching band.

   b. **Describe the event represented by the complement of E.**

   A randomly selected student is not majoring in engineering.

   c. **Describe the event M and E (M intersect E).**

   A randomly selected student is majoring in engineering and is in the marching band.

   d. **Describe the event M or E (M union E).**

   A randomly selected student is majoring in engineering or is in the marching band.

The union can be a challenge for students to describe or to identify. Start with one of the joint cells of the two-way table. Ask if the joint cell includes students majoring in engineering or students in the marching band. If either description applies, then that cell is part of the union. Work through each of the joint cells in this way to identify all of the cells that compose the union.

4. Based on the completed two-way frequency table, determine the following, and explain how you got your answer:

a. The probability that a randomly selected student is in the marching band

$\frac{160}{805} \approx 0.199$

I compared the number of students in the marching band to the total number of students.

b. The probability that a randomly selected student is an engineering major

$\frac{175}{805} \approx 0.217$

I compared the number of engineering majors to the total number of students

c. The probability that a randomly selected student is in the marching band and an engineering major

$\frac{40}{805} \approx 0.05$

I found the number of students who are in the band and are engineering majors and compared it to the total number of students.

d. The probability that a randomly selected student is in the marching band and not an engineering major

$\frac{120}{805} \approx 0.149$

I found the number of students who are in the band and are NOT engineering majors and compared it to the total number of students.

5. Indicate if the following conditional probabilities would be calculated using the rows or the columns of the two-way frequency table:

a. A randomly selected student is majoring in engineering. What is the probability this student is in the marching band?

This probability is based on the row Engineering Major.

b. A randomly selected student is not in the marching band. What is the probability that this student is majoring in engineering?

This probability is based on the column Not in the Marching Band.

6. Based on the two-way frequency table, determine the following conditional probabilities:

a. A randomly selected student is majoring in engineering. What is the probability that this student is in the marching band?

$\frac{40}{175} \approx 0.229$

b. A randomly selected student is not majoring in engineering. What is the probability that this student is in the marching band?

$\frac{120}{630} \approx 0.190$

7. **The claim that started this investigation was that students majoring in engineering are more likely to be in the marching band than students from other majors. Describe the conditional probabilities that would be used to determine if this claim is accurate.**

Given a randomly selected student is an engineering major, what is the probability the student is in the marching band. Also, given a randomly selected student is not an engineering major, what is the probability the student is in the marching band.

8. **Based on the two-way frequency table, calculate the conditional probabilities identified in Problem 7.**

The probabilities were calculated in Problem 6. Approximately 0.229 (or 22.9%) of the engineering students are in the marching band. Approximately 0.190 (or 19.0%) of the students not majoring in engineering are in the marching band.

9. **Do you think the claim that students majoring in engineering are more likely to be in the marching band than students for other majors is accurate? Explain your answer.**

The claim is accurate based on the conditional probabilities.

10. **There are 40 students at Oostburg College majoring in computer science. Computer science is not considered an engineering major. Calculate an estimate of the number of computer science majors you think are in the marching band. Explain how you calculated your estimate.**

The probability that a randomly selected student who is not majoring in engineering is in the marching band is 0.190. As a result, you would estimate that 19% of the 40 computer science majors are in the marching band. Since $40(0.190) = 7.6$, I would expect that 7 or 8 computer science majors are in the marching band.

Lesson 4: Calculating Conditional Probabilities and Evaluating Independence Using Two-Way Tables

Student Outcomes

- Students use a hypothetical 1000 two-way table to calculate probabilities of events.
- Students calculate conditional probabilities given a two-way data table or using a hypothetical 1000 two-way table.
- Students use two-way tables (data tables or hypothetical 1000 two-way tables) to determine if two events are independent.
- Students interpret probabilities, including conditional probabilities, in context.

Lesson Notes

This lesson builds on students' previous work with conditional probabilities to define independent events. In previous lessons, conditional probabilities were used to investigate whether or not there is a connection between two events. This lesson formalizes this idea and introduces the concept of independence.

Classwork

Exercise 1 (5 minutes)

Consider asking students to respond to each question independently in writing and then share answers as a class. Allow multiple student responses for each question.

Exercise 1

In previous lessons, conditional probabilities were used to investigate whether or not there is a connection between two events. This lesson formalizes this idea and introduces the concept of *independence*.

1. **Several questions are posed below. Each question is about a possible connection between two events. For each question, identify the two events, and indicate whether or not you think that there would be a connection. Explain your reasoning.**
   a. **Are high school students whose parents or guardians set a midnight curfew less likely to have a traffic violation than students whose parents or guardians have not set such a curfew?**

      Responses vary.

      The two events are "parents set a midnight curfew" and "students have a traffic violation." Anticipate that students may indicate either that students with a curfew are less likely to have a traffic violation or that there is no connection. Either answer is acceptable as long as students provide an explanation.

Scaffolding:

Consider focusing on only one of these examples. Also, consider revising examples to be more representative of students' experiences and interests. Supplying a visual to accompany the situations may also increase accessibility.

b. **Are left-handed people more likely than right-handed people to be interested in the arts?**

Responses vary.

The two events are "people are right-handed" and "people are interested in the arts." Students may argue that there is a connection or that there is not a connection between being right-handed and being interested in the arts. Look for some reasoning that indicates that if one of the descriptions is met (selecting a right-handed person), it is more (or less) likely this person is also interested in the arts. Students may indicate that they do not think that there is a connection, which is acceptable as long as they explain their reasoning.

c. **Are students who regularly listen to classical music more likely to be interested in mathematics than students who do not regularly listen to classical music?**

Responses vary.

The two events are "students regularly listen to classical music" and "students are interested in mathematics." Students may argue that there is a connection or there is not a connection between classical music and an interest in mathematics. Look for some reasoning that indicates that if one of the descriptions is met, it is more likely the other description will or will not occur.

d. **Are people who play video games more than 10 hours per week more likely to select football as their favorite sport than people who do not play video games more than 10 hours per week?**

Responses vary.

The two events are "people who play video games more than 10 hours per week" and "people whose favorite sport is football." Students may argue that there is a connection or there is not a connection between video games and an interest in football, which is acceptable as long as they explain their reasoning.

Ask students to write down the answer to the question below. How can conditional probabilities be used to tell if two events are independent or not independent? Then, ask students to share their conjectures with a partner.

MP.3

Two events are independent when knowing that one event has occurred does not change the likelihood that the second event has occurred. How can conditional probabilities be used to tell if two events are independent or not independent?

Exercises 2–6 (15 minutes)

A two-way frequency table is used in these exercises in order to calculate probabilities, which helps students answer the types of questions posed in Exercise 1. The conditional probabilities are used to develop an understanding of the concept of independence. The data and several of the calculations in these exercises were first introduced in Lesson 3. This is intentional because the following exercises formalize ideas developed in students' previous work.

Exercises 2–6

Recall the hypothetical 1000 two-way frequency table that was used to classify students at Rufus King High School according to gender and whether or not they participated in an after-school athletics program.

Table 1: Participation of Male and Female Students in an After-School Athletics Program

	Participate in an After-School Athletics Program	Do Not Participate in an After-School Athletics Program	Total
Female	232	348	580
Male	168	252	420
Total	400	600	$1,000$

2. **For each of the following, indicate whether the probability described is one that can be calculated using the values in Table 1. Also indicate whether or not it is a conditional probability.**

   a. **The probability that a randomly selected student participates in an after-school athletics program**

   This probability can be calculated from the values in the table. It is not a conditional probability. The probability is based on the entire school population.

   b. **The probability that a randomly selected student who is female participates in an after-school athletics program**

   This probability can be calculated from the values in the table. It is a conditional probability because it is a probability based on only the female students at the school.

   c. **The probability that a randomly selected student who is male participates in an after-school athletics program**

   This probability can be calculated from the values in the table. It is a conditional probability because it is based on only the male students at the school.

3. **Use Table 1 to calculate each of the probabilities described in Exercise 2.**

   a. **The probability that a randomly selected student participates in an after-school athletics program**

   $\frac{400}{1000} = 0.400$ *(This means that* 40% *of all students participate in after-school athletics programs.)*

   b. **The probability that a randomly selected student who is female participates in an after-school athletics program**

   $\frac{232}{580} = 0.400$ *(This means that* 40% *of female students participate in after-school athletics programs.)*

   c. **The probability that a randomly selected student who is male participates in an after-school athletics program**

   $\frac{168}{420} = 0.400$ *(This means that* 40% *of male students participate in after-school athletics programs.)*

Exercise 4 is an important question as it provides students an understanding of the meaning of independent events. MP.2 Students are expected to connect the concept of conditional probabilities with the definition of independent events. This exercise also provides students an opportunity to express their reasoning abstractly and quantitatively. Students use numerical probabilities to make inferences about real-world situations. Allow students to work individually on this question for a few minutes, and then discuss as a whole group. Use this question to define two independent events.

- Two events are independent when knowing that one event has occurred does not change the likelihood that the second event has occurred. How can conditional probabilities be used to tell if two events are independent or not independent?
  - *Allow for multiple responses. Then, read through the definition, which is also provided for students, so that they can use it to complete this lesson.*

4. **Would your prediction of whether or not a student participates in an after-school athletics program change if you knew the gender of the student? Explain your answer.**

No. Based on the conditional probabilities, the probability that a student participates in an after-school athletics program is 0.400 for both male and female students.

Two events are *independent* if knowing that one event has occurred does not change the probability that the other event has occurred. For example, consider the following two events:

F: The event that a randomly selected student is female

S: The event that a randomly selected student participates in an after-school athletics program

Events F and S would be independent if the probability that a randomly selected student participates in an after-school athletics program is equal to the probability that a randomly selected student who is female participates in an after-school athletics program. If this was the case, knowing that a randomly selected student is female does not change the probability that the selected student participates in an after-school athletics program. Then, F and S would be independent.

Exercise 5 and the definition of independence provide the opportunity to have students explore independence using the conditional probabilities based on columns. Exercise 6 is designed for students to observe that the probability that a randomly selected student who participates in an after-school athletics program is female is equal to the probability that a randomly selected student (from the entire school population) is female. The probability that a randomly selected student who participates in an after-school athletics program is male is equal to the probability that a randomly selected student (from the entire school population) is male. Point out that when this happens, then the events are independent. Also point out that conditional probabilities in either row or column can be used to decide if events are independent.

Scaffolding:

Students may have learned the term *independence* or *independent* in other contexts (such as social studies) and, as such, may need opportunities to add to their understandings. A Frayer model diagram may be useful for this purpose.

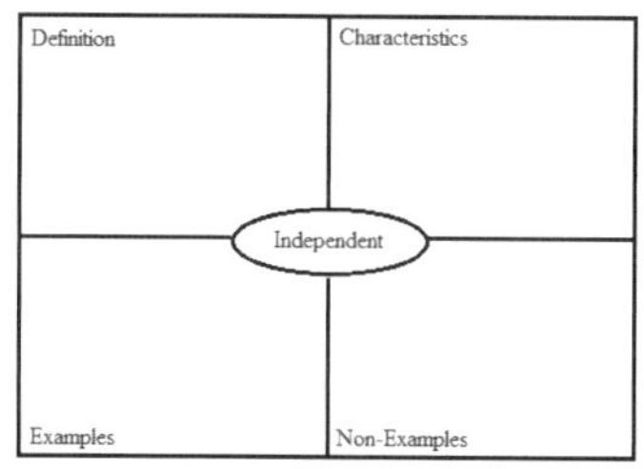

5. **Based on the definition of independence, are the events "randomly selected student is female" and "randomly selected student participates in an after-school athletics program" independent? Explain.**

Yes, they are independent because knowing that a randomly selected student is female does not change the probability that the selected student participates in an after-school athletics program. The probability that a randomly selected student participates in an after-school athletics program is 0.400, and the probability that a randomly selected student who is female participates is also 0.400.

6. **A randomly selected student participates in an after-school athletics program.**

   a. **What is the probability this student is female?**

   $\frac{232}{400} = 0.58$

   This is equal to the probability that a randomly selected student is female.

   b. **Using only your answer from part (a), what is the probability that this student is male? Explain how you arrived at your answer.**

   The probability is 0.42.

   The probability in part (a) can be interpreted as 58% of the students who participate in after-school athletics programs are female. The rest must be male, so the probability that a randomly selected student who participates in an after-school athletics program is male is 0.42. $1 - 0.58 = 0.42$

Exercises 7–11 (15 minutes)

The following exercises also provide examples of events that are *not independent*. Students may have already expressed this idea in their previous work. They now have a more formal way to express the relationship between two events.

Exercises 7–11

Consider the data below.

	No Household Member Smokes	At Least One Household Member Smokes	Total
Student Has Asthma	69	113	182
Student Does Not Have Asthma	473	282	755
Total	542	395	937

7. **You are asked to determine if the two events "a randomly selected student has asthma" and "a randomly selected student has a household member who smokes" are independent. What probabilities could you calculate to answer this question?**

   Students could indicate that the probability of selecting a student who has a household member who smokes is the same for students who have asthma as for those who do not have asthma. Or students could indicate that the probability of selecting a student who has a household member who smokes from the students who have asthma would need to be equal to the probability of selecting a student who has a household member who smokes from all of the students. Or students could indicate that the probability of selecting a student who has a household member who smokes from the students who do not have asthma would need to be equal to the probability of selecting a student who has a household member who smokes from all of the students.

   Students could also indicate that conditional probabilities based on the columns would have to be equal for the events to be independent. The probability of selecting a student who has asthma from the students who have no household member who smokes would need to be equal to the probability of selecting a student who has asthma from the students who have at least one household member who smokes.

8. **Calculate the probabilities you described in Exercise 7.**

   The row conditional probabilities described are as follows:

   $\frac{69}{182} \approx 0.379$

   $\frac{473}{755} \approx 0.626$

   The column conditional probabilities described are as follows:

   $\frac{69}{542} \approx 0.127$

   $\frac{113}{395} \approx 0.286$

9. **Based on the probabilities you calculated in Exercise 8, are these two events independent or not independent? Explain.**

   No. The conditional probabilities need to be equal for the events to be independent.

10. **Is the probability that a randomly selected student who has asthma and who has a household member who smokes the same as or different from the probability that a randomly selected student who does not have asthma but does have a household member who smokes? Explain your answer.**

The probabilities are different as the events are not independent. The probability of a randomly selected student who has asthma having a household member who smokes is $\frac{113}{182}$, *which is approximately* 0.62. *The probability of a randomly selected student who does not have asthma having a household member who smokes is* $\frac{282}{755}$, *which is approximately* 0.37.

11. **A student is selected at random. The selected student indicates that he has a household member who smokes. What is the probability that the selected student has asthma?**

$\frac{113}{395} = 0.29$ *This is the conditional probability that a randomly selected student who has a household family member who smokes has asthma.*

Closing (5 minutes)

Ask students to summarize the main ideas of the lesson in writing or with a neighbor. Use this as an opportunity to informally assess comprehension of the lesson. The Lesson Summary below offers some important ideas that should be included.

It is important that students do not conclude that just because two events are not independent, there is a relationship between the events that indicates one event causes the other. Students often make this mistake. (Causation is covered in Algebra I.)

Explain to students that when two events are not independent, it is important to remember that this does not mean that one event causes the other (*causation*). The asthma research was not a statistical experiment. There may be other possible explanations for why the events "has asthma" and "has a household member who smokes" might not be independent.

- If you know the probability that a randomly selected student from your school plans to attend a college or university after graduation, and you also know the probability that a randomly selected student from your school has a job, what would it mean for these two events to be independent?
  - *Students should indicate that knowing one event has occurred (for example, the selected student plans to attend college) does not change the probability that the second event occurred (for example, the selected student has a job).*

Lesson Summary

Data organized in a two-way frequency table can be used to calculate conditional probabilities.

Two events are independent if knowing that one event has occurred does not change the probability that the second event has occurred.

Probabilities calculated from two-way frequency tables can be used to determine if two events are independent or not independent.

Exit Ticket (5 minutes)

Name ______________________________ Date ______________

Lesson 4: Calculating Conditional Probabilities and Evaluating Independence Using Two-Way Tables

Exit Ticket

1. The following hypothetical 1000 two-way table was introduced in the previous lesson:

	Plan to Vacation in New York Within the Next Year	Do Not Plan to Vacation in New York Within the Next Year	Total
Watched the Online Ad	300	450	750
Did Not Watch the Online Ad	100	150	250
Total	400	600	1,000

Are the events "a randomly selected person watched the online ad" and "a randomly selected person plans to vacation in New York within the next year" independent or not independent? Justify your answer using probabilities calculated from information in the table.

2. A survey conducted at a local high school indicated that 30% of students have a job during the school year. If having a job and being in the eleventh grade are not independent, what do you know about the probability that a randomly selected student who is in the eleventh grade would have a job? Justify your answer.

3. Eighty percent of the dogs at a local kennel are in good health. If the events "a randomly selected dog at this kennel is in good health" and "a randomly selected dog at this kennel weighs more than 30 pounds" are independent, what do you know about the probability that a randomly selected dog that weighs more than 30 pounds will be in good health? Justify your answer.

Exit Ticket Sample Solutions

1. **The following hypothetical 1000 two-way table was introduced in the previous lesson:**

	Plan to Vacation in New York Within the Next Year	Do Not Plan to Vacation in New York Within the Next Year	Total
Watched the Online Ad	300	450	750
Did Not Watch the Online Ad	100	150	250
Total	400	600	$1{,}000$

Are the events "a randomly selected person watched the online ad" and "a randomly selected person plans to vacation in New York within the next year" independent or not independent? Justify your answer using probabilities calculated from information in the table.

The conditional probabilities that could be used to evaluate if the events are independent are the probability that given the selected member watched the online ad, the member plans to vacation in New York, and the probability that given the selected member did not watch the ad, the member plans to vacation in New York. Because the probabilities are equal, the events are independent.

$\frac{300}{750} = 0.40$ *and* $\frac{100}{250} = 0.40$

2. **A survey conducted at a local high school indicated that 30% of students have a job during the school year. If having a job and being in the eleventh grade are not independent, what do you know about the probability that a randomly selected student who is in the eleventh grade would have a job? Justify your answer.**

The probability that a student selected from the eleventh grade has a job would not be equal to 0.30. Not independent means that knowing the selected student is in the eleventh grade changes the probability that the student has a job.

3. **Eighty percent of the dogs at a local kennel are in good health. If the events "a randomly selected dog at this kennel is in good health" and "a randomly selected dog at this kennel weighs more than 30 pounds" are independent, what do you know about the probability that a randomly selected dog that weighs more than 30 pounds will be in good health? Justify your answer.**

As the events are independent, knowing that the selected dog weighs more than 30 pounds does not change the probability that the dog is in good health. This means that the probability that a large dog is in good health is also 0.80.

Problem Set Sample Solutions

1. **Consider the following questions:**
   a. **A survey of the students at a Midwest high school asked the following questions:**

      "Do you use a computer at least 3 times a week to complete your schoolwork?"
      "Are you taking a mathematics class?"

      Do you think the events "a randomly selected student uses a computer at least 3 times a week" and "a randomly selected student is taking a mathematics class" are independent or not independent? Explain your reasoning.

      Anticipate students indicate that using a computer at least 3 times per week and taking a mathematics class are not independent. However, it is also acceptable for students to make a case for independence. Examine the explanations students provide to see if they understand the meaning of independence.

b. The same survey also asked students the following:

"Do you participate in any extracurricular activities at your school?"
"Do you know what you want to do after high school?"

Do you think the events "a randomly selected student participates in extracurricular activities" and "a randomly selected student knows what she wants to do after completing high school" are independent or not independent? Explain your reasoning.

Answers will vary. Anticipate students indicate that students involved in extracurricular activities are often students who want to attend college. It is likely the events are not independent.

c. People attending a professional football game in 2013 completed a survey that included the following questions:

"Is this the first time you have attended a professional football game?"
"Do you think football is too violent?"

Do you think the events "a randomly selected person who completed the survey is attending a professional football game for the first time" and "a randomly selected person who completed the survey thinks football is too violent" are independent or not independent? Explain your reasoning.

Answers will vary. Anticipate that students indicate that people who attend football more often are more likely to not think the game is too violent. It is likely the events are not independent. Again, examine the explanations students provide if they indicate the events are not independent.

2. Complete the table below in a way that would indicate the two events "uses a computer" and "is taking a mathematics class" are independent.

	Uses a Computer at Least 3 Times a Week for Schoolwork	Does Not Use a Computer at Least 3 Times a Week for Schoolwork	Total
In a Mathematics Class	420	280	700
Not in a Mathematics Class	180	120	300
Total	600	400	$1,000$

The values in the table were based on 0.60 of the students not in mathematics use a computer at least 3 times a week for school (0.60×300). Also, 0.60 of the students in mathematics use a computer at least 3 times a week for school (0.60×700).

3. Complete the following hypothetical 1000 table. Are the events "participates in extracurricular activities" and "know what I want to do after high school" independent or not independent? Justify your answer.

	Participates in Extracurricular Activities	Does Not Participate in Extracurricular Activities	Total
Know What I Want to Do After High School	550	250	800
Do Not Know What I Want to Do After High School	50	150	200
Total	600	400	$1,000$

The events "student participates in extracurricular activities" and "student knows what I want to do after high school" are not independent. Students could indicate that the events are not independent in several ways. For example, $\frac{50}{200}$ (the probability that a randomly selected student who does not know what he wants to do after high school participates in extracurricular activities) does not equal $\frac{550}{800}$ (the probability that a randomly selected student who does know what he wants to do after high school participates in extracurricular activities).

4. The following hypothetical 1000 table is from Lesson 2:

	No Household Member Smokes	At Least One Household Member Smokes	Total
Student Has Asthma	73	120	193
Student Does Not Have Asthma	506	301	807
Total	579	421	$1,000$

The actual data from the entire population are given in the table below.

	No Household Member Smokes	At Least One Household Member Smokes	Total
Student Has Asthma	69	113	182
Student Does Not Have Asthma	473	282	755
Total	542	395	937

a. Based on the hypothetical 1000 table, what is the probability that a randomly selected student who has asthma has at least one household member who smokes?

$$\frac{120}{193} \approx 0.622$$

b. Based on the actual data, what is the probability that a randomly selected student who has asthma has at least one household member who smokes (round your answer to 3 decimal places)?

$$\frac{113}{182} \approx 0.621$$

c. Based on the hypothetical 1000 table, what is the probability that a randomly selected student who has no household member who smokes has asthma?

$$\frac{73}{579} \approx 0.126$$

d. Based on the actual data, what is the probability that a randomly selected student who has no household member who smokes has asthma?

$$\frac{69}{542} \approx 0.127$$

e. What do you notice about the probabilities calculated from the actual data and the probabilities calculated from the hypothetical 1000 table?

The conditional probabilities differ only due to rounding in constructing the hypothetical 1000 table from probability information based on the actual data. When an actual data table is available, it can be used to calculate probabilities. When only probability information is available, constructing a hypothetical 1000 table from that information and using it to compute other probabilities will give the same answers as if the actual data were available.

5. **As part of the asthma research, the investigators wondered if students who have asthma are less likely to have a pet at home than students who do not have asthma. They asked the following two questions:**

"Do you have asthma?"
"Do you have a pet at home?"

Based on the responses to these questions, you would like to set up a two-way table that you could use to determine if the following two events are independent or not independent:

Event 1: A randomly selected student has asthma.
Event 2: A randomly selected student has a pet at home.

a. **How would you label the rows of the two-way table?**

Anticipate students indicate for the rows "Has Asthma" and "Does Not Have Asthma." Students might use these labels for the columns rather than the rows, which is also acceptable.

b. **How would you label the columns of the two-way table?**

Anticipate students indicate for the columns "Has a Pet" and "Does Not Have a Pet."

c. **What probabilities would you calculate to determine if Event 1 and Event 2 are independent?**

Answers may vary. Row conditional probabilities or column conditional probabilities would have to be equal if the events are independent. For column conditional probabilities (based on the definition of rows and columns above), this would mean that the probability that a randomly selected student who has a pet has asthma is equal to the probability that a randomly selected student who does not have a pet has asthma.

Lesson 5: Events and Venn Diagrams

Student Outcomes

- Students represent events by shading appropriate regions in a Venn diagram.
- Given a chance experiment with equally likely outcomes, students calculate counts and probabilities by adding or subtracting given counts or probabilities.
- Students interpret probabilities in context.

Lesson Notes

This lesson introduces Venn diagrams to represent the sample space and various events and sets the stage for the two lessons that follow, which introduce students to probability formulas. The purpose is to provide a bridge between using the two-way table approach and using formulas to calculate probabilities. Venn diagrams also provide an opportunity to visually represent the population needed to understand what is requested in the exercises.

Classwork

Opening (4 minutes)

While blank Venn diagrams are supplied for most exercises, use the examples to informally assess students. As students become proficient with Venn diagrams, consider asking them to solve without providing the diagrams.

Draw the following Venn diagram on the board:

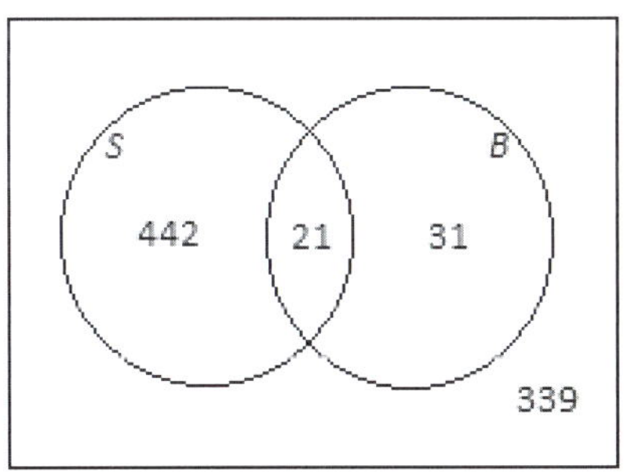

Discuss the following descriptions of a certain high school:

- 442 students participate in organized sports but do not play in the band,
- 31 students play in the band but do not participate in organized sports,
- 21 students participate in organized sports and play in the band, and
- 339 students neither participate in organized sports nor play in the band.

> *Scaffolding:*
>
> For students working above grade level, consider giving the image of the Venn diagram and asking students to describe a situation that could be modeled by this Venn diagram.
>
> For students working below grade level, consider beginning class by creating a Venn diagram from information about the class (for example, which students take chemistry, which students take the bus to school, and which students take chemistry *and* take the bus to school). Basing the Venn diagram on a concrete situation may increase accessibility.

Indicate to students (especially if this is their first time working with Venn diagrams) that the diagram and the descriptions of a certain high school are connected. Ask students what they think the outer rectangle represents. If necessary to continue this discussion, point out that the rectangle is a visual representation of all of the student *population* of the school (emphasize *population*). Also, ask students to explain what they think the circle labeled S represents and what the circle labeled B represents. Allow students to develop a description of circle S as high school students who participate in sports and circle B as students who play in the band. Ask them to also explain why the circles overlap and what the overlapping part of the circles represents in this school. As students begin to make sense of this diagram with the numbers provided about this high school, ask them the following questions:

MP.1 & MP.2

- How many students participate in organized sports?
  - $442 + 21 = 463$
- How many students play in the band?
  - $31 + 21 = 52$
- How many students do not participate in organized sports?
  - $31 + 339 = 370$
- How many students participate in organized sports *or* play in the band? (Explain that *or* always includes the possibility of *both*.)
  - $442 + 21 + 31 = 494$

Connecting the numbers in the Venn diagram to probability questions is a focus of this lesson.

Example 1 (5 minutes): Shading Regions of a Venn Diagram

Here students are introduced to Venn diagrams and are shown the process of shading appropriate regions. Work through each part as a class.

Example 1: Shading Regions of a Venn Diagram

At a high school, some students play soccer, and some do not. Also, some students play basketball, and some do not. This scenario can be represented by a Venn diagram, as shown below. The circle labeled S represents the students who play soccer, the circle labeled B represents the students who play basketball, and the rectangle represents all the students at the school.

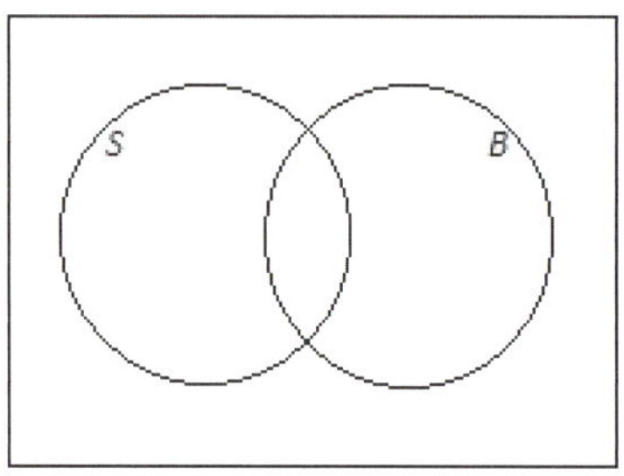

On the Venn diagrams provided, shade the region representing the following instances:

a. The students who play soccer

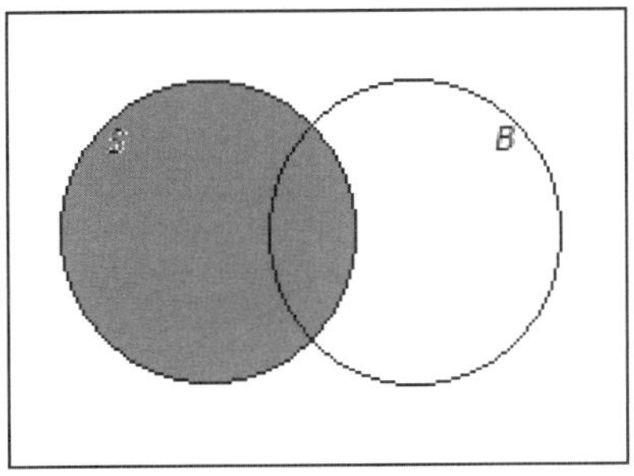

b. The students who do not play soccer

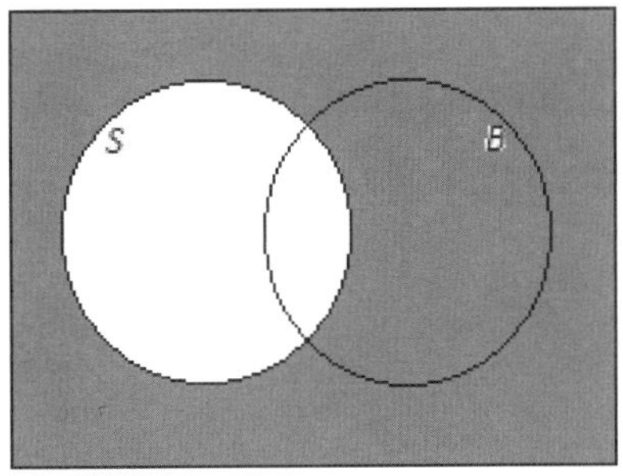

c. The students who play soccer and basketball

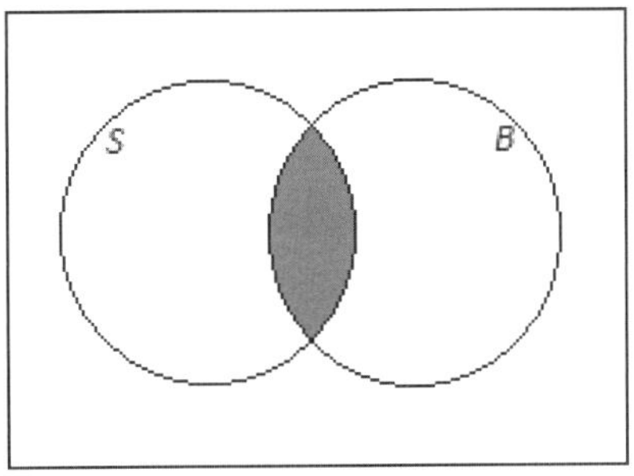

d. The students who play soccer or basketball

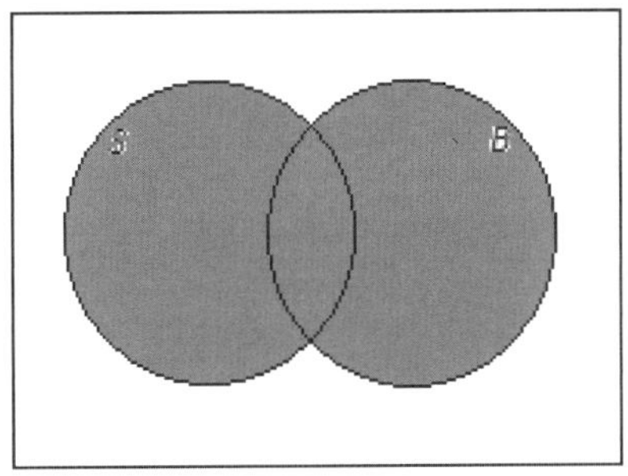

Exercise 1 (5 minutes)

Let students work individually shading regions of a Venn diagram. Then, allow them to compare answers with a neighbor.

Exercise 1

An online bookstore offers a large selection of books. Some of the books are works of fiction, and some are not. Also, some of the books are available as e-books, and some are not. Let F be the set of books that are works of fiction, and let E be the set of books that are available as e-books. On the Venn diagrams provided, shade the regions representing the following instances.

a. Books that are available as e-books

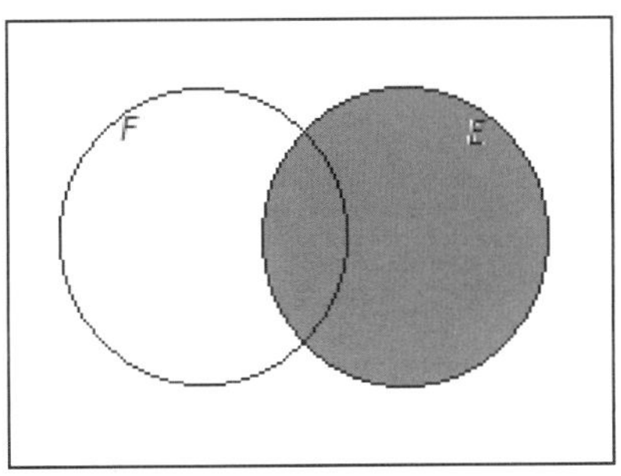

b. Books that are not works of fiction

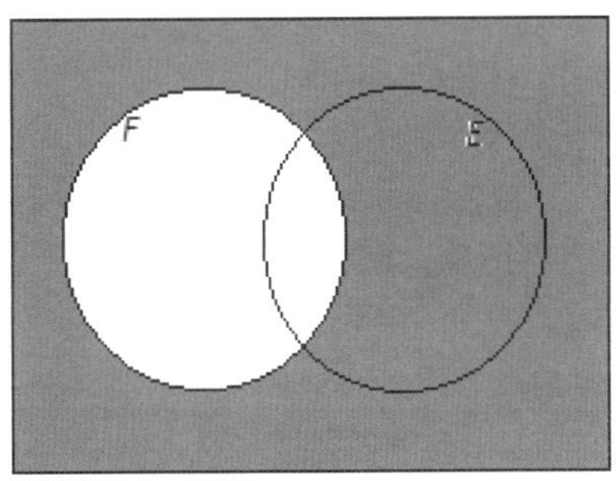

c. **Books that are works of fiction and available as e-books**

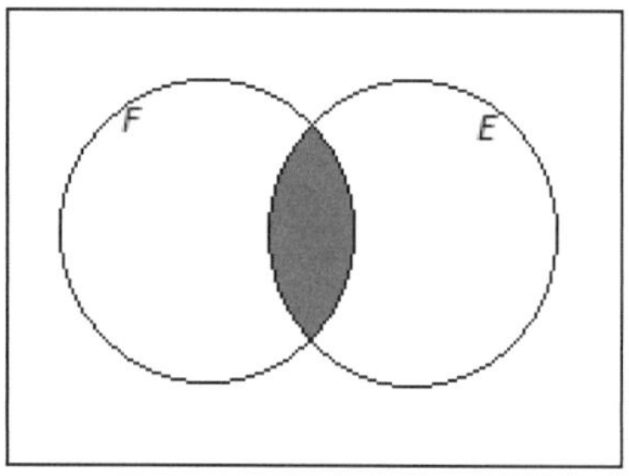

d. **Books that are works of fiction or available as e-books**

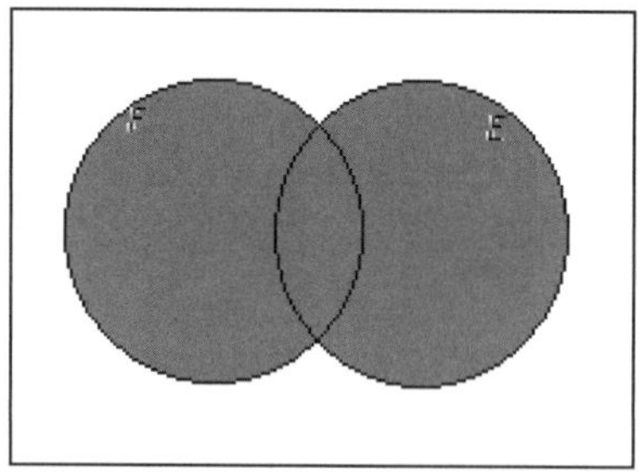

e. **Books that are neither works of fiction nor available as e-books**

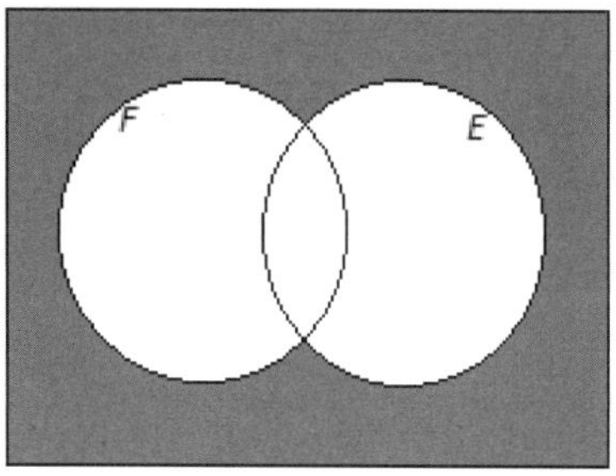

f. **Books that are works of fiction that are not available as e-books**

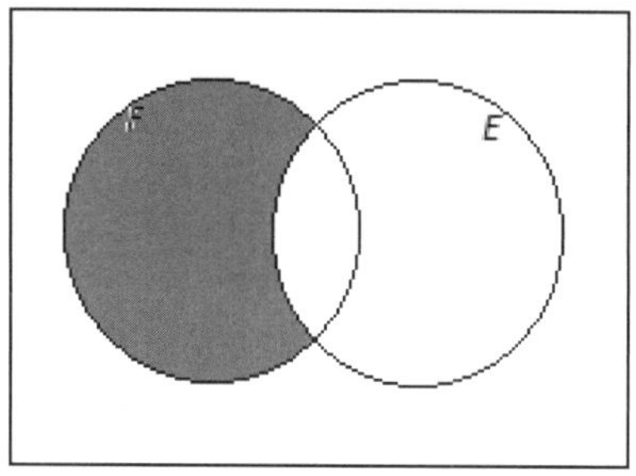

Example 2 (6 minutes): Showing Numbers of Possible Outcomes (and Probabilities) in a Venn Diagram

Students are introduced to the use of Venn diagrams to display numbers of possible outcomes and probabilities and how these numbers or probabilities can be added or subtracted.

Note that the technique of showing the number (or probability) associated with an entire circle in the Venn diagram uses a line drawn to the circle. This is the only way to add this information to the diagram without introducing confusion.

Each of the probabilities in the final part of the question is calculated by dividing the equivalent number in the first Venn diagram by 500. It is important to point out to students that the four probabilities in the diagram sum to 1, as this fact is used in the exercises that follow:

Example 2: Showing Numbers of Possible Outcomes (and Probabilities) in a Venn Diagram

Think again about the school introduced in Example 1. Suppose that 230 students play soccer, 190 students play basketball, and 60 students play both sports. There are a total of 500 students at the school.

a. **Complete the Venn diagram below by writing the numbers of students in the various regions of the diagram.**

Answer:

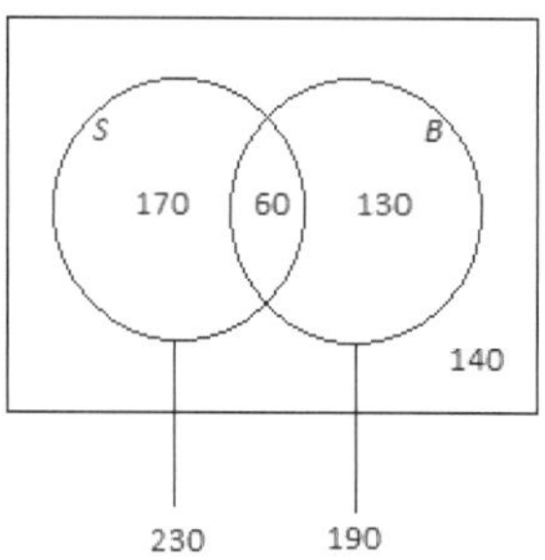

b. How many students play basketball but not soccer?

$190 - 60 = 130$

c. Suppose that a student will be selected at random from the school.

i. What is the probability that the selected student plays both sports?

$\frac{60}{500} = 0.12$

ii. Complete the Venn diagram below by writing the probabilities associated with the various regions of the diagram.

Answer:

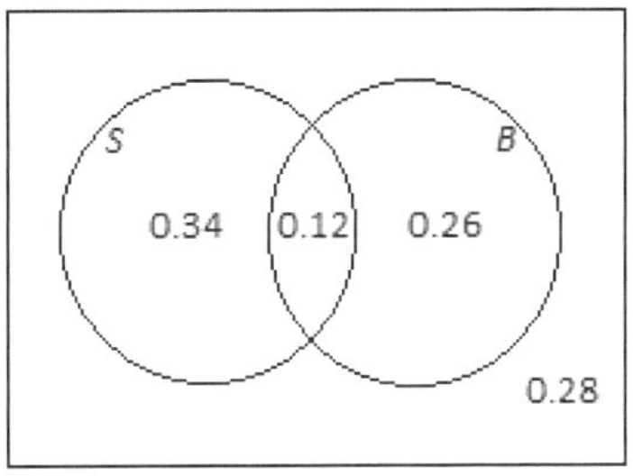

Example 3 (8 minutes): Adding and Subtracting Probabilities

Students are introduced to problems where probabilities (not counts) are given and to more challenging additions and subtractions than in Example 2. The proportions are given as percentages in the questions, but the solution should be expressed entirely in terms of decimals.

Part (b) is designed to demonstrate that students are doing the same work as in the previous lessons but expressing it in a different way (Venn diagrams instead of hypothetical 1000 tables). However, the process of transcribing the probabilities from the Venn diagram to the table is a relatively straightforward one, so this part of the question can be omitted if there is not enough time. Nonetheless, it is important that students be aware that both approaches are valid and ultimately lead to the same answer.

Example 3: Adding and Subtracting Probabilities

Think again about the online bookstore introduced in Exercise 1, and suppose that 62% of the books are works of fiction, 47% are available as e-books, and 14% are available as e-books but are not works of fiction. A book will be selected at random.

a. **Using a Venn diagram, find the following probabilities:**

i. **The book is a work of fiction and available as an e-book.**

$P(F \text{ and } E) = 0.33$

ii. **The book is neither a work of fiction nor available as an e-book.**

$P(\text{neither } F \text{ nor } E) = 0.24$

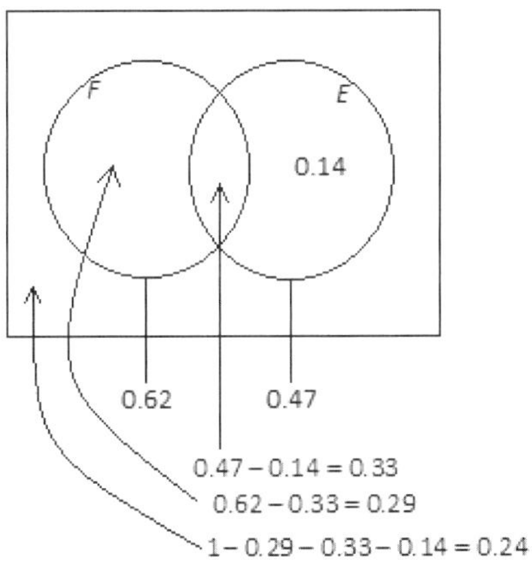

b. Return to the information given at the beginning of the question: 62% of the books are works of fiction, 47% are available as e-books, and 14% are available as e-books but are not works of fiction.

i. How would this information be shown in a hypothetical 1000 table? (Show your answers in the table provided below.)

	Fiction	Not Fiction	Total
Available as E-Book	330	140	470
Not Available as E-Book	290	240	530
Total	620	380	$1,000$

ii. Complete the hypothetical 1000 table given above.

iii. Complete the table below showing the probabilities of the events represented by the cells in the table.

	Fiction	Not Fiction	Total
Available as E-Book	0.33	0.14	0.47
Not Available as E-Book	0.29	0.24	0.53
Total	0.62	0.38	1

iv. How do the probabilities in your table relate to the probabilities you calculated in part (a)?

The probabilities are the same.

Scaffolding:

Consider asking students to generate their own tables rather than completing one that is already constructed.

Exercise 2 (5 minutes)

MP.2 Students practice the approaches introduced in the preceding examples. In this exercise, students are given an opportunity to reason abstractly and quantitatively by using a Venn diagram to represent the given information and answering questions about probability.

Let students work with a partner and confirm answers as a class. In this exercise, students could be permitted to omit part (c) if they might be short of time.

Exercise 2

When a fish is selected at random from a tank, the probability that it has a green tail is 0.64, the probability that it has red fins is 0.25, and the probability that it has both a green tail and red fins is 0.19.

a. Draw a Venn diagram to represent this information.

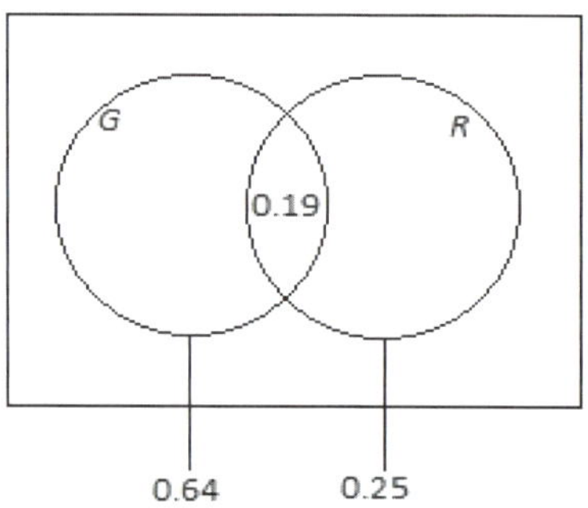

b. Find the following probabilities:

i. The fish has red fins but does not have a green tail.

$0.25 - 0.19 = 0.06$

ii. The fish has a green tail but not red fins.

$0.64 - 0.19 = 0.45$

iii. The fish has neither a green tail nor red fins.

$1 - 0.45 - 0.19 - 0.06 = 0.30$

c. Complete the table below showing the probabilities of the events corresponding to the cells of the table.

	Green Tail	Not Green Tail	Total
Red Fins	0.19	0.06	0.25
Not Red Fins	0.45	0.30	0.75
Total	0.64	0.36	1.00

Exercise 3 (5 minutes)

This exercise is slightly more challenging and provides additional practice with the process of adding and subtracting probabilities using a Venn diagram. The proportions are given in the question as percentages, but the solution should be expressed entirely in terms of decimals, as probabilities are typically expressed on a scale from 0 to 1.

Let students continue to work with a partner and then confirm the answer as a class.

Exercise 3

In a company, 43% of the employees have access to a fax machine, 38% have access to a fax machine and a scanner, and 24% have access to neither a fax machine nor a scanner. Suppose that an employee will be selected at random. Using a Venn diagram, calculate the probability that the randomly selected employee will not have access to a scanner. (Note that Venn diagrams and probabilities use decimals or fractions, not percentages.) Explain how you used the Venn diagram to determine your answer.

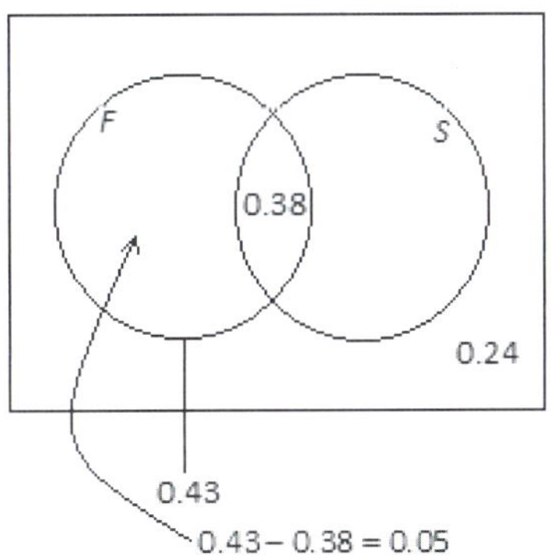

$P(\text{not } S) = 0.05 + 0.24 = 0.29$

I can see from the Venn diagram that 5% of the employees have access to a fax machine but not a scanner, and 24% do not have access to either. So, I combined the two to find the probability that a randomly selected employee will not have access to a scanner.

Closing (2 minutes)

If time allows, consider introducing the mathematical symbols for *and*, *or*, and *not*. This terminology and the notation for intersections, unions, and complements are introduced in Problem 4 of the Problem Set, but if there is time, consider doing this problem in class.

- The *intersection* of the set A and the set B is written as $A \cap B$ and is read as A intersect B. It consists of the elements that are in both A and B.
- The *union* of the set A and the set B is written as $A \cup B$ and is read as A union B. It consists of the elements that are in either A or B or both.
- The *complement* of the set A is written as A^C and is read as A complement. It consists of the elements that are not in A. Other common notations for the A complement are A' and $\bar{A}$.

Note that it may be useful for students to have access to these symbols in a graphic organizer or as a visual on a poster. This should include the symbol, what it is called, and what it means.

Ask students to summarize the main ideas of the lesson in writing or with a neighbor. Use this as an opportunity to informally assess comprehension of the lesson. The Lesson Summary below offers some important concepts that should be included.

> **Lesson Summary**
>
> **In a probability experiment, the events can be represented by circles in a Venn diagram.**
>
> **Combinations of events using *and*, *or*, and *not* can be shown by shading the appropriate regions of the Venn diagram.**
>
> **The number of possible outcomes can be shown in each region of the Venn diagram; alternatively, probabilities may be shown. The number of outcomes in a given region (or the probability associated with it) can be calculated by adding or subtracting the known numbers of possible outcomes (or probabilities).**

Exit Ticket (5 minutes)

Name ______________________________ Date ______________

Lesson 5: Events and Venn Diagrams

Exit Ticket

1. At a high school, some students take Spanish, and some do not. Also, some students take an arts subject, and some do not. Let S be the set of students who take Spanish and A be the set of students who take an arts subject. On the Venn diagrams given, shade the region representing the following instances:

   a. Students who take Spanish and an arts subject

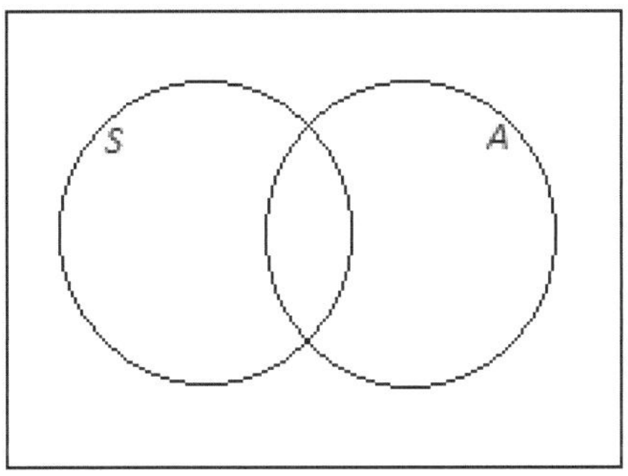

   b. Students who take Spanish or an arts subject

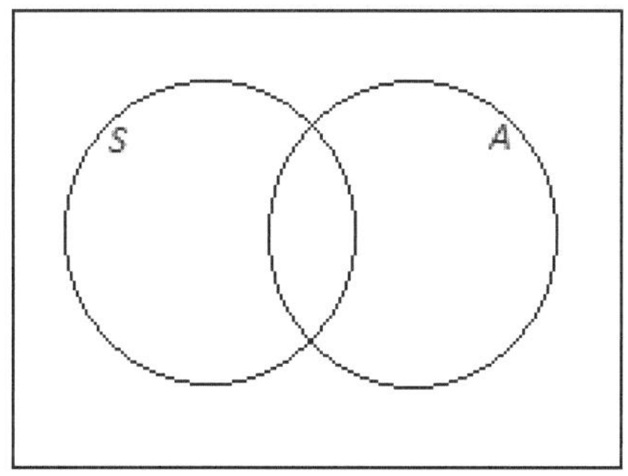

   c. Students who take Spanish but do not take an arts subject

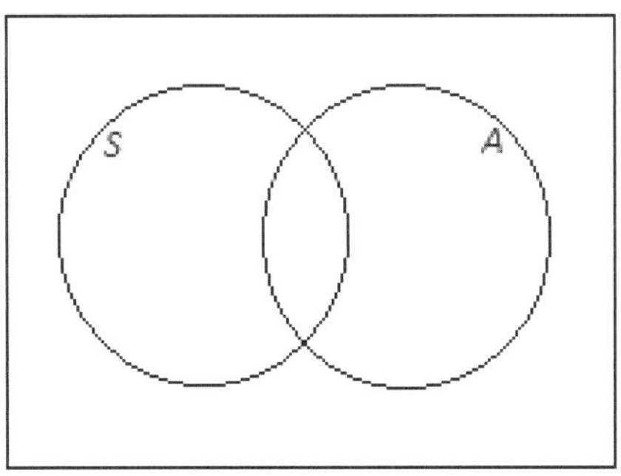

   d. Students who do not take an arts subject

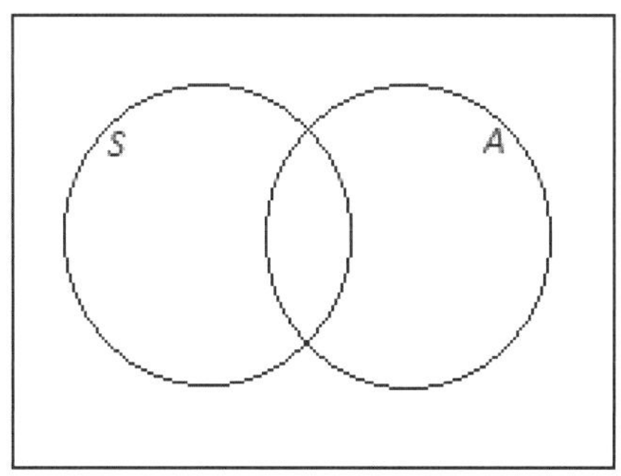

2. When a player is selected at random from a high school boys' baseball team, the probability that he is a pitcher is 0.35, the probability that he is right-handed is 0.79, and the probability that he is a right-handed pitcher is 0.26. Let P be the event that a player is a pitcher, and let R be the event that a player is right-handed. A Venn diagram is provided below.

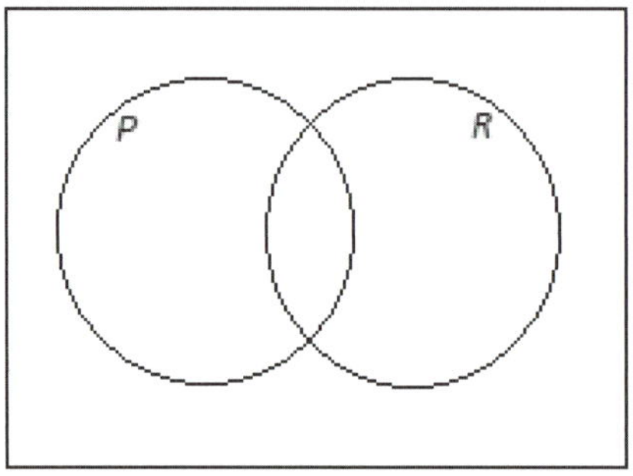

Use the Venn diagram to calculate the probability that a randomly selected player is each of the following. Explain how you used the Venn diagram to determine your answer.

a. Right-handed but not a pitcher

b. A pitcher but not right-handed

c. Neither right-handed nor a pitcher

Exit Ticket Sample Solutions

1. At a high school, some students take Spanish, and some do not. Also, some students take an arts subject, and some do not. Let S be the set of students who take Spanish and A be the set of students who take an arts subject. On the Venn diagrams given, shade the region representing the following instances:

   a. Students who take Spanish and an arts subject

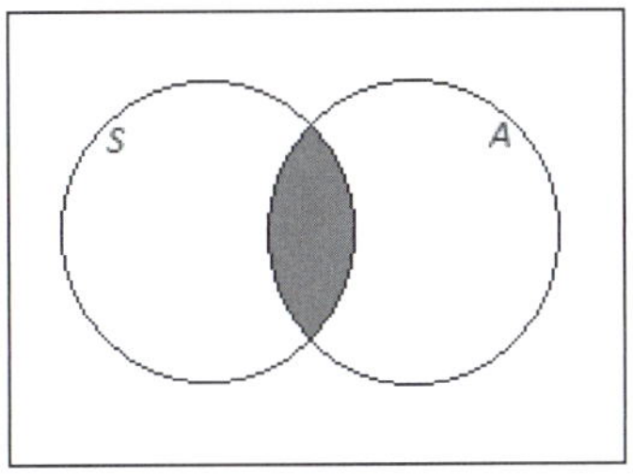

   b. Students who take Spanish or an arts subject

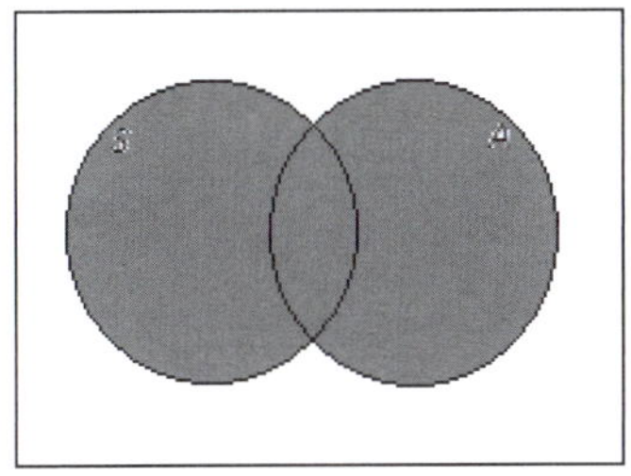

   c. Students who take Spanish but do not take an arts subject

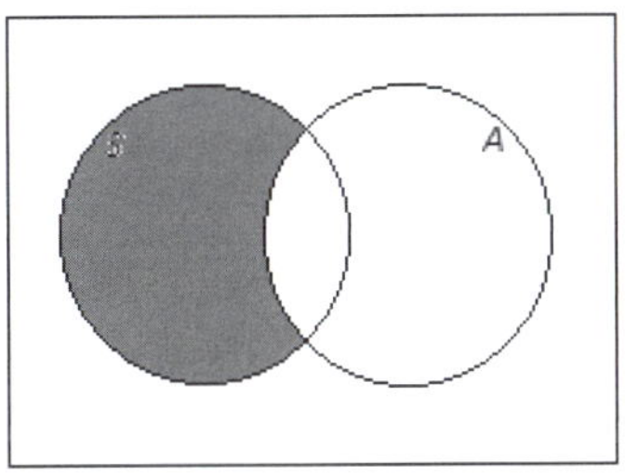

   d. Students who do not take an arts subject

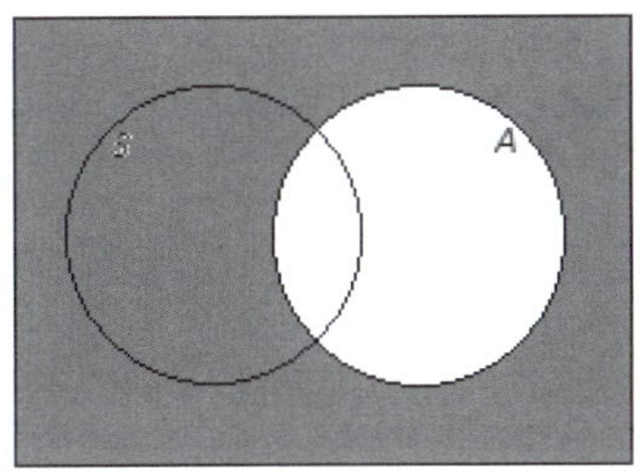

2. When a player is selected at random from a high school boys' baseball team, the probability that he is a pitcher is 0.35, the probability that he is right-handed is 0.79, and the probability that he is a right-handed pitcher is 0.26. Let P be the event that a player is a pitcher, and let R be the event that a player is right-handed. A Venn diagram is provided below.

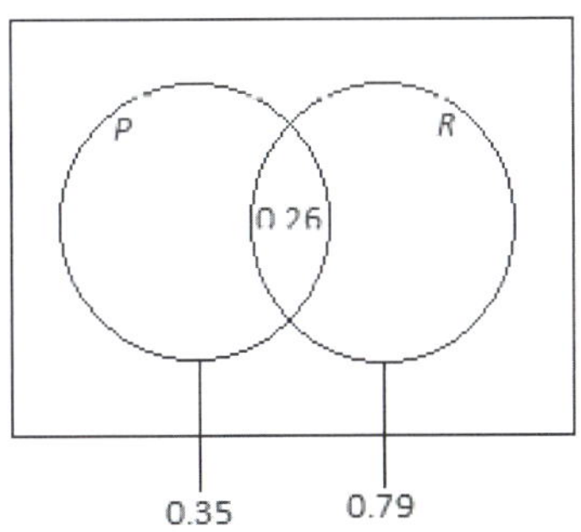

   Use the Venn diagram to calculate the probability that a randomly selected player is each of the following. Explain how you used the Venn diagram to determine your answer.

   a. Right-handed but not a pitcher

   $0.79 - 0.26 = 0.53$

   I found the probability that the player was right-handed (0.79) *and subtracted the probability that the player was also a pitcher* (0.26).

b. **A pitcher but not right-handed**

$0.35 - 0.26 = 0.09$

I found the probability that the player was a pitcher (0.35) *and subtracted the probability that the player was also right-handed* (0.26).

c. **Neither right-handed nor a pitcher**

$1 - 0.09 - 0.26 - 0.53 = 0.12$

I used the Venn diagram to determine all of the probabilities associated with the player being right-handed and/or a pitcher. Then, I subtracted each from 1 *to find the probability that the player was not either.*

Problem Set Sample Solutions

1. **On a flight, some of the passengers have frequent-flier status, and some do not. Also, some of the passengers have checked baggage, and some do not. Let the set of passengers who have frequent-flier status be F and the set of passengers who have checked baggage be C. On the Venn diagrams provided, shade the regions representing the following instances:**

a. **Passengers who have frequent-flier status and have checked baggage**

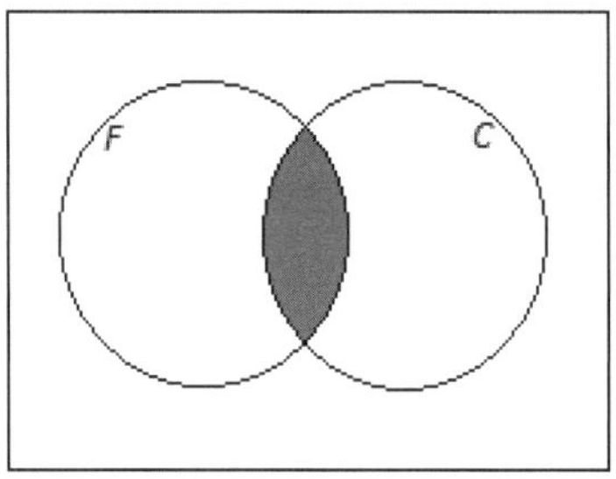

b. **Passengers who have frequent-flier status or have checked baggage**

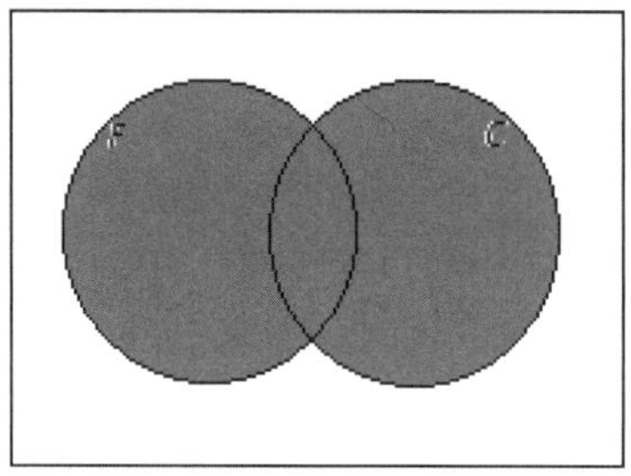

c. **Passengers who do not have both frequent-flier status and checked baggage**

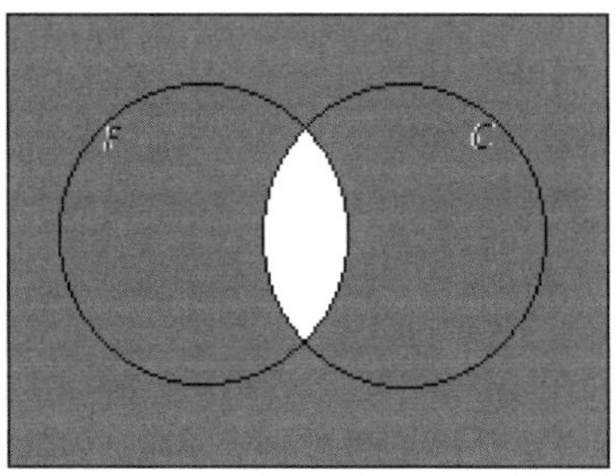

d. **Passengers who have frequent-flier status or do not have checked baggage**

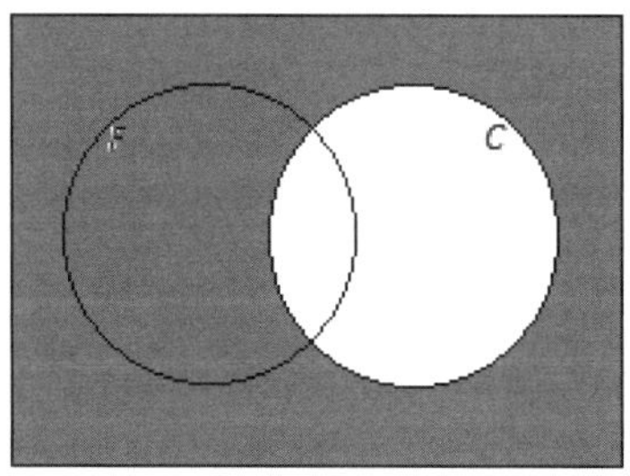

2. For the scenario introduced in Problem 1, suppose that, of the 400 people on the flight, 368 have checked baggage, 228 have checked baggage but do not have frequent-flier status, and 8 have neither frequent-flier status nor checked baggage.

a. Using a Venn diagram, calculate the following:

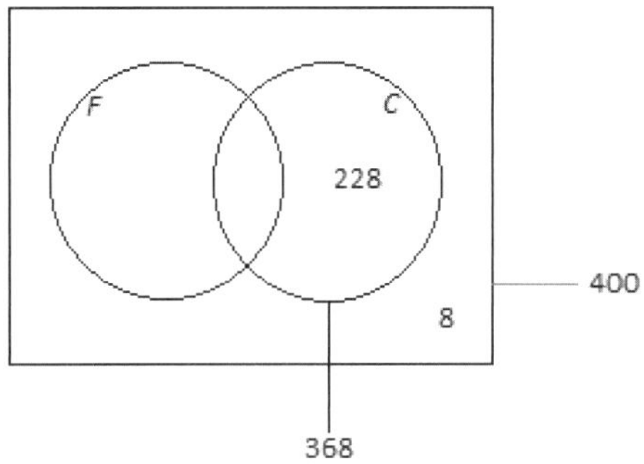

i. The number of people on the flight who have frequent-flier status and have checked baggage

Number of passengers with frequent-flier status and checked baggage:

$$368 - 228 = 140$$

ii. The number of people on the flight who have frequent-flier status

Number of passengers with frequent-flier status:

$$400 - 8 - 228 = 164$$

b. In the Venn diagram provided below, write the probabilities of the events associated with the regions marked with a star (*).

Answer:

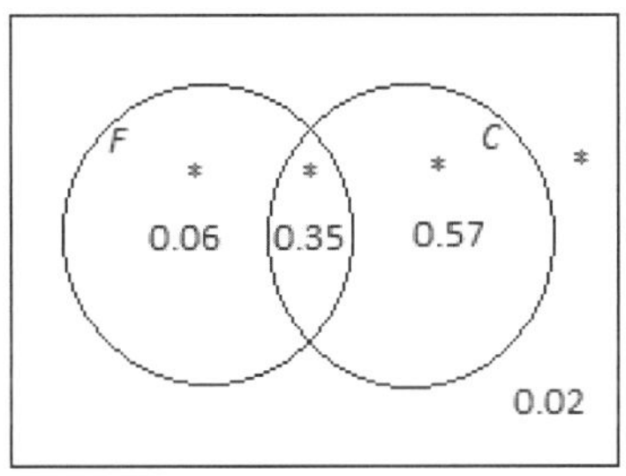

3. When an animal is selected at random from those at a zoo, the probability that it is North American (meaning that its natural habitat is in the North American continent) is 0.65, the probability that it is both North American and a carnivore is 0.16, and the probability that it is neither American nor a carnivore is 0.17.

a. Using a Venn diagram, calculate the probability that a randomly selected animal is a carnivore.

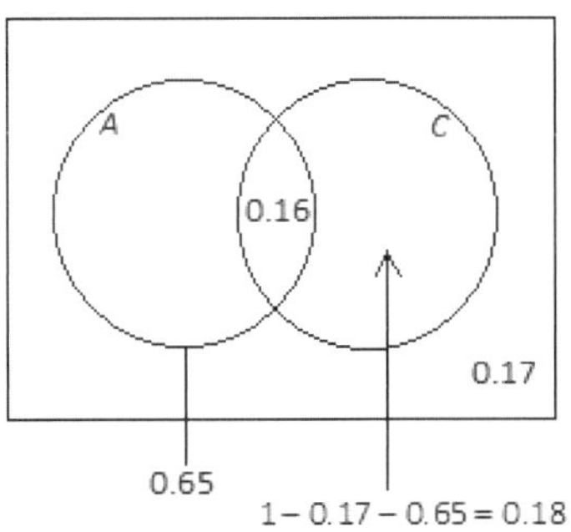

$$P(C) = 0.16 + 0.18 = 0.34$$

b. Complete the table below showing the probabilities of the events corresponding to the cells of the table.

	North American	Not North American	Total
Carnivore	0.16	0.18	0.34
Not Carnivore	0.49	0.17	0.66
Total	0.65	0.35	1.00

4. This question introduces the mathematical symbols for *and*, *or*, and *not*.

Considering all the people in the world, let A be the set of Americans (citizens of the United States), and let B be the set of people who have brothers.

- The set of people who are Americans and have brothers is represented by the shaded region in the Venn diagram below.

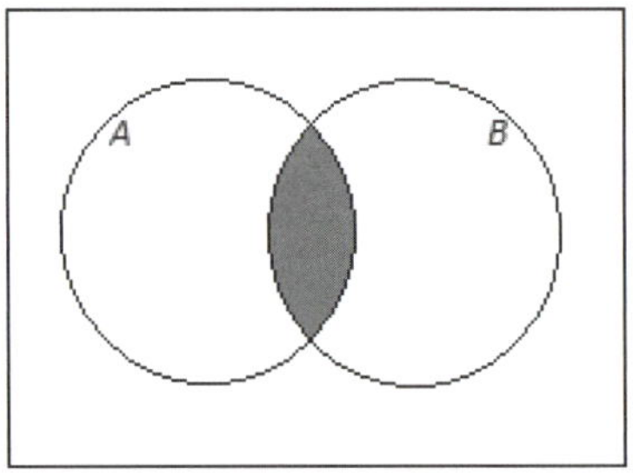

This set is written $A \cap B$ (read A intersect B), and the probability that a randomly selected person is American and has a brother is written $P(A \cap B)$.

- The set of people who are Americans or have brothers is represented by the shaded region in the Venn diagram below.

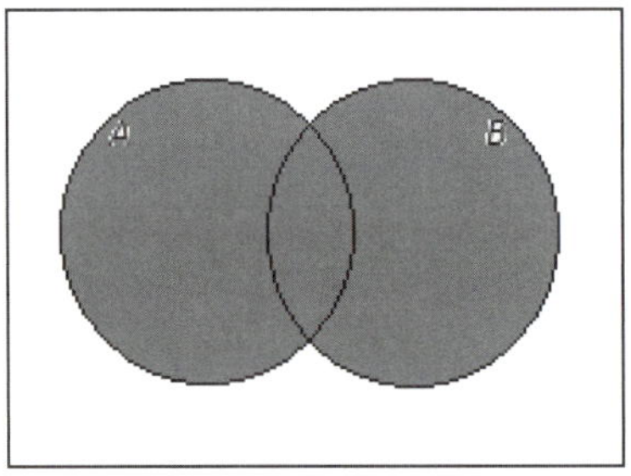

This set is written $A \cup B$ (read A union B), and the probability that a randomly selected person is American or has a brother is written $P(A \cup B)$.

- The set of people who are not Americans is represented by the shaded region in the Venn diagram below.

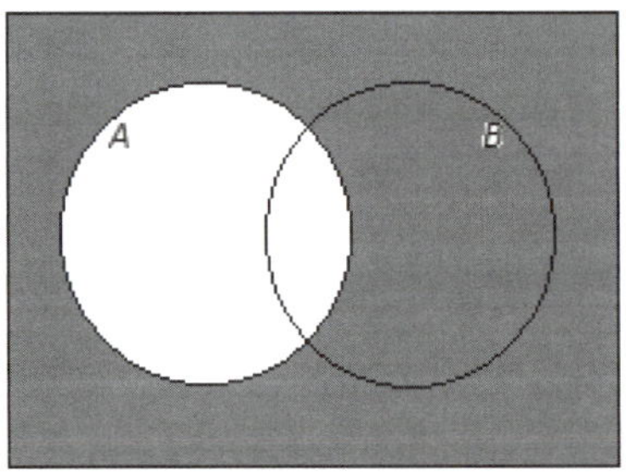

This set is written A^C (read A complement), and the probability that a randomly selected person is not American is written $P(A^C)$.

Now, think about the cars available at a dealership. Suppose a car is selected at random from the cars at this dealership. Let the event that the car has manual transmission be denoted by M, and let the event that the car is a sedan be denoted by S. The Venn diagram below shows the probabilities associated with four of the regions of the diagram.

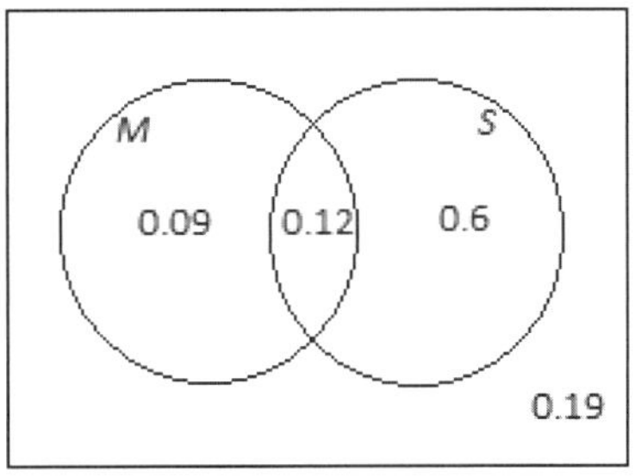

a. **What is the value of $P(M \cap S)$?**

0.12

b. **Complete this sentence using *and* or *or*:**

$P(M \cap S)$ is the probability that a randomly selected car has a manual transmission ___*and*___ **is a sedan.**

c. **What is the value of $P(M \cup S)$?**

$0.09 + 0.12 + 0.60 = 0.81$

d. **Complete this sentence using *and* or *or*:**

$P(M \cup S)$ is the probability that a randomly selected car has a manual transmission ___*or*___ **is a sedan.**

e. **What is the value of $P(S^C)$?**

$1 - (0.6 + 0.12) = 0.28$

f. **Explain the meaning of $P(S^C)$.**

$P(S^C)$ is the probability that a randomly selected car is not a sedan.

Lesson 6: Probability Rules

Student Outcomes

- Students use the complement rule to calculate the probability of the complement of an event and the multiplication rule for independent events to calculate the probability of the intersection of two independent events.
- Students recognize that two events A and B are independent if and only if $P(A \text{ and } B) = P(A)P(B)$ and interpret independence of two events A and B as meaning that the conditional probability of A given B is equal to $P(A)$.
- Students use the formula for conditional probability to calculate conditional probabilities and interpret probabilities in context.

Lesson Notes

This lesson introduces the formulas for calculating the probability of the complement of an event, the probability of an intersection when events are independent, and conditional probabilities. Because these concepts have already been presented using the more intuitive approach of earlier lessons, students should readily understand why the formulas are true.

Classwork

Opening (4 minutes)

Begin this lesson by asking students the following questions:

- There are 300 students at a certain school. All students indicated they were either right-handed or left-handed but not both. Fifty of the students are left-handed. How many students are right-handed? What is the probability of randomly selecting a right-handed student at this school? How did you determine this?
- The United States Census Bureau indicates there are $19{,}378{,}102$ people in the State of New York. If $4{,}324{,}929$ are under the age of 18, how many are 18 and older? What is the probability of randomly selecting a person from New York who is 18 or older? How did you determine this?
- It is estimated that approximately 20% of the people in the United States have asthma and severe allergies. What is the probability that a randomly selected person does not have asthma or a severe allergy? How did you determine this?

Scaffolding:

For students who are struggling with this concept, consider displaying a visual of the numerical representation of each problem (e.g., $300 - 50$, $19\,378\,102 - 4\,324\,929$, $100\% - 20\%$) and asking them, "What do all of these expressions have in common?"

Discuss with students how these examples, or similar examples, involve finding the probability of the *complement of a given event.* Work with students in deriving the probabilities by thinking of the probability of people who are not in the given probability and connecting that probability to the fact that the two probabilities (or the probability of the given event and the probability of the complement) add up to 1.00 or 100%. Although students have previously worked with this idea, this lesson formalizes their understanding of complement and the probability of complementary events.

MP.8 Use this as an opportunity to have students look for a general method or rule for determining the probability of the complement of an event. Encourage students to try to write a rule, and ask them to share it with their neighbor.

- Based on the three examples we looked at, what is a general rule we could write to determine the probability of the complement of an event?

Scaffolding:

- Clarify that the term *complement* is not the same as *compliment*. Point out the difference in spelling between the terms, as well as their meanings.
- The *complement* of any event is the event that does not occur.
- A *compliment* is an expression of respect or admiration.
- Consider using visual displays and repeated choral readings to reinforce the mathematical meaning of the word.
- A Frayer model may also be used.

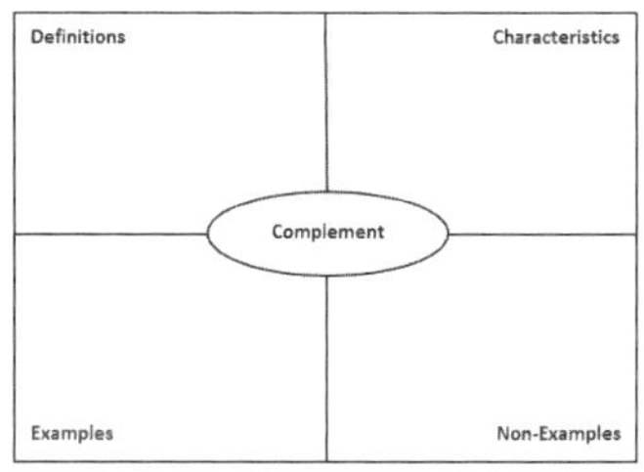

Example 1 (3 minutes): The Complement Rule

At this point, students have informally used the complement rule, but in this example it is given as a formula. The example serves as a quick illustration of the rule. Read through the example as a class, and then give students a moment to calculate the probability presented at the end of the example.

Example 1: The Complement Rule

In previous lessons, you have seen that to calculate the probability that an event *does not happen,* you can subtract the probability of the event from 1. If the event is denoted by A, then this rule can be written:

$$P(\text{not } A) = 1 - P(A).$$

For example, suppose that the probability that a particular flight is on time is 0.78. What is the probability that the flight is not on time?

$$P(\text{not on time}) = 1 - 0.78 = 0.22$$

Example 2 (6 minutes): Formula for Conditional Probability

The purpose of this example is to introduce the formula for conditional probability. A conditional probability is first calculated using a hypothetical 1000 table (as in previous lessons), and then the formula is shown to produce the same result.

When working through part (c) with the class, it would be helpful to illustrate the division with the aid of a Venn diagram so that students get a visual idea of what is being divided by what. (The probability of the intersection is being divided by the probability of B.)

Consider asking students to try solving the entire problem or parts and then discussing the results as a class.

Scaffolding:

- In addition to using Venn diagrams, if students have trouble identifying the intersection of two events, point out real-world examples like intersecting streets to connect students to the concept.
- Students may confuse the term *conditional* with *conditioner* or *condition*. Point out the difference in spelling and meaning.
- Again, the use of visuals and repeated choral readings helps reinforce these words.

Additionally, point out that $\frac{0.38}{0.43}$ is exactly the same thing as $\frac{380}{430}$ (the numerator and the denominator have been divided by $1{,}000$). By seeing this, students should see why the conditional probability formula is valid. (This is the main point of this example.)

Example 2: Formula for Conditional Probability

When a room is randomly selected in a downtown hotel, the probability that the room has a king-sized bed is 0.62, the probability that the room has a view of the town square is 0.43, and the probability that it has a king-sized bed *and* a view of the town square is 0.38. Let A be the event that the room has a king-sized bed, and let B be the event that the room has a view of the town square.

a. **What is the meaning of $P(A \text{ given } B)$ in this context?**

$P(A \text{ given } B)$ is the probability that a room known to have a view of the town square also has a king-sized bed.

b. **Use a hypothetical 1000 table to calculate $P(A \text{ given } B)$.**

	A (room has a king-sized bed)	Not A (room does not have a king-sized bed)	Total
B (room has a view of the town square)	380	50	430
Not B (room does not have a view of the town square)	240	330	570
Total	620	380	1,000

$$P(A \text{ given } B) = \frac{380}{430} \approx 0.884$$

c. **There is also a formula for calculating a conditional probability. The formula for conditional probability is**

$$P(A \text{ given } B) = \frac{P(A \text{ and } B)}{P(B)}.$$

Use this formula to calculate $P(A \text{ given } B)$, where the events A and B are as defined in this example.

$$P(A \text{ given } B) = \frac{P(A \text{ and } B)}{P(B)} = \frac{0.38}{0.43} \approx 0.884$$

d. **How does the probability you calculated using the formula compare to the probability you calculated using the hypothetical 1000 table?**

The probabilities are the same.

Exercise 1 (15 minutes)

This exercise provides practice using the conditional probability formula. Additionally, in part (e), students are asked to compare a conditional and an unconditional probability and to provide an interpretation. In part (f), students are asked to recall the definition of independence in terms of equality of the conditional and unconditional probabilities.

Let students work with a partner and then confirm answers as a class, spending about 10 minutes total on the exercise. After confirming the answer to part (f), present the multiplication rule for independent events (approximately 5 minutes). Use this as an opportunity to informally assess student understanding of the lesson's outcomes.

Exercise 1

A credit card company states that 42% of its customers are classified as long-term cardholders, 35% pay their bills in full each month, and 23% are long-term cardholders who also pay their bills in full each month. Let the event that a randomly selected customer is a long-term cardholder be L and the event that a randomly selected customer pays his bill in full each month be F.

a. **What are the values of $P(L)$, $P(F)$, and $P(L \text{ and } F)$?**

$P(L) = 0.42$, $P(F) = 0.35$, $P(L \text{ and } F) = 0.23$

b. **Draw a Venn diagram, and label it with the probabilities from part (a).**

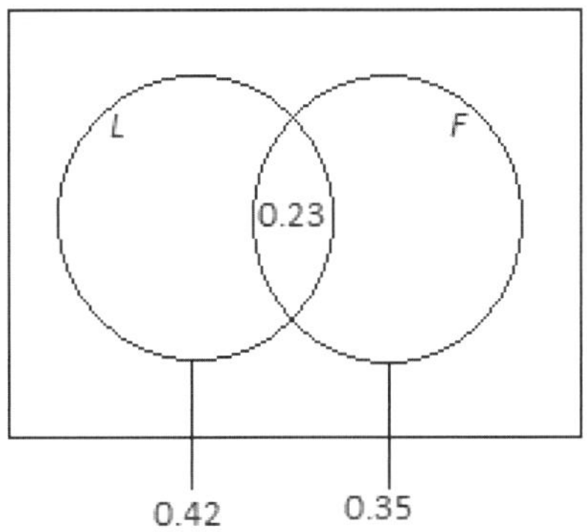

c. **Use the conditional probability formula to calculate $P(L \text{ given } F)$. (Round your answer to the nearest thousandth.)**

$$P(L \text{ given } F) = \frac{P(L \text{ and } F)}{P(F)} = \frac{0.23}{0.35} \approx 0.657$$

d. **Use the conditional probability formula to calculate $P(F \text{ given } L)$. (Round your answer to the nearest thousandth.)**

$$P(F \text{ given } L) = \frac{P(F \text{ and } L)}{P(L)} = \frac{0.23}{0.42} \approx 0.548$$

e. **Which is greater, $P(F \text{ given } L)$ or $P(F)$? Explain why this is relevant.**

$P(F \text{ given } L) \approx 0.548$, and $P(F) = 0.35$; therefore, $P(F \text{ given } L)$ is larger than $P(F)$. This tells us that long-term cardholders are more likely to pay their bills in full each month than customers in general.

f. **Remember that two events A and B are said to be independent if $P(A \text{ given } B) = P(A)$. Are the events F and L independent? Explain.**

Note that there are several ways to answer this question. Here are three possibilities:

No, because $P(F \text{ given } L) \neq P(F)$.

No, because $P(L \text{ given } F) \neq P(L)$.

No, because $(L \text{ and } F) \neq P(L)P(F)$.

After confirming answers to part (f), introduce the multiplication rule for independent events.

- Events A and B are independent if and only if $P(A \text{ and } B) = P(A)P(B)$.

This means that if we know the events A and B are independent, then we can conclude that $P(A \text{ and } B) = P(A)P(B)$, and if we know that $P(A \text{ and } B) = P(A)P(B)$, then we can also conclude that the events A and B are independent. Consider asking students to state the meaning of this rule in their own words. The easiest way to explain this is as follows:

> *Scaffolding:*
>
> For students working above grade level, consider asking the following:
> "Explain why this statement makes sense using conditional probabilities."
>
> (The table below offers a sample explanation.)

If the events A and B are independent, then we know:	$P(A \text{ given } B) = P(A)$
Use the formula for conditional probability to replace $P(A \text{ given } B)$: $$P(A \text{ given } B) = \frac{P(A \text{ and } B)}{P(B)}$$	$\frac{P(A \text{ and } B)}{P(B)} = P(A)$
Now, isolate $P(A \text{ and } B)$ and conclude that:	$P(A \text{ and } B) = P(A)P(B)$

Example 3 (5 minutes): Using the Multiplication Rule for Independent Events

This is an example of the use of the multiplication rule for independent events. There are two ways of telling whether events are independent. Either it is obvious from the description of the problem (as in part (a)), or the question tells students that the events are independent (as in part (b)). Consider asking students to attempt to solve independently or with a neighbor, informally assessing understanding and offering guidance as necessary. When tackling part (a):

- Explain to students that the result for the number cube cannot possibly affect the result for the coin, and so the two events are independent.
- Explain that the result of any one roll of the number cube cannot have an effect on the results of the other two rolls.

Example 3: Using the Multiplication Rule for Independent Events

A number cube has faces numbered 1 through 6, and a coin has two sides, heads and tails.

The number cube will be rolled, and the coin will be flipped. Find the probability that the cube shows a 4 and the coin lands on heads. Because the events are independent, we can use the multiplication rules we just learned.

$$\left(\frac{1}{6}\right)\left(\frac{1}{2}\right) = \frac{1}{12}$$

If you toss the coin five times, what is the probability you will see a head on all five tosses?

$(0.5)(0.5)(0.5)(0.5)(0.5) = 0.03125$

If you tossed the coin five times and got five heads, would you think that this coin is a fair coin? Why or why not?

Although getting five heads is possible (about 3% of the time you would expect this), it is not likely; therefore, you would suspect that the coin is not fair.

If you roll the number cube three times, what is the probability that it will show 4 on all three throws?

$\left(\frac{1}{6}\right)\left(\frac{1}{6}\right)\left(\frac{1}{6}\right) = \frac{1}{216} \approx 0.005$

If you rolled the number cube three times and got a 4 on all three rolls, would you think that this number cube is fair? Why or why not?

The probability of getting a 4 on all three rolls is very small. As a result, you suspect the number cube is not fair.

Suppose that the credit card company introduced in Exercise 1 states that when a customer is selected at random, the probability that the customer pays his bill in full each month is 0.35, the probability that the customer makes regular online purchases is 0.83, and these two events are independent. What is the probability that a randomly selected customer pays his bill in full each month *and* makes regular online purchases?

$(0.35)(0.83) = 0.2905$

Exercise 2 (5 minutes)

MP.4

This exercise provides practice with use of the multiplication rule for independent events. Students should think about *why* the events given are independent (in this case, because it is obvious from the problem description). Particularly in part (c), students have an opportunity to interpret their results in the context of the question and to reflect on whether the results make sense. Consider asking students to work independently on this exercise.

Exercise 2

A spinner has a pointer, and when the pointer is spun, the probability that it stops in the red section of the spinner is 0.25.

a. **If the pointer is spun twice, what is the probability that it will stop in the red section on both occasions?**

$(0.25)(0.25) = 0.0625$

b. **If the pointer is spun four times, what is the probability that it will stop in the red section on all four occasions? (Round your answer to the nearest thousandth.)**

$(0.25)(0.25)(0.25)(0.25) \approx 0.004$

c. **If the pointer is spun five times, what is the probability that it never stops on red? (Round your answer to the nearest thousandth.)**

$(0.75)(0.75)(0.75)(0.75)(0.75) \approx 0.237$

Closing (2 minutes)

Ask students to summarize the main ideas of the lesson in writing or with a neighbor. Use this as an opportunity to informally assess comprehension of the lesson. The Lesson Summary below offers some important ideas that should be included.

- Does it make sense to consider the two events in Example 3, part (b), to be independent? (Events are that customer "pays bill in full each month" and "makes regular online purchases.")
  - *Yes, it seems feasible that long-term cardholders might be as likely to make regular online purchases as customers in general.*

> **Lesson Summary**
>
> For any event A, $P(\text{not } A) = 1 - P(A)$.
>
> For any two events A and B, $P(A \text{ given } B) = \frac{P(A \text{ and } B)}{P(A)}$.
>
> Events A and B are independent if and only if $P(A \text{ and } B) = P(A)P(B)$.

Exit Ticket (5 minutes)

Name ______________________________ Date ______________

Lesson 6: Probability Rules

Exit Ticket

1. Of the light bulbs available at a store, 42% are fluorescent, 23% are labeled as long life, and 12% are fluorescent *and* long life.
   a. A light bulb will be selected at random from the light bulbs at this store. Rounding your answer to the nearest thousandth where necessary, find the probability that
      i. The selected light bulb is not fluorescent.

      ii. The selected light bulb is fluorescent given that it is labeled as long life.

   b. Are the events "fluorescent" and "long life" independent? Explain.

2. When a person is selected at random from a very large population, the probability that the selected person is right-handed is 0.82. If three people are selected at random, what is the probability that
   a. They are all right-handed?

   b. None of them is right-handed?

Exit Ticket Sample Solutions

1. Of the light bulbs available at a store, 42% are fluorescent, 23% are labeled as long life, and 12% are fluorescent *and* long life.

   a. A light bulb will be selected at random from the light bulbs at this store. Rounding your answer to the nearest thousandth where necessary, find the probability that

      i. The selected light bulb is not fluorescent.

      $1 - 0.42 = 0.58$

      ii. The selected light bulb is fluorescent given that it is labeled as long life.

      $$P(\text{fluorescent given long life}) = \frac{P(\text{fluorescent and long life})}{P(\text{long life})} = \frac{0.12}{0.23} \approx 0.522$$

   b. Are the events "fluorescent" and "long life" independent? Explain.

   No. $P(\text{fluorescent given long life}) \neq P(\text{fluorescent})$

2. When a person is selected at random from a very large population, the probability that the selected person is right-handed is 0.82. If three people are selected at random, what is the probability that

   a. They are all right-handed?

   $(0.82)(0.82)(0.82) \approx 0.551$

   b. None of them is right-handed?

   $(0.18)(0.18)(0.18) \approx 0.006$

Problem Set Sample Solutions

1. When an avocado is selected at random from those delivered to a food store, the probability that it is ripe is 0.12, the probability that it is bruised is 0.054, and the probability that it is ripe and bruised is 0.019.

   a. Rounding your answers to the nearest thousandth where necessary, find the probability that an avocado randomly selected from those delivered to the store is

      i. Not bruised.

      $1 - 0.054 = 0.946$

      ii. Ripe given that it is bruised.

      $$P(\text{ripe given bruised}) = \frac{P(\text{ripe and bruised})}{P(\text{bruised})} = \frac{0.019}{0.054} \approx 0.352$$

      iii. Bruised given that it is ripe.

      $$P(\text{bruised given ripe}) = \frac{P(\text{bruised and ripe})}{P(\text{ripe})} = \frac{0.019}{0.12} \approx 0.158$$

   b. Which is larger, the probability that a randomly selected avocado is bruised given that it is ripe or the probability that a randomly selected avocado is bruised? Explain in words what this tells you.

   $P(\text{bruised given ripe}) = 0.158$ *and* $P(\text{bruised}) = 0.054$*. Therefore* $P(\text{bruised given ripe})$ *is greater than* $P(\text{bruised})$*, which tells you that ripe avocados are more likely to be bruised than avocados in general.*

c. Are the events "ripe" and "bruised" independent? Explain.

No, because $P(\text{bruised given ripe})$ *is different from* $P(\text{bruised})$.

2. Return to the probability information given in Problem 1. Complete the hypothetical 1000 table given below, and use it to find the probability that a randomly selected avocado is bruised given that it is not ripe. (Round your answer to the nearest thousandth.)

	Ripe	Not Ripe	Total
Bruised	19	35	54
Not Bruised	101	845	946
Total	120	880	1,000

$P(\text{bruised given not ripe}) = \frac{35}{880} \approx 0.040$

3. According to the U.S. census website (www.census.gov), based on the U.S. population in 2010, the probability that a randomly selected man is 65 or older is 0.114, and the probability that a randomly selected woman is 65 or older is 0.146. In the questions that follow, round your answers to the nearest thousandth:

a. If a man is selected at random and a woman is selected at random, what is the probability that both people selected are 65 or older? (Hint: Use the multiplication rule for independent events.)

$(0.114)(0.146) \approx 0.017$

b. If two men are selected at random, what is the probability that both of them are 65 or older?

$(0.114)(0.114) \approx 0.013$

c. If two women are selected at random, what is the probability that neither of them is 65 or older?

If one woman is selected at random, the probability that she is not 65 or older is $1 - 0.146 = 0.854$.
So, if two women are selected at random, the probability that neither of them is 65 *or older is* $(0.854)(0.854) \approx 0.729$.

4. In a large community, 72% of the people are adults, 78% of the people have traveled outside the state, and 11% are adults who have not traveled outside the state.

a. Using a Venn diagram or a hypothetical 1000 table, calculate the probability that a randomly selected person from the community is an adult and has traveled outside the state.

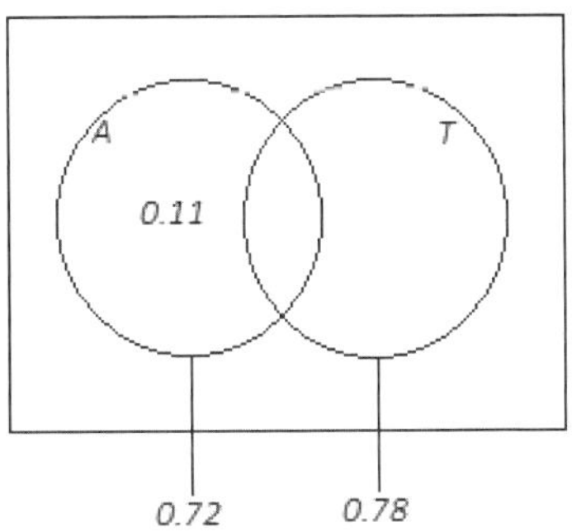

$P(\text{adult and traveled out of state}) = 0.72 - 0.11 = 0.61$

b. Use the multiplication rule for independent events to decide whether the events "is an adult" and "has traveled outside the state" are independent.

$P(\text{adult and traveled out of state}) = 0.61$

$P(\text{adult})P(\text{traveled out of state}) = (0.72)(0.78) = 0.5616$

Since these two quantities are not equal, the two events are not independent.

5. In a particular calendar year, 10% of the registered voters in a small city are called for jury duty. In this city, people are selected for jury duty at random from all registered voters in the city, and the same individual cannot be called more than once during the calendar year.

a. What is the probability that a registered voter is not called for jury duty during a particular year?

0.90

b. What is the probability that a registered voter is called for jury duty two years in a row?

$(0.10)(0.10) = 0.01$

6. A survey of registered voters in a city in New York was carried out to assess support for a new school tax. 51% of the respondents supported the school tax. Of those with school-age children, 56% supported the school tax, while only 45% of those who did not have school-age children supported the school tax.

a. If a person who responded to this survey is selected at random, what is the probability that

i. The person selected supports the school tax?

0.51

ii. The person supports the school tax given that she does not have school-age children?

0.45

b. Are the two events "has school-age children" and "supports the school tax" independent? Explain how you know this.

These two events are not independent because the probability of support given no school-age children is not the same as the probability of support.

c. Suppose that 35% of those responding to the survey were over the age of 65 and that 10% of those responding to the survey were both over age 65 and supported the school tax. What is the probability that a randomly selected person who responded to this survey supported the school tax given that she was over age 65?

$$\begin{aligned} P(\text{support given over age 65}) &= \frac{P(\text{support and over age 65})}{P(\text{over age 65})} \\ &= \frac{0.10}{0.35} \\ &\approx 0.286 \end{aligned}$$

Lesson 7: Probability Rules

Student Outcomes

- Students use the addition rule to calculate the probability of a union of two events.
- Students interpret probabilities in context.

Lesson Notes

This lesson builds off of the probability rules presented in Lesson 6 and introduces the addition rule for calculating the probability of the union of two events. The general form of the rule is considered, as well as the special cases for disjoint and independent events. The use of Venn diagrams is encouraged throughout the lesson to illustrate problems.

Classwork

Opening (3 minutes)

Revisit the high school considered in the opening discussion of Lesson 5. Encourage students to work independently in finding the answer to the central question, "What is the number of students in the band or in organized sports?"

- 442 students participate in organized sports but do not play in the band,
- 31 students play in the band but do not participate in organized sports,
- 21 students participate in organized sports *and* play in the band, and
- 339 students neither participate in organized sports nor play in the band.

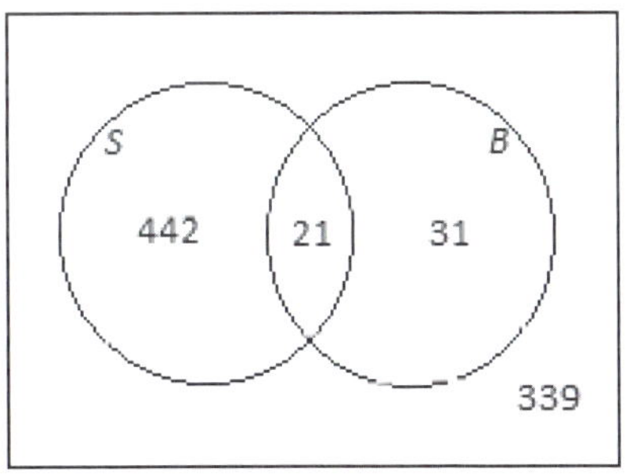

Use the Venn diagram to highlight the pieces involved in answering the following questions, as the answers to each of these questions are used to determine the number of students in the band *or* in organized sports:

How would you find the number of students in sports?

How would you find the number of students in band?

Point out to students that to answer each of the above questions, they had to add the 21 students involved in sports *and* band (the intersection) to find the total number of students in sports or to find the total number of students in band.

Discuss how the number of students in band or sports would be calculated if the summary of the school was presented differently. In particular, discuss with students how they would determine the number of students in band or sports if the description of the school was the following:

463 students are in sports,

52 students are in the band, and

21 students are in both sports and band.

The number of students in sports or the number of students in band (called the *union*) is $463 + 52 - 21$. The piece representing students in band and sports (or the intersection) is part of the total number of students in band, and it is also part of the total number of students in sports. As a result, if the number of students in band (52) and the number of students in sports (463) are added together, the 21 students in both band and sports are counted twice. As indicated, it is necessary to subtract the 21 students in both band and sports to make sure that these students are counted only once. Generalizing this as a probability of the union of two overlapping events is the focus of this lesson.

Scaffolding:

Ask students working above grade level to draw a Venn diagram to illustrate this scenario and determine how many students are in sports or in the band.

For students working below grade level, use the Venn diagram below to illustrate this scenario.

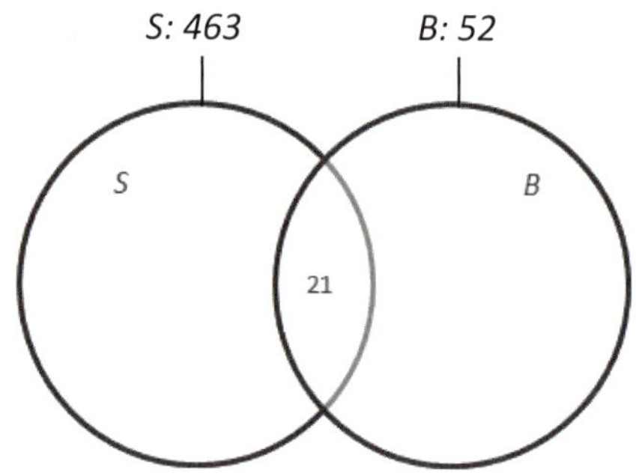

Exercise 1 (9 minutes)

Introduce the following addition rule to students. (This rule was informally illustrated with the above example and in several questions in the earlier lessons with two-way frequency tables.) The addition rule states that for any two events A and B,

$$P(A \text{ or } B) = P(A) + P(B) - P(A \text{ and } B).$$

To illustrate, draw a Venn diagram, denoting the probabilities of events as shown.

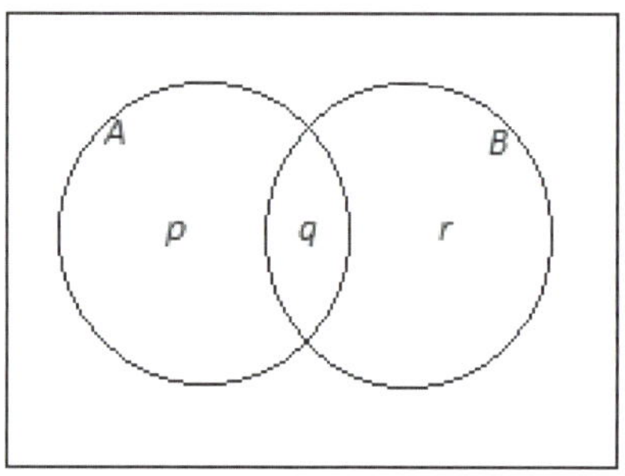

(Note that $p = P(A \text{ and not } B)$, $q = P(A \text{ and } B)$, and $r = P(B \text{ and not } A)$, but the labeling of the Venn diagram should be sufficient to communicate this.)

Therefore, $P(A) + P(B) - P(A \text{ and } B) = (p + q) + (q + r) - q$

$$= p + q + r$$

$$= P(A \text{ or } B).$$

Scaffolding:

For students who may be struggling with this concept, consider displaying and discussing several concrete examples in conjunction with the abstract representation.

For example, using the case of a coin flip, with A represents the event of heads and B represents the event of tails:

$$P(A \text{ or } B) = P(A) + P(B) - P(A \text{ and } B)$$
$$1 = 0.5 + 0.5 - 0$$

Discuss the meaning of each component of the equation in context (e.g., 1 makes sense in this situation because it is certain that the coin either lands on heads or tails; 0 makes sense because it could not land on both heads and tails in one flip).

For students working above grade level, consider encouraging them to determine the addition rule independently by asking, "What is a general rule for determining $P(A \text{ or } B)$? Use the example from the Opening to determine your answer."

Indicate to students that $P(A \text{ or } B)$ is $p + q + r$ using the Venn diagram directly. Note that when the probability of the events A or B were added together, the probability of q was added twice; therefore, the addition rule indicates that q (the intersection) is subtracted from the sum of the two events to make sure it is not added twice.

Exercise 1 is a straightforward application of the addition rule. Since Exercises 1 and 2 are students' first experience using the addition rule, consider having students work in pairs. Use this as an opportunity to informally assess student understanding of the addition rule.

Exercise 1

When a car is brought to a repair shop for a service, the probability that it will need the transmission fluid replaced is 0.38, the probability that it will need the brake pads replaced is 0.28, and the probability that it will need both the transmission fluid and the brake pads replaced is 0.16. Let the event that a car needs the transmission fluid replaced be T and the event that a car needs the brake pads replaced be B.

a. **What are the values of the following probabilities?**

i. $P(T)$ $\quad 0.38$

ii. $P(B)$ $\quad 0.28$

iii. $P(T \text{ and } B)$ $\quad 0.16$

b. **Use the addition rule to find the probability that a randomly selected car needs the transmission fluid or the brake pads replaced.**

$P(T \text{ or } B) = P(T) + P(B) - P(T \text{ and } B) = 0.38 + 0.28 - 0.16 = 0.5$

Exercise 2 (5 minutes)

Here students are asked to use the addition rule in conjunction with the multiplication rule for independent events.

Exercise 2

Josie will soon be taking exams in math and Spanish. She estimates that the probability she passes the math exam is 0.9, and the probability that she passes the Spanish exam is 0.8. She is also willing to assume that the results of the two exams are independent of each other.

a. **Using Josie's assumption of independence, calculate the probability that she passes both exams.**

$P(\text{passes both}) = (0.9)(0.8) = 0.72$

b. **Find the probability that Josie passes at least one of the exams. (Hint: Passing at least one of the exams is passing math *or* passing Spanish.)**

$P(\text{passes math or Spanish}) = P(\text{passes math}) + P(\text{passes Spanish}) - P(\text{passes both})$
$= 0.9 + 0.8 - 0.72 = 0.98$

Example 1 (7 minutes): Use of the Addition Rule for Disjoint Events

Introduce the idea of *disjoint events* and how the addition rule for disjoint events is different from the addition rule for events that had an intersection. While discussing the following examples with students, ask them how these events differ from the previous examples:

- An animal hospital has 5 dogs and 3 cats out of 10 animals in the hospital. What is the probability that an animal selected at random is a dog or a cat?

- At a certain high school, 100 students are involved in an after-school community service program. Students can only sign up for one project. Currently, 25 students are involved in cleaning up nearby parks, 20 students are tutoring elementary students in mathematics, and the rest of the students are working at helping out at a community recreational center. What is the probability that a randomly selected student is involved in cleaning up nearby parks or tutoring elementary students in mathematics?

The above examples are different in that they do not have any students in the intersection. Students would indicate that the addition rule of two events would not have a piece that needs to be subtracted. The probability of randomly selecting a dog or a cat is $\frac{5}{10} + \frac{3}{10}$. The probability of randomly selecting a student involved in cleaning a nearby park or tutoring elementary students in mathematics is $\frac{25}{100} + \frac{20}{100}$.

Summarize the following with students:

- Two events are said to be *disjoint* if they have no outcomes in common. So, if the events A and B are disjoint, the Venn diagram looks like this.

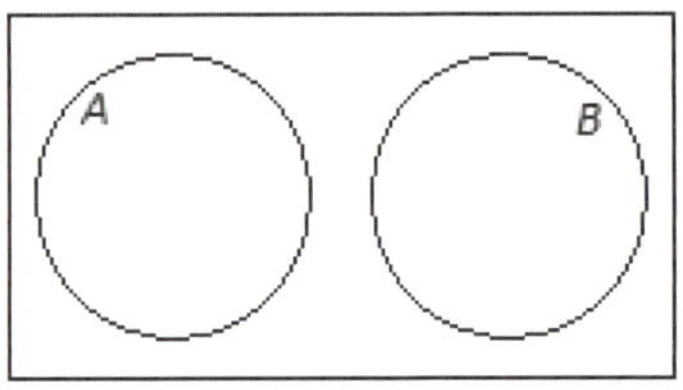

Another way of describing disjoint events is by saying that they cannot both happen at the same time. Continue discussing with students other examples.

If a number cube has faces numbered 1–6, and the number cube is rolled once, then the events "the result is even" and "the result is a 5" are disjoint (since *even* and 5 cannot both happen on a single roll), but the events "the result is even" and "the result is greater than 4" are not (since getting a 6 results in both events occurring).

It would be a good idea at this stage to provide some other examples of disjoint and non-disjoint events so that students get used to the meaning of the term.

Scaffolding:

Point out to students that the meaning of *disjoint* can be found by examining the prefix *dis-* and the word *joint*. Remind students that *dis-* means *not*. The stem *joint* has several other meanings that might need to be explained or explored.

If A and B are disjoint, then $P(A \text{ and } B) = 0$. So, the addition rule for disjoint events can be written as

$$P(A \text{ or } B) = P(A) + P(B).$$

Now, work through the example presented in the lesson as a class. This is a straightforward application of the addition rule for disjoint events.

Example 1: Use of the Addition Rule for Disjoint Events

A set of 40 cards consists of the following:

- 10 black cards showing squares
- 10 black cards showing circles
- 10 red cards showing X's
- 10 red cards showing diamonds

A card will be selected at random from the set. Find the probability that the card is black or shows a diamond.

The events "is black" and "shows a diamond" are disjoint since there are no black cards that show diamonds. So,

$$\begin{aligned} P(\text{black or diamond}) &= P(\text{black}) + P(\text{diamond}) \\ &= \frac{20}{40} + \frac{10}{40} \\ &= \frac{30}{40} = \frac{3}{4}. \end{aligned}$$

Example 2 (4 minutes): Combining Use of the Multiplication and Addition Rules

The addition rule for disjoint events is often used in conjunction with the multiplication rule for independent events. This example illustrates this.

When tackling part (b), point out to students that the three events—red shows 6 and blue shows 5, red shows 5 and blue shows 6, red shows 6 and blue shows 6—are disjoint. This is why the probabilities are added together.

Example 2: Combining Use of the Multiplication and Addition Rules

A red cube has faces labeled 1 through 6, and a blue cube has faces labeled in the same way. The two cubes are rolled. Find the probability of each event.

a. Both cubes show 6's.

$$P(\text{red shows 6 and blue shows 6}) = P(\text{red shows 6})P(\text{blue shows 6}) = \frac{1}{6}\cdot\frac{1}{6} = \frac{1}{36}$$

b. The total score is at least 11.

$$P(\text{total is at least 11}) = P(\text{red shows 6 and blue shows 5}) + P(\text{red shows 5 and blue shows 6}) + P(\text{red shows 6 and blue shows 6})$$

$$P(\text{total is at least 11}) = \frac{1}{6}\cdot\frac{1}{6} + \frac{1}{6}\cdot\frac{1}{6} + \frac{1}{6}\cdot\frac{1}{6} = \frac{1}{36} + \frac{1}{36} + \frac{1}{36} = \frac{1}{12}$$

Exercise 3 (7 minutes)

This exercise provides additional practice with the ideas introduced in Example 2. Part (c) is a little more complex than any part of Example 2, and part (d) requires students to recall the complement rule. Have students first work the solutions independently. Then, have students compare and discuss solutions with a partner. Ask students to describe and compare with each other how they found their solutions. This is an opportunity for students to show persistence in solving a problem.

MP.1

Exercise 3

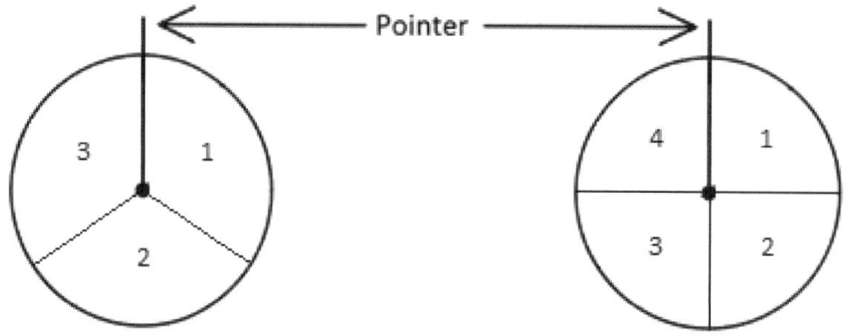

The diagram above shows two spinners. For the first spinner, the scores 1, 2, and 3 are equally likely, and for the second spinner, the scores 1, 2, 3, and 4 are equally likely. Both pointers will be spun. Writing your answers as fractions in lowest terms, find the probability of each event.

a. **The total of the scores on the two spinners is 2.**

$$P(\text{total} = 2) = P(1,1) = \frac{1}{3} \cdot \frac{1}{4} = \frac{1}{12}$$

b. **The total of the scores on the two spinners is 3.**

$$P(\text{total} = 3) = P(1,2) + P(2,1) = \frac{1}{3} \cdot \frac{1}{4} + \frac{1}{3} \cdot \frac{1}{4} = \frac{1}{12} + \frac{1}{12} = \frac{2}{12} = \frac{1}{6}$$

c. **The total of the scores on the two spinners is 5.**

$$\begin{aligned} P(\text{total} = 5) &= P(1,4) + P(2,3) + P(3,2) \\ &= \frac{1}{3} \cdot \frac{1}{4} + \frac{1}{3} \cdot \frac{1}{4} + \frac{1}{3} \cdot \frac{1}{4} \\ &= \frac{1}{12} + \frac{1}{12} + \frac{1}{12} \\ &= \frac{3}{12} \\ &= \frac{1}{4} \end{aligned}$$

d. **The total of the scores on the two spinners is not 5.**

$$P(\text{total is not } 5) = 1 - P(\text{total is } 5) = 1 - \frac{1}{4} = \frac{3}{4}$$

Closing (3 minutes)

Review the difference between independent and disjoint events.

- Define independent and disjoint events.
  - *Two events are independent if knowing that one event has occurred does not change the probability that the other event has occurred. Two events are said to be disjoint if they have no outcomes in common.*

Ask students to summarize the main ideas of the lesson with a neighbor or in writing. Use this as an opportunity to informally assess comprehension of the lesson. The Lesson Summary below offers some important ideas that should be included.

Lesson Summary

The addition rule states that for any two events A and B, $P(A \text{ or } B) = P(A) + P(B) - P(A \text{ and } B)$.

The addition rule can be used in conjunction with the multiplication rule for independent events: Events A and B are independent if and only if $P(A \text{ and } B) = P(A)P(B)$.

Two events are said to be *disjoint* if they have no outcomes in common. If A and B are disjoint events, then $P(A \text{ or } B) = P(A) + P(B)$.

The addition rule for disjoint events can be used in conjunction with the multiplication rule for independent events.

Exit Ticket (7 minutes)

Name ______________________________ Date______________

Lesson 7: Probability Rules

Exit Ticket

1. When a call is received at an airline's call center, the probability that it comes from abroad is 0.32, and the probability that it is to make a change to an existing reservation is 0.38.

   a. Suppose that you are told that the probability that a call is both from abroad and is to make a change to an existing reservation is 0.15. Calculate the probability that a randomly selected call is either from abroad or is to make a change to an existing reservation.

   b. Suppose now that you are *not* given the information in part (a), but you are told that the events "the call is from abroad" and "the call is to make a change to an existing reservation" are independent. What is the probability that a randomly selected call is either from abroad or is to make a change to an existing reservation?

2. A golfer will play two holes of a course. Suppose that on each hole the player will score 3, 4, 5, 6, or 7, with these five scores being equally likely. Find the probability, and explain how the answer was determined that the player's total score for the two holes will be

   a. 14.

   b. 12.

Exit Ticket Sample Solutions

1. When a call is received at an airline's call center, the probability that it comes from abroad is 0.32, and the probability that it is to make a change to an existing reservation is 0.38.

   a. Suppose that you are told that the probability that a call is both from abroad and is to make a change to an existing reservation is 0.15. Calculate the probability that a randomly selected call is either from abroad or is to make a change to an existing reservation.

   $$\begin{aligned} P(\text{abroad or change}) &= P(\text{abroad}) + P(\text{change}) - P(\text{abroad and change}) \\ &= 0.32 + 0.38 - 0.15 \\ &= 0.55 \end{aligned}$$

   b. Suppose now that you are *not* given the information in part (a), but you are told that the events "the call is from abroad" and "the call is to make a change to an existing reservation" are independent. What is the probability that a randomly selected call is either from abroad or is to make a change to an existing reservation?

   $$\begin{aligned} P(\text{abroad and change}) &= P(\text{abroad})P(\text{change}) \\ &= (0.32)(0.38) \\ &= 0.1216 \end{aligned}$$

   So,

   $$\begin{aligned} P(\text{abroad or change}) &= P(\text{abroad}) + P(\text{change}) - P(\text{abroad and change}) \\ &= 0.32 + 0.38 - 0.1216 \\ &= 0.5784. \end{aligned}$$

2. A golfer will play two holes of a course. Suppose that on each hole the player will score 3, 4, 5, 6, or 7, with these five scores being equally likely. Find the probability, and explain how the answer was determined that the player's total score for the two holes will be

   a. 14.

   The only way to have a total score of 14 is if the player scores 7 on each hole, which is $P(7) \cdot P(7)$. Each of the scores is equally likely, so $P(7) = \frac{1}{5} = 0.2$.

   $P(\text{total} = 14) = P(7) \cdot P(7) = (0.2)(0.2) = 0.04$

   b. 12.

   There are three ways to have a total score of 12. The player could score 7 on the first hole and 5 on the second. The player could score 6 on the first hole and 6 on the second. Finally, the player could score 5 on the first hole and 7 on the second. Again, all of the scores are equally likely, so $P(5) = P(6) = P(7) = \frac{1}{5} = 0.2$.

   $$\begin{aligned} P(\text{total} = 12) &= P(7,5) + P(6,6) + P(5,7) \\ &= (0.2)(0.2) + (0.2)(0.2) + (0.2)(0.2) \\ &= 0.04 + 0.04 + 0.04 \\ &= 0.12 \end{aligned}$$

Problem Set Sample Solutions

1. Of the works of art at a large gallery, 59% are paintings, and 83% are for sale. When a work of art is selected at random, let the event that it is a painting be A and the event that it is for sale be B.

   a. What are the values of $P(A)$ and $P(B)$?

   $P(A) = 0.59$

   $P(B) = 0.83$

   b. Suppose you are told that $P(A \text{ and } B) = 0.51$. Find $P(A \text{ or } B)$.

   $P(A \text{ or } B) = P(A) + P(B) - P(A \text{ and } B) = 0.59 + 0.83 - 0.51 = 0.91$

   c. Suppose now that you are not given the information in part (b), but you are told that the events A and B are independent. Find $P(A \text{ or } B)$.

   $$P(A \text{ and } B) = P(A)P(B) = (0.59)(0.83) = 0.4897$$

   So, $P(A \text{ or } B) = P(A) + P(B) - P(A \text{ and } B) = 0.59 + 0.83 - 0.4897 = 0.9303.$

2. A traveler estimates that, for an upcoming trip, the probability of catching malaria is 0.18, the probability of catching typhoid is 0.13, and the probability of catching neither of the two diseases is 0.75.

   a. Draw a Venn diagram to represent this information.

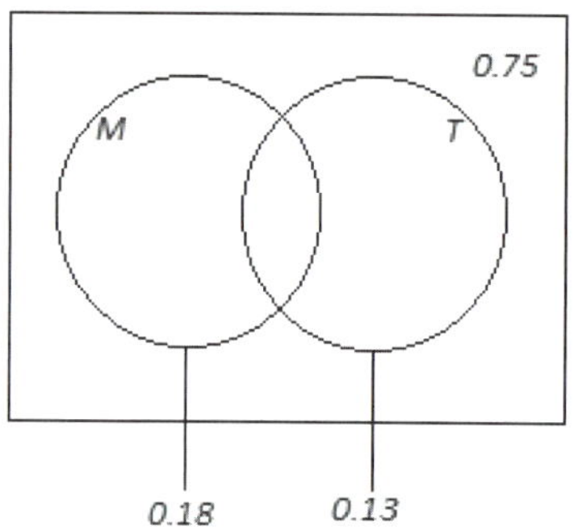

   b. Calculate the probability of catching both of the diseases.

   $$P(M \text{ or } T) = 1 - 0.75 = 0.25$$

   By the addition rule:

   $$P(M \text{ or } T) = P(M) + P(T) - P(M \text{ and } T)$$
   $$0.25 = 0.18 + 0.13 - P(M \text{ and } T)$$
   $$0.25 = 0.31 - P(M \text{ and } T)$$
   $$P(M \text{ and } T) = 0.06$$

   c. Are the events "catches malaria" and "catches typhoid" independent? Explain your answer.

   $P(M \text{ and } T) = 0.06$

   $P(M)P(T) = (0.18)(0.13) = 0.0234$

   Since these quantities are different, the two events are not independent.

3. A deck of 40 cards consists of the following:
   - 10 black cards showing squares, numbered 1–10
   - 10 black cards showing circles, numbered 1–10
   - 10 red cards showing X's, numbered 1–10
   - 10 red cards showing diamonds, numbered 1–10

   A card will be selected at random from the deck.

   a. i. Are the events "the card shows a square" and "the card is red" disjoint? Explain.

   Yes. There is no red card that shows a square.

   ii. Calculate the probability that the card will show a square or will be red.

   $$P(\text{square or red}) = P(\text{square}) + P(\text{red})$$
   $$= \frac{10}{40} + \frac{20}{40} = \frac{30}{40} = \frac{3}{4}$$

   b. i. Are the events "the card shows a 5" and "the card is red" disjoint? Explain.

   No. There are red fives in the deck.

   ii. Calculate the probability that the card will show a 5 or will be red.

   $$P(5 \text{ or red}) = P(5) + P(\text{red}) - P(5 \text{ and red})$$
   $$= \frac{4}{40} + \frac{20}{40} - \frac{2}{40} = \frac{22}{40} = \frac{11}{20}$$

4. The diagram below shows a spinner. When the pointer is spun, it is equally likely to stop on 1, 2, or 3. The pointer will be spun three times. Expressing your answers as fractions in lowest terms, find the probability, and explain how the answer was determined that the total of the values from all thee spins is

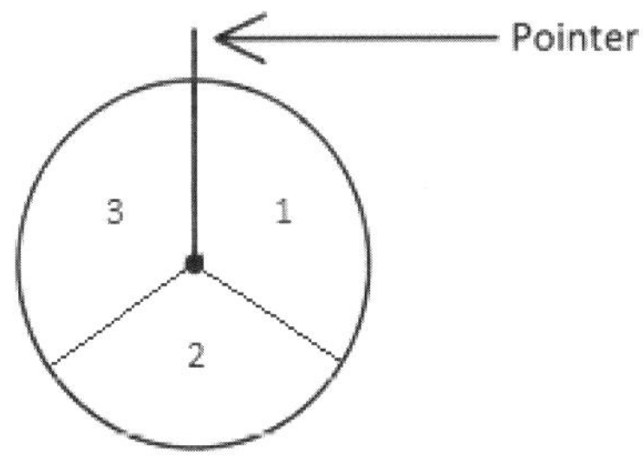

   a. 9.

   The only way to get a total of 9 is to spin a 3, 3 times. Since the probability of spinning a 3 is $\frac{1}{3}$,

   $$P(\text{total is } 9) = P(3,3,3) = \frac{1}{3} \cdot \frac{1}{3} \cdot \frac{1}{3} = \frac{1}{27}.$$

   b. 8.

   There are 3 ways to get a total of 8. Since the probability of spinning a 1, 2, and 3 are all equally likely $\left(\frac{1}{3}\right)$:

   $$P(\text{total is } 8) = P(3,3,2) + P(3,2,3) + P(2,3,3)$$
   $$= \frac{1}{3} \cdot \frac{1}{3} \cdot \frac{1}{3} + \frac{1}{3} \cdot \frac{1}{3} \cdot \frac{1}{3} + \frac{1}{3} \cdot \frac{1}{3} \cdot \frac{1}{3}$$
   $$= \frac{1}{27} + \frac{1}{27} + \frac{1}{27}$$
   $$= \frac{3}{27}$$
   $$= \frac{1}{9}$$

c. 7.

There are 6 ways to get a total of 7. Since the probability of spinning a 1, 2, and 3 are all equally likely $\left(\frac{1}{3}\right)$:

$$\begin{aligned} P(\text{total is } 7) &= P(3,3,1) + P(3,1,3) + P(1,3,3) + P(3,2,2) + P(2,3,2) + P(2,2,3) \\ &= 6\left(\frac{1}{3}\cdot\frac{1}{3}\cdot\frac{1}{3}\right) \\ &= \frac{6}{27} \\ &= \frac{2}{9} \end{aligned}$$

5. A number cube has faces numbered 1 through 6, and a coin has two sides—heads and tails. The number cube will be rolled once, and the coin will be flipped once. Find the probabilities of the following events. (Express your answers as fractions in lowest terms.)

a. The number cube shows a 6.

$$\frac{1}{6}$$

b. The coin shows heads.

$$\frac{1}{2}$$

c. The number cube shows a 6, and the coin shows heads.

$$\frac{1}{6}\cdot\frac{1}{2}=\frac{1}{12}$$

d. The number cube shows a 6, or the coin shows heads.

$$\begin{aligned} P(6 \text{ or heads}) &= P(6) + P(\text{heads}) - P(6 \text{ and heads}) \\ &= \frac{1}{6}+\frac{1}{2}-\frac{1}{12}=\frac{2}{12}+\frac{6}{12}-\frac{1}{12}=\frac{7}{12} \end{aligned}$$

6. Kevin will soon be taking exams in math, physics, and French. He estimates the probabilities of his passing these exams to be as follows:

- Math: 0.9
- Physics: 0.8
- French: 0.7

Kevin is willing to assume that the results of the three exams are independent of each other. Find the probability of each event.

a. Kevin will pass all three exams.

$$(0.9)(0.8)(0.7) = 0.504$$

b. Kevin will pass math but fail the other two exams.

$$(0.9)(0.2)(0.3) = 0.054$$

EUREKA MATH™

c. **Kevin will pass exactly one of the three exams.**

$$\begin{aligned} P(\text{passes exactly one}) = {} & P(\text{passes math, fails physics, fails French}) \\ & + P(\text{fails math, passes physics, fails French}) \\ & + P(\text{fails math, fails physics, passes French}) \\ & = (0.9)(0.2)(0.3) + (0.1)(0.8)(0.3) + (0.1)(0.2)(0.7) \\ & = 0.092 \end{aligned}$$

Topic B

Modeling Data Distributions

S-ID.A.4

Focus Standard:	S-ID.A.4	Use the mean and standard deviation of a data set to fit it to a normal distribution and to estimate population percentages. Recognize that there are data sets for which such a procedure is not appropriate. Use calculators, spreadsheets, and tables to estimate areas under the normal curve.
Instructional Days:	4	
Lesson 8:	Distributions—Center, Shape, and Spread (P)[1]	
Lesson 9:	Using a Curve to Model a Data Distribution (P)	
Lessons 10–11:	Normal Distributions (P,P)	

This topic introduces students to the idea of using a smooth curve to model a data distribution, eventually leading to using the normal distribution to model data distributions that are bell shaped and symmetric. Many naturally occurring variables, such as arm span, weight, reaction times, and standardized test scores, have distributions that are well described by a normal curve.

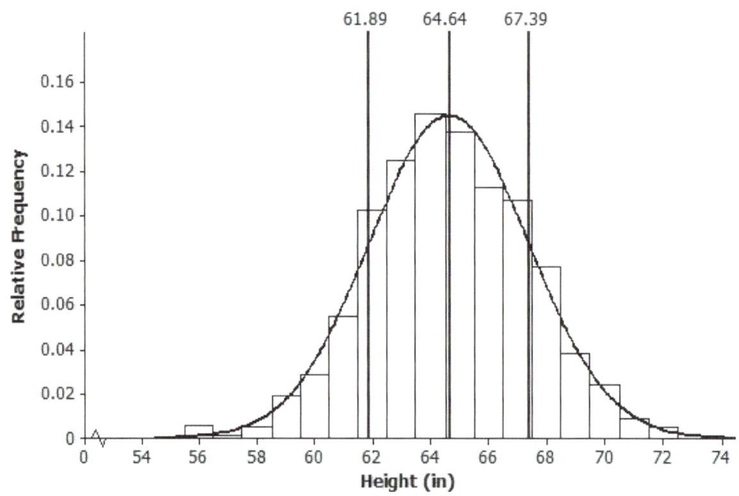

[1]Lesson Structure Key: **P**-Problem Set Lesson, **M**-Modeling Cycle Lesson, **E**-Exploration Lesson, **S**-Socratic Lesson

Students begin by reviewing their previous work with shape, center, and variability. Students use the mean and standard deviation to describe center and variability for a data distribution that is approximately symmetric. This provides a foundation for selecting an appropriate normal distribution to model a given data distribution.

Students learn to draw a smooth curve that could be used to model a given data distribution. A smooth curve is first used to model a relative frequency histogram, which shows that the area under the curve represents the approximate proportion of data falling in a given interval. Properties of the normal distribution are introduced by asking students to distinguish between reasonable and unreasonable data distributions for using a normal distribution model. Students use tables and technology to calculate normal probabilities. They work with graphing calculators, tables of normal curve areas, and spreadsheets to calculate probabilities in the examples and exercises provided (**S-ID.A.4**).

Lesson 8: Distributions—Center, Shape, and Spread

Student Outcomes

- Students describe data distributions in terms of shape, center, and variability.
- Students use the mean and standard deviation to describe center and variability for a data distribution that is approximately symmetric.

Lesson Notes

In this lesson, students review key ideas developed in Grades 6 and 7 and Algebra I. In particular, this lesson revisits distribution shapes (approximately symmetric, mound shaped, and skewed) and the use of the mean and standard deviation to describe center and variability for distributions that are approximately symmetric. The steps to calculate standard deviation are not included in the student edition since it is recommended that students use technology to calculate standard deviation. If students are not familiar with standard deviation, the steps to calculate this value are included under the notes for Example 1. The major emphasis should be on the interpretation of the mean and standard deviation rather than on calculations.

Classwork

Opening (5 minutes): Review of Distribution Shapes and Standard Deviation

Prior to beginning Example 1, students may need to review the interpretation of the *mean* of a distribution. The mean is the "fair share" value. It also represents the balance point of the distribution (the point where the sum of the deviations to the left of the mean is equal to the sum of the deviations to the right of the mean). The mean can be interpreted as an average or a typical value for a data distribution.

Show students the following data on the amount of time (in minutes) spent studying for a math test for 10 students:

34	38	48	41	38	61	29	46	45	40

For this data set, the mean is 42 minutes, and this would be interpreted as a typical amount of time spent studying for this group of students. Also, consider showing the dot plot of the data shown below, and point out that the data values are centered at 42 minutes.

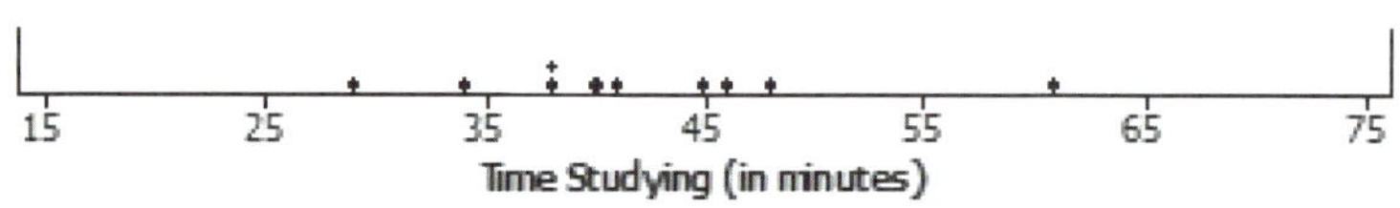

In addition to discussing the definition of *mean*, it may be necessary to review the definition of *standard deviation*.

Scaffolding:

The word *mean* has multiple definitions (including from different parts of speech). Although students have been exposed to the term *mean* in Grades 6 and 7 and Algebra I, those new to the curriculum or English language learners may struggle with its meaning. Teachers should clarify the term for these students.

Consider using visual displays and repeated choral readings to reinforce the mathematical meaning of this word. A Frayer diagram may be used.

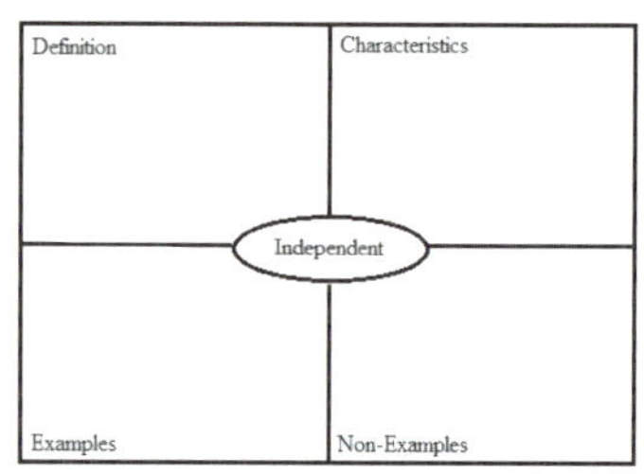

Remind students that the *standard deviation* is one measure of *spread* or *variability* in a data distribution. The standard deviation describes variation in terms of deviation from the mean. The standard deviation can be interpreted as a typical deviation from the mean. A small standard deviation indicates that the data points tend to be very close to the mean, and a large standard deviation indicates that the data points are spread out over a large range of values.

The formula for the standard deviation of a sample is $s = \sqrt{\frac{\Sigma(x-\overline{x})^2}{n-1}}$, where $\overline{x}$ is the sample mean.

The standard deviation, s, of the study time data is approximately 8.77, which would be interpreted as a typical difference from the mean number of minutes this group of students spent studying. Consider showing the following dot plots, and ask students if they think that the standard deviation would be less than or greater than 8.77 for each of Data Sets A, B, and C. (The standard deviation is less than 8.77 for Data Set B and greater for Data Sets A and C.)

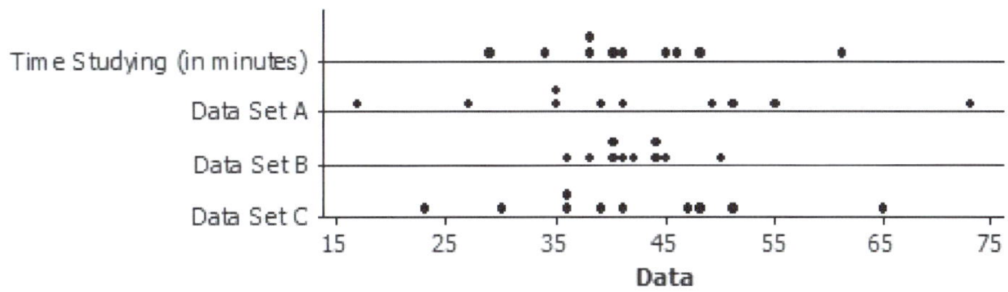

To calculate the standard deviation,

1. Find the mean, $\overline{x}$.
2. Find the difference between each data point and the mean. These values are called *deviations* from the mean, and a deviation is denoted by $(x - \overline{x})$.
3. Square each of the deviations.
4. Find the sum of the squared deviations.
5. Divide the sum of the squared deviations by $n - 1$.
6. Determine the square root.

To find the standard deviation of data from an entire population (as opposed to a sample), divide by n instead of $n - 1$.

Review the different distribution shapes: *symmetric* and *skewed*. A distribution is approximately *symmetric* if the left side and the right side of the distribution are roughly mirror images of each other (if students were to fold the distribution in half, the two sides would be a close match). A distribution is *skewed* if it has a noticeably longer tail on one side than the other. A distribution with a longer tail to the right is considered *skewed to the right*, whereas a distribution with a longer tail on the left side is *skewed to the left*. Display the distributions shown in the Lesson Summary, and discuss the shape of each distribution. Also, point out that a distribution that is approximately symmetric with a single peak is often described as mound shaped.

Example 1 (10 minutes): Center, Shape, and Spread

All of the histograms in this lesson are relative frequency histograms. Relative frequency histograms were introduced in earlier grades. If necessary, remind students that the heights of the bars in a relative frequency histogram represent the proportion of the observations within each interval and not the number of observations (the frequency) within the interval. Prior to students beginning to work on Exercises 1 through 6, discuss the scales on the battery life histogram. Address the following points using the histogram of battery life:

- What is the width of each bar?
  - 5 *hours*
- What does the height of each bar represent?
  - *The proportion of all batteries with a life in the interval corresponding to the bar. For example, approximately* 5% *of the batteries lasted between* 85 *and* 90 *hours.*

Example 1: Center, Shape, and Spread

Have you ever noticed how sometimes batteries seem to last a long time, and other times the batteries seem to last only a short time?

The histogram below shows the distribution of battery life (hours) for a sample of 40 batteries of the same brand. When studying a distribution, it is important to think about the shape, center, and spread of the data.

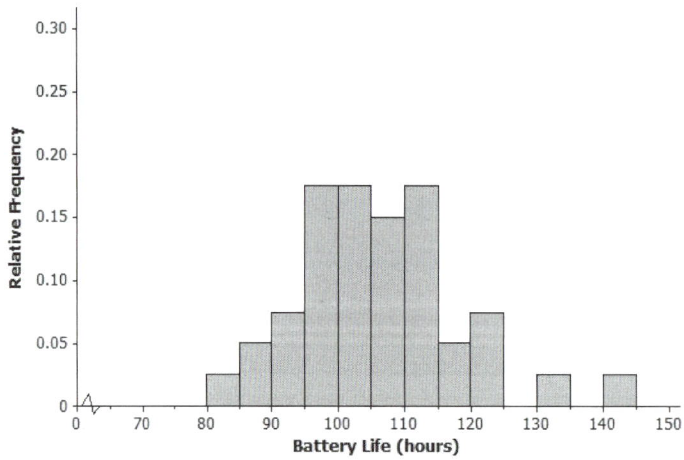

Exercises 1–6 (10 minutes)

Have students work independently and confirm answers with a partner or in a small group. Discuss answers as needed.

MP.2 Before beginning Exercise 7, review the interpretation of standard deviation. Have students describe how they made their estimates for the standard deviation in Exercises 3 and 6. This allows students to practice reasoning abstractly and quantitatively.

In the next several exercises, students are asked to estimate and interpret the standard deviation. Standard deviation can be challenging for students. Often in the process of focusing on how it is calculated, students lose sight of what it indicates about the data distribution. The calculation of the standard deviation can be done using technology or by following a precise sequence of steps; understanding what it indicates about the data should be the focus of the exercises.

The standard deviation is a measure of variability that is based on how far observations in a data set fall from the mean. It can be interpreted as a typical or an average distance from the mean. Various rules and shortcuts are often used to estimate a standard deviation, but for this lesson, keep the focus on understanding the standard deviation as a value that describes a typical distance from the mean. Students should observe that a typical distance is one for which some distances would be less than this value and some would be greater. When students estimate a standard deviation, ask them whether that value is representative of the collection of distances. Is the estimate a reasonable value for the average distance of observations from the mean? If the estimate is a value that is less than most of the distances, then it is not a good estimate and is probably too small. If the estimate is a value that is greater than most of the distances, then the estimate is probably too large. This understanding is developed in several of the following exercises:

Exercises 1–9

1. **Would you describe the distribution of battery life as approximately symmetric or as skewed? Explain your answer.**

   The distribution is approximately symmetric. The right and left halves of the distribution are similar.

Indicate that because this distribution is approximately symmetric, the standard deviation is a reasonable way to describe the variability of the data. From students' previous work in Grade 6 and Algebra I, they should recall that for a data distribution that is skewed rather than symmetric, the interquartile range (IQR) would be used to describe variability.

2. **Is the mean of the battery life distribution closer to 95, 105, or 115 hours? Explain your answer.**

   The mean of the battery life distribution is closer to 105 hours because the data appear to center around 105.

The next exercise provides an opportunity to discuss a typical distance from the mean. If students struggle with understanding this question, for each estimate of the standard deviation, ask if it would be a good estimate of a typical or an average distance of observations from the mean. If 5 was the standard deviation, how many of the distances would be greater than and how many less than this value? Students should be able to see that most of the data values are more than 5 units from the mean. If 25 was the standard deviation, how many of the distances would be greater than and how many less than this value? Again, students should see that most of the data values are less than 25 units away from the mean. The estimate of 10 is more reasonable as an estimate of the average distance from the mean and is a better estimate of the standard deviation.

3. **Consider 5, 10, or 25 hours as an estimate of the standard deviation for the battery life distribution.**

   a. **Consider 5 hours as an estimate of the standard deviation. Is it a reasonable description of a typical distance from the mean? Explain your answer.**

      Most of the distances from the mean are greater than 5 hours. It is not a good estimate of the standard deviation.

   b. **Consider 10 hours as an estimate of the standard deviation. Is it a reasonable description of a typical distance from the mean? Explain your answer.**

      It looks like 10 is a reasonable estimate of a typical distance from the mean. It is a reasonable estimate of the standard deviation.

   c. **Consider 25 hours as an estimate of the standard deviation. Is it a reasonable description of a typical distance from the mean? Explain your answer.**

      Nearly all of the data values are less than 25 hours from the mean of 105. It is not a good estimate of the standard deviation.

The histogram below shows the distribution of the greatest drop (in feet) for 55 major roller coasters in the United States.

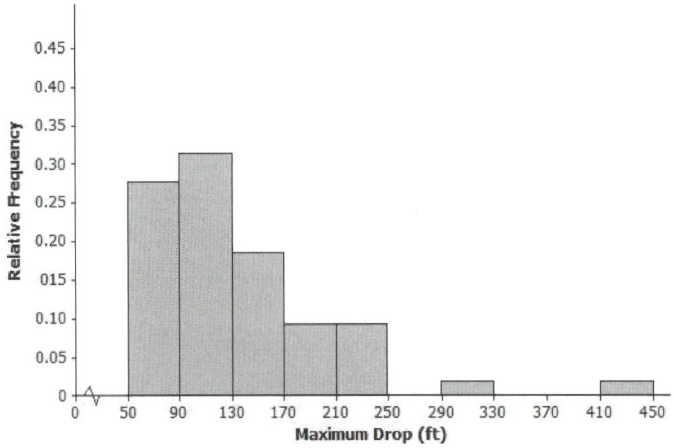

4. **Would you describe this distribution of roller coaster maximum drop as approximately symmetric or as skewed? Explain your answer.**

   The distribution is skewed to the right because there is a long tail on the right side.

5. **Is the mean of the maximum drop distribution closer to 90, 135, or 240 feet? Explain your answer.**

   The mean is closer to 135 feet because 90 is too small and 240 is too large to be considered a typical value for this data set.

In the same way that students estimated the standard deviation for battery life, the following exercise asks students to select an estimate for the drop data. Here again, students should consider each estimate separately and determine which one is the most typical of the distances from the estimated mean. As this is a skewed distribution, estimating a typical distance from the mean is a challenge and requires careful thought.

6. **Is the standard deviation of the maximum drop distribution closer to 40, 70, or 100 hours? Explain your answer.**

   It seems that 70 is about right for a typical distance from the mean. A deviation of 40 would be too small, and 100 would be too large to be considered a typical distance from the mean for this data set. Most of the data values differ from the estimated mean of 135 by more than 40, which means that 40 is not a reasonable estimate of the standard deviation. Most of the data values differ from the estimated mean of 135 by less than 100, which means that 100 is not a reasonable estimate of the standard deviation. This can be illustrated for students using the following picture:

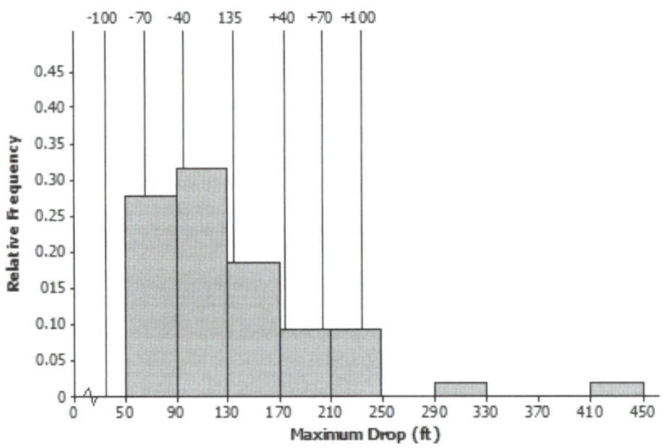

Exercises 7–9 (10 minutes)

Encourage students to work in pairs on Exercises 7, 8, and 9. Then, discuss and confirm the answers.

7. **Consider the following histograms: Histogram 1, Histogram 2, Histogram 3, and Histogram 4. Descriptions of four distributions are also given. Match the description of a distribution with the appropriate histogram.**

Histogram	Distribution
1	B
2	A
3	C
4	D

Description of distributions:

Distribution	Shape	Mean	Standard Deviation
A	Skewed to the right	100	10
B	Approximately symmetric, mound shaped	100	10
C	Approximately symmetric, mound shaped	100	40
D	Skewed to the right	100	40

Histograms:

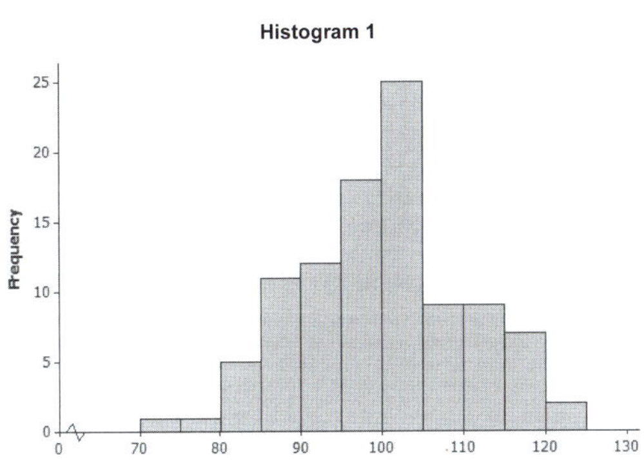

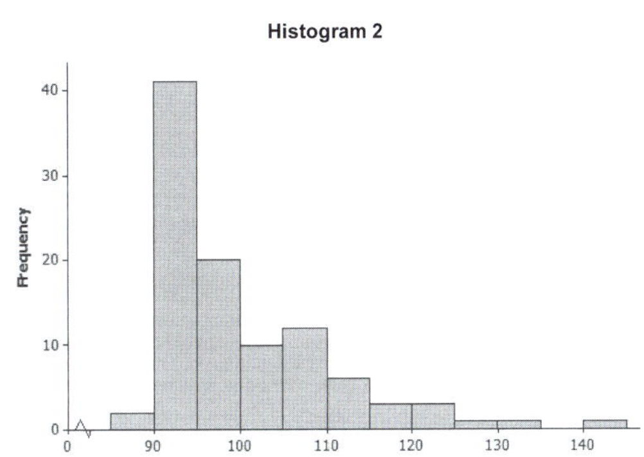

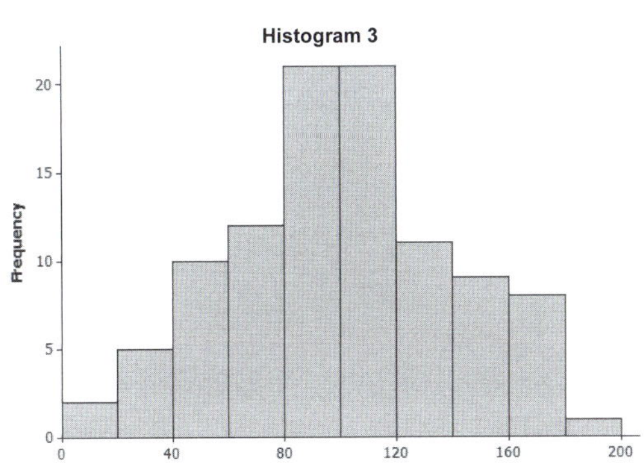

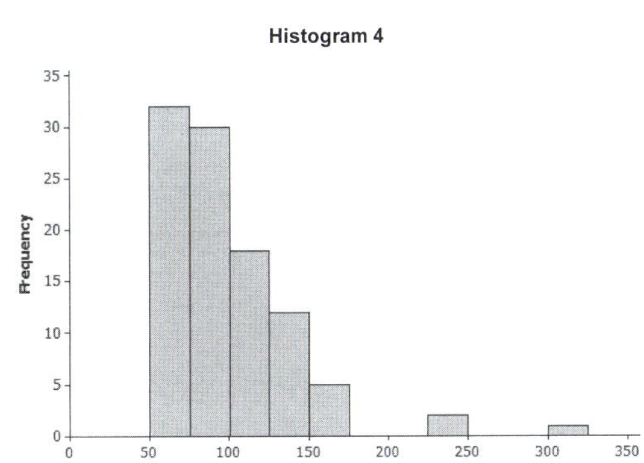

8. **The histogram below shows the distribution of gasoline tax per gallon for the 50 states and the District of Columbia in 2010. Describe the shape, center, and spread of this distribution.**

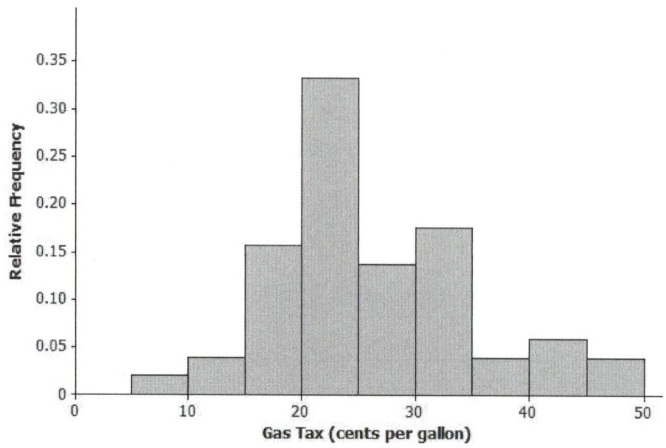

The distribution shape is skewed to the right. Answers for center and spread will vary, but the center is approximately 25, and the standard deviation is approximately 10.

9. **The histogram below shows the distribution of the number of automobile accidents per year for every $1,000$ people in different occupations. Describe the shape, center, and spread of this distribution.**

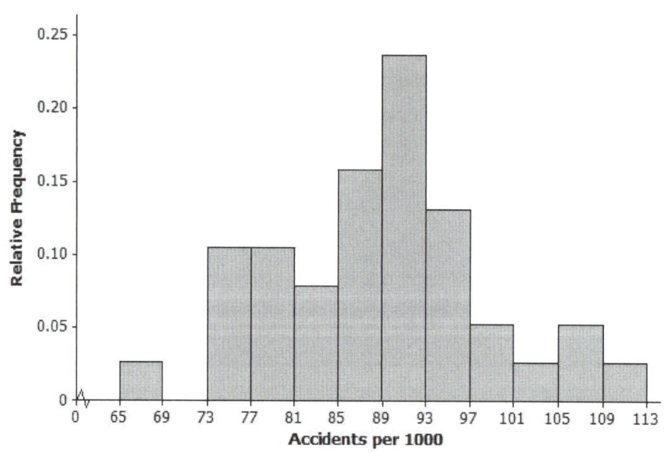

The shape of the distribution is approximately symmetric. Answers for center and spread will vary, but the center is approximately 89, and the standard deviation is approximately 10.

Closing (5 minutes)

Ask students to summarize one of the histograms presented in class in terms of center, shape, and spread. Allow them to select any one of the many examples presented in this lesson. Call on a representative group of students to present descriptions of the histogram they selected.

Ask students to summarize the main concepts of the lesson in writing or with a neighbor. Use this as an opportunity to informally assess comprehension of the lesson. The Lesson Summary below offers some important concepts that should be included.

Lesson Summary

Distributions are described by the shape (symmetric or skewed), the center, and the spread (variability) of the distribution.

A distribution that is approximately symmetric can take different forms.

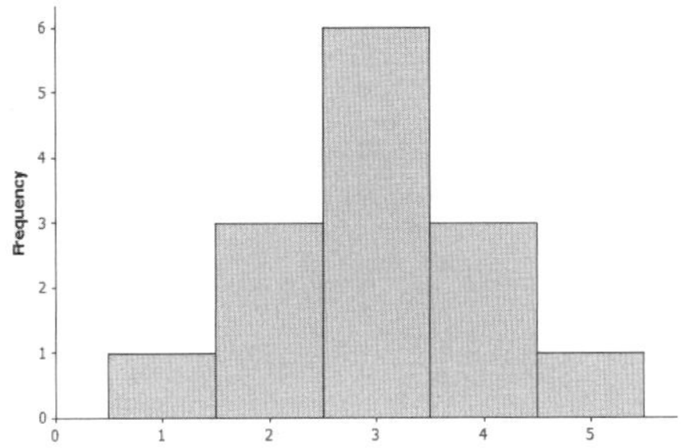

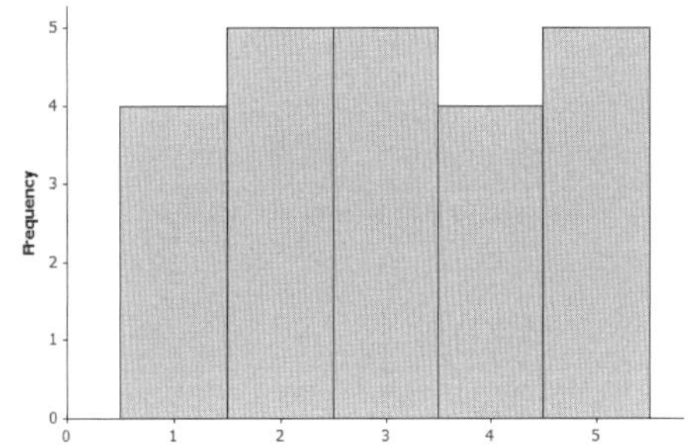

A distribution is described as *mound shaped* if it is approximately symmetric and has a single peak.

A distribution is *skewed to the right* or *skewed to the left* if one of its tails is longer than the other.

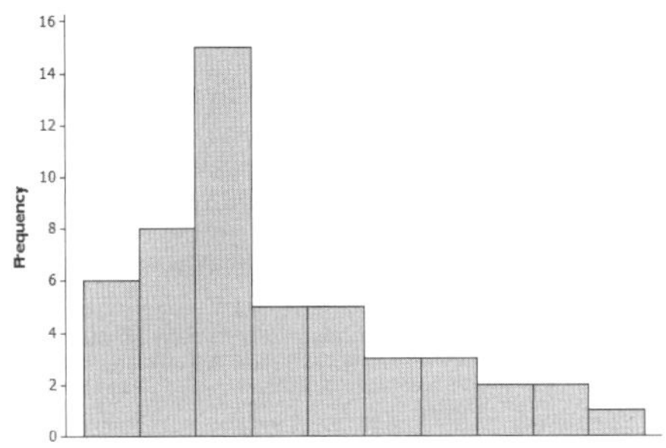

Skewed to the Right

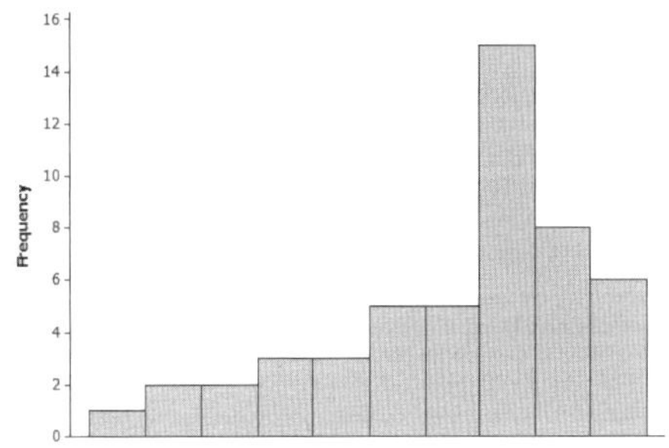

Skewed to the Left

The *mean of a distribution* is interpreted as a typical value and is the average of the data values that make up the distribution.

The *standard deviation* is a value that describes a typical distance from the mean.

Exit Ticket (5 minutes)

Name ___________________________________ Date ________________

Lesson 8: Distributions—Center, Shape, and Spread

Exit Ticket

A local utility company wanted to gather data on the age of air conditioners that people have in their homes. The company took a random sample of 200 residents of a large city and asked if the residents had an air conditioner, and if they did, how old it was. Below is the distribution in the reported ages of the air conditioners.

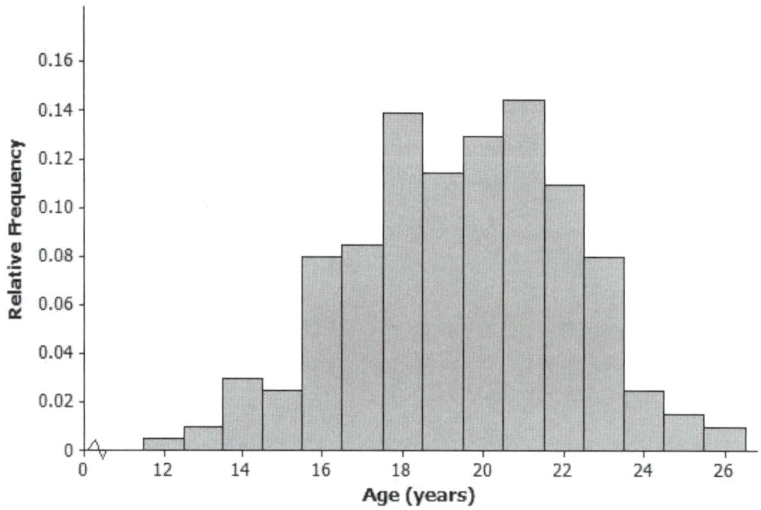

1. Would you describe this distribution of air conditioner ages as approximately symmetric or as skewed? Explain your answer.

2. Is the mean of the age distribution closer to 15, 20, or 25 years? Explain your answer.

3. Is the standard deviation of the age distribution closer to 3, 6, or 9 years? Explain your answer.

Exit Ticket Sample Solutions

A local utility company wanted to gather data on the age of air conditioners that people have in their homes. The company took a random sample of 200 residents of a large city and asked if the residents had an air conditioner, and if they did, how old it was. Below is the distribution in the reported ages of the air conditioners.

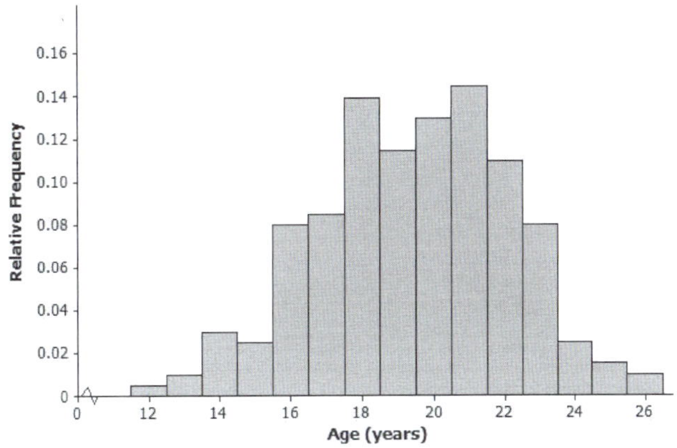

1. Would you describe this distribution of air conditioner ages as approximately symmetric or as skewed? Explain your answer.

   The distribution is approximately symmetric. The left and right sides of the distribution are similar. This distribution would also be described as mound shaped.

2. Is the mean of the age distribution closer to 15, 20, or 25 years? Explain your answer.

   The mean of the age distribution is closer to 20 years because the distribution is centered at about 20.

3. Is the standard deviation of the age distribution closer to 3, 6, or 9 years? Explain your answer.

   A reasonable estimate of an average distance from the mean would be 3, so the standard deviation of the age distribution would be about 3.

Problem Set Sample Solutions

1. For each of the following histograms, describe the shape, and give estimates of the mean and standard deviation of the distributions:

   a. Distribution of head circumferences (mm)

   Shape: Approximately symmetric and mound shaped

   Mean: Approximately 560

   Standard Deviation: Approximately 25

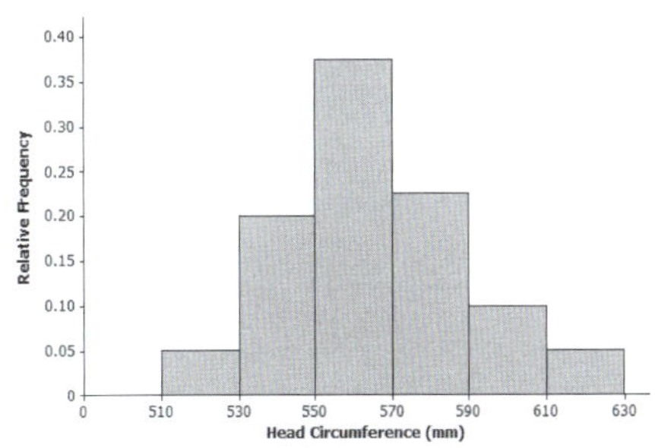

b. Distribution of NBA arena seating capacity

Shape: Approximately symmetric

Center: Approximately $19,000$

Spread: The standard deviation is approximately $1,000$

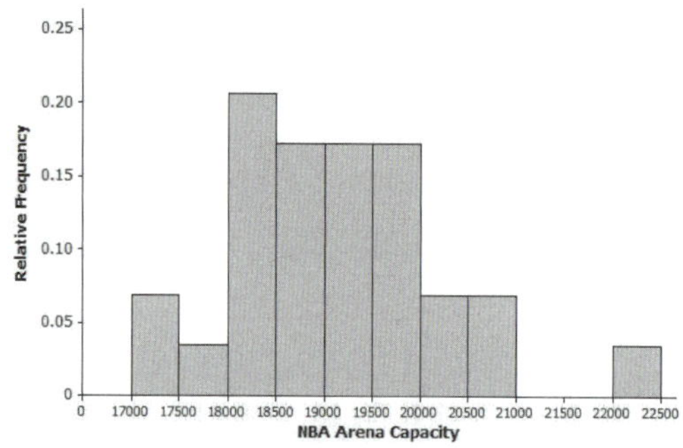

2. For the each of the following, match the description of each distribution with the appropriate histogram:

Histogram	Distribution
1	C
2	B
3	A
4	D

Description of distributions:

Distribution	Shape	Mean	Standard Deviation
A	Approximately symmetric, mound shaped	50	5
B	Approximately symmetric, mound shaped	50	10
C	Approximately symmetric, mound shaped	30	10
D	Approximately symmetric, mound shaped	30	5

Histogram 1

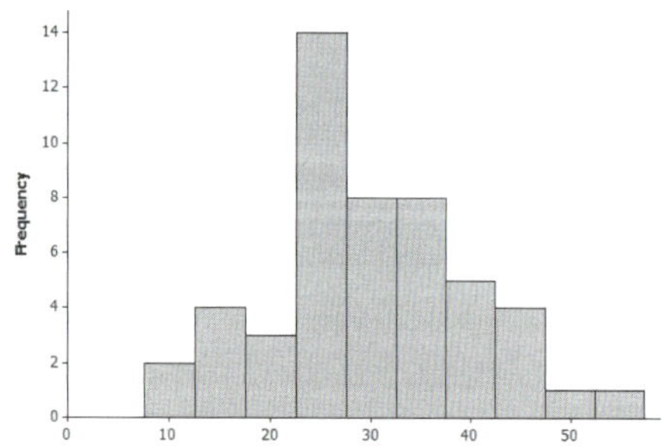

Histogram 2

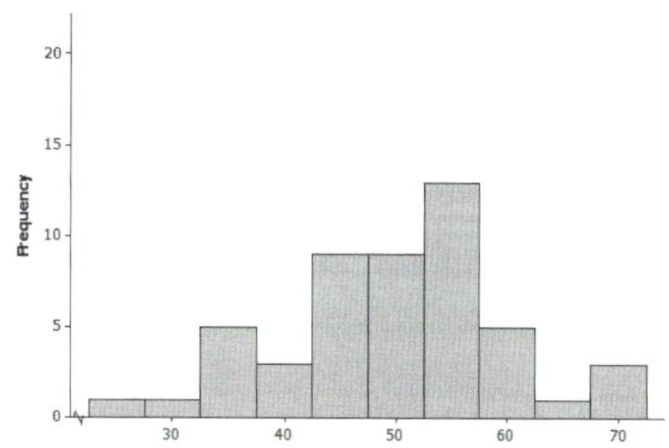

Histogram 3

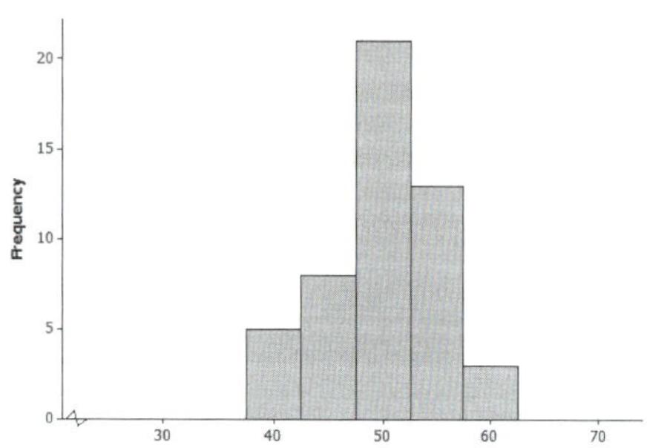

Histogram 4

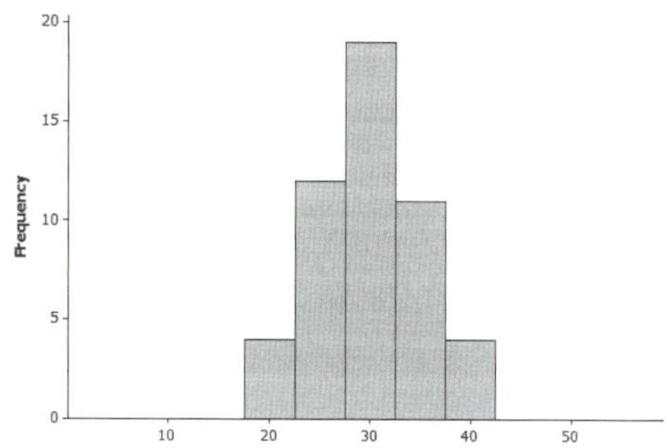

3. Following are the number of calories in a basic hamburger (one meat patty with no cheese) at various fast food restaurants around the country:

$380, 790, 680, 460, 725, 1130, 240, 260, 930, 331, 710, 680, 1080, 612, 1180, 400, 866, 700, 1060, 270, 550, 380, 940, 280, 940, 550, 549, 937, 820, 870, 250, 740$

a. Draw a dot plot on the scale below.

Completed dot plot:

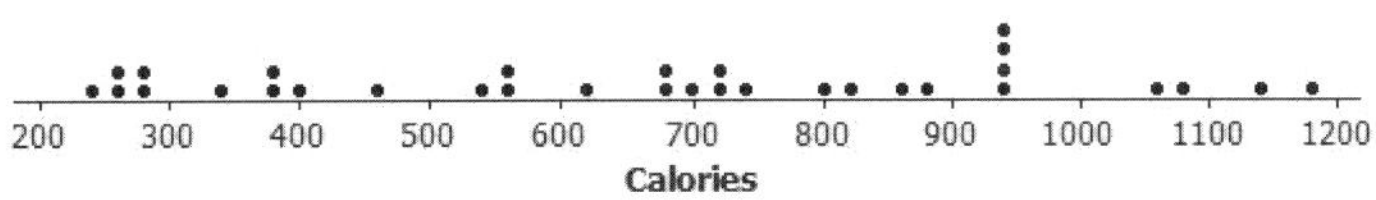

b. Describe the shape of the calorie distribution.

Approximately symmetric

c. Using technology, find the mean and standard deviation of the calorie data.

Mean: 665

Standard deviation: 283

d. Why do you think there is a lot of variability in the calorie data?

Answers will vary. Possible answers include that the size (number of ounces) of the hamburger patty may be different, the type of bun may vary, and burgers may have different types and amounts of various condiments.

Lesson 9: Using a Curve to Model a Data Distribution

Student Outcomes

- Students draw a smooth curve that could be used as a model for a given data distribution.
- Students recognize when it is reasonable and when it is not reasonable to use a normal curve as a model for a given data distribution.

Lesson Notes

This lesson introduces the concept of using a curve to model a data distribution. A smooth curve is used to model a relative frequency histogram, and the idea of an area under the curve representing the approximate proportion of data falling in a given interval is introduced. When data are approximated with a smooth curve, meaningful information can be learned about the distribution. The normal curve (a smooth curve that is bell shaped and symmetric) is introduced. Examples of data distributions that could reasonably be modeled using a normal curve and data distributions that cannot reasonably be modeled by a normal curve are both used in the lesson. In Lessons 10 and 11, students calculate the area under a normal curve and interpret the associated probabilities in context.

Classwork

Example 1 (5 minutes): Heights of Dinosaurs and the Normal Curve

All of the histograms in this lesson are relative frequency histograms. Consider reviewing the meaning of relative frequency prior to having students work on the first exercises. Relative frequency histograms were introduced in Grade 6 and Algebra I. If necessary, discuss how the height of each bar of the histogram is interpreted as the proportion of the data values that fall in the corresponding interval rather than the number of the data values (the frequency) in the interval. The relative frequency can be expressed as either a decimal or a percent.

In many of the exercises, students are asked to find an approximate percent of the data that are within one standard deviation of the mean. Students should base their estimates on the relative frequency that can be found by adding the heights of the bars within one standard deviation of the mean. When the mark for the standard deviation falls within a bar, have students round to the nearest edge of the bar.

MP.4 In several exercises, students model with mathematics when they draw a smooth curve that could be used to model the distribution. Suggest to students that if the distribution is approximately normal, the curve they draw should be bell shaped and roughly passing through the midpoints of bars and the peak in the center of the distribution. When students draw the curve, allow some leeway on the appearance of the curve. This section is the first introduction to modeling a distribution with a curve.

> *Scaffolding:*
>
> For students working below grade level, consider using Exercises 1–7 as an opportunity for teacher modeling or as an activity for mixed-ability groups.
>
> Consider asking students working above grade level to develop their own plans for answering the question, "How tall was a *compy* dinosaur?" Allow them to perform calculations and create their own data displays to answer this question.

To develop motivation for the activities, consider using a discussion question, such as the following:

- Imagine you are a scientist studying dinosaurs that lived millions of years ago. What are some questions you might try to answer about these dinosaurs?
  - *Expect multiple responses such as how much they weighed, average life span, how fast they were, etc.*

In this example, the question to be answered is "How tall was a *compy* dinosaur?" Display the table of data showing the heights of 660 compy dinosaurs. Ask students what each column represents, and emphasize the meaning of relative frequency.

Example 1: Heights of Dinosaurs and the Normal Curve

A paleontologist studies prehistoric life and sometimes works with dinosaur fossils. The table below shows the distribution of heights (rounded to the nearest inch) of 660 procompsognathids, otherwise known as compys.

The heights were determined by studying the fossil remains of the compys.

Height (cm)	Number of Compys	Relative Frequency
26	1	0.002
27	5	0.008
28	12	0.018
29	22	0.033
30	40	0.061
31	60	0.091
32	90	0.136
33	100	0.152
34	100	0.152
35	90	0.136
36	60	0.091
37	40	0.061
38	22	0.033
39	12	0.018
40	5	0.008
41	1	0.002
Total	660	1.000

Exercises 1–8 (15 minutes)

Let students work independently on Exercises 1–8. Then, discuss answers as a class. Some students may have a slightly different answer for the percent within one standard deviation. Since students are approximating an answer, results will vary. Ask students to explain how they arrived at their answers (percent). In addition, ask students to share how they drew the smooth curve for Exercise 6.

Exercises 1–8

The following is a relative frequency histogram of the compy heights:

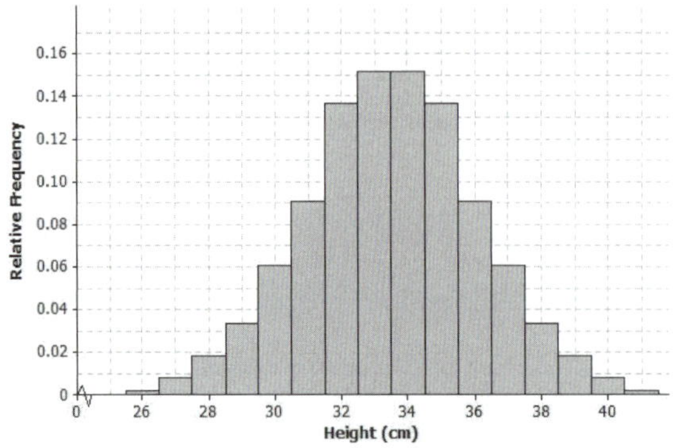

1. **What does the relative frequency of 0.136 mean for the height of 32 cm?**

   13.6% *of the* 660 *compys were* 32 cm *tall.*

2. **What is the width of each bar? What does the height of the bar represent?**

   Each bar has a width of 1 cm. *The height of the bar is the relative frequency for the corresponding compy height.*

3. **What is the area of the bar that represents the relative frequency for compys with a height of 32 cm?**

   The area of the bar is equal to the relative frequency of 0.136.

4. **The mean of the distribution of compy heights is 33.5 cm, and the standard deviation is 2.56 cm. Interpret the mean and standard deviation in this context.**

   The mean of 33.5 cm *is the average height of the compys in the sample. It can be interpreted as a typical compy height.*

   The standard deviation of 2.56 cm *is the typical distance that a compy height is from the mean height.*

5. **Mark the mean on the graph, and mark *one* deviation above and below the mean.**

   a. **Approximately what percent of the values in this data set are within one standard deviation of the mean (i.e., between $33.5\text{ cm} - 2.56\text{ cm} = 30.94\text{ cm}$ and $33.5\text{ cm} + 2.56\text{ cm} = 36.06\text{ cm}$)?**

   $0.091 + 0.136 + 0.152 + 0.152 + 0.136 + 0.091 = 0.758 = 75.8\%$

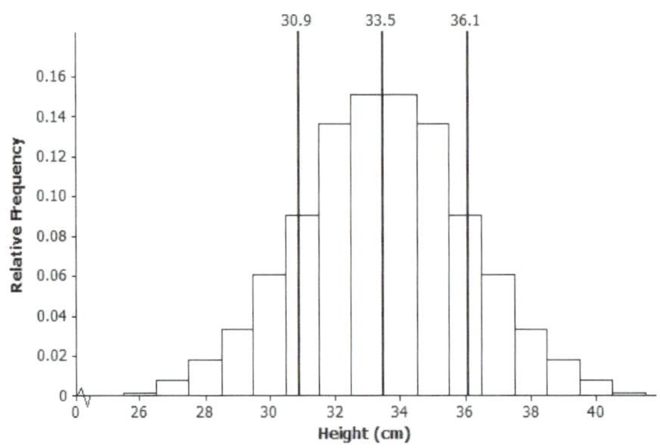

b. **Approximately what percent of the values in this data set are within *two* standard deviations of the mean?**

Between 29 and 38, about 94%

6. **Draw a smooth curve that comes reasonably close to passing through the midpoints of the tops of the bars in the histogram. Describe the shape of the distribution.**

The curve is bell shaped and approximately symmetric and mound shaped.

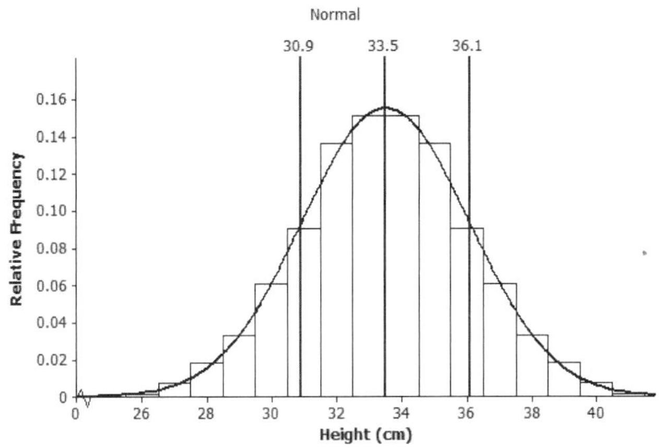

7. **Shade the area of the histogram that represents the proportion of heights that are within one standard deviation of the mean.**

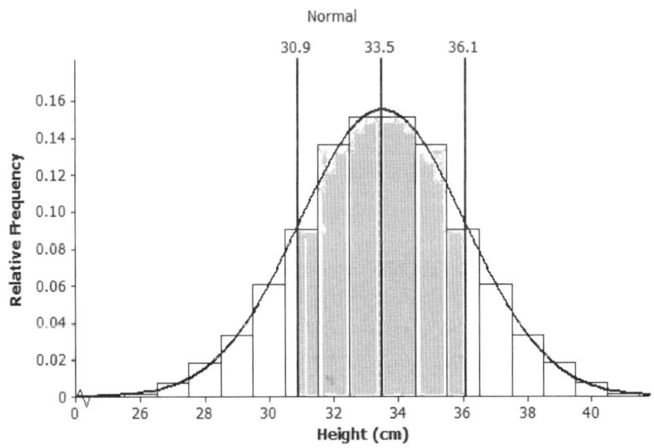

8. **Based on our analysis, how would you answer the question, "How tall was a compy?"**

Answers will vary. It is likely that students will say the height was between 31 cm and 36 cm.

Example 2 (5 minutes): Gas Mileage and the Normal Distribution

Read through the example as a class.

Example 2: Gas Mileage and the Normal Distribution

A normal curve is a smooth curve that is symmetric and bell shaped. Data distributions that are mound shaped are often modeled using a normal curve, and we say that such a distribution is approximately normal. One example of a distribution that is approximately normal is the distribution of compy heights from Example 1. Distributions that are approximately normal occur in many different settings. For example, a salesman kept track of the gas mileage for his car over a 25-week span.

The mileages (miles per gallon rounded to the nearest whole number) were

$23, 27, 27, 28, 25, 26, 25, 29, 26, 27, 24, 26, 26, 24, 27, 25, 28, 25, 26, 25, 29, 26, 27, 24, 26.$

Exercise 9 (10 minutes)

MP.5 Students are asked to find the mean and standard deviation using technology. Ask students to indicate how technology helps them make sense of the data. If using a graphing calculator similar to the TI-84, the mileages are entered into L1 and the frequency into L2. To find the mean and standard deviation, select 1-Var Stats and type *L1, L2*. After the 1-Var Stats entry, select Enter. Consult an appropriate manual or similar resource if using a different type of calculator or if using a statistical software package that is different from the program described above.

Remind students that when they construct the histogram, they should center the mileage in the middle of each bar.

Let students work with a partner or in a small group based on available access to technology.

Exercise 9

9. Consider the following:

a. Use technology to find the mean and standard deviation of the mileage data. How did you use technology to assist you?

Mean $= 26.04$ mpg
Standard deviation $= 1.54$ mpg

The graphing calculator does several tedious calculations for me. I entered the data into lists and was able to indicate what calculations I wanted done by writing an expression using lists. I did not have to set up the organization to find the standard deviation and perform the rather messy calculations.

b. Calculate the relative frequency of each of the mileage values. For example, the mileage of 26 mpg has a frequency of 7. To find the relative frequency, divide 7 by 25, the total number of mileages recorded. Complete the following table:

Mileage (mpg)	Frequency	Relative Frequency
23	1	0.04
24	3	0.12
25	5	0.20
26	7	0.28
27	5	0.20
28	2	0.08
29	2	0.08
Total	25	1.00

c. **Construct a relative frequency histogram using the scale below.**

Completed histogram:

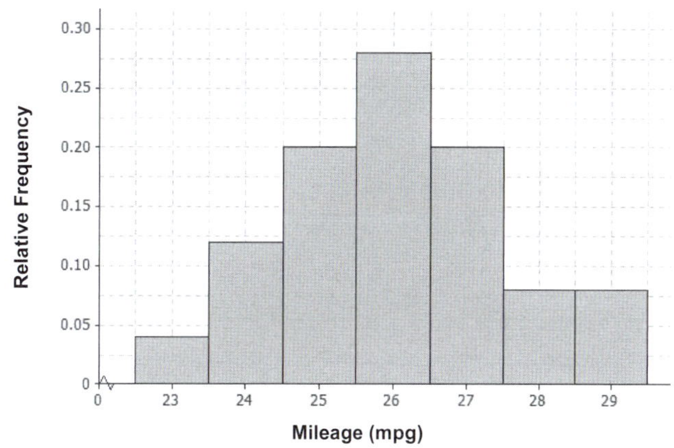

d. **Describe the shape of the mileage distribution. Draw a smooth curve that comes reasonably close to passing through the midpoints of the tops of the bars in the histogram. Is this approximately a normal curve?**

The shape is approximately normal. See the graph in part (e).

e. **Mark the mean on the histogram. Mark one standard deviation to the left and right of the mean. Shade the area of the histogram that represents the proportion of mileages that are within one standard deviation of the mean. Find the proportion of the data within one standard deviation of the mean.**

One standard deviation to the left (or below) the mean: $26.04 - 1.54 = 24.5$

One standard deviation to the right (or above) the mean: $26.04 + 1.54 = 27.58$

The proportion of the data within one standard deviation of the mean is approximately 0.68 *(which is the sum of* $0.20 + 0.28 + 0.20$*).*

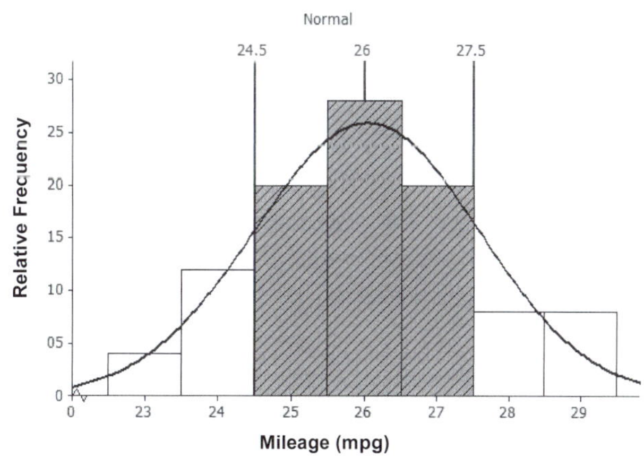

Closing (2 minutes)

Ask students to summarize the main ideas of the lesson with a neighbor or in writing. Use this as an opportunity to informally assess comprehension of the lesson. The Lesson Summary below offers some important ideas that should be included.

- Is the mean of a distribution that is approximately normal located near where the curve is the highest?
  - *Yes*
- Is the mean of a skewed distribution located near where the curve is the highest? Why does this happen?
  - *No. In a skewed distribution, the mean will be pulled toward the values in the tail of the distribution.*

Lesson Summary

- **A normal curve is symmetric and bell shaped. The mean of a normal distribution is located in the center of the distribution. Areas under a normal curve can be used to estimate the proportion of the data values that fall within a given interval.**

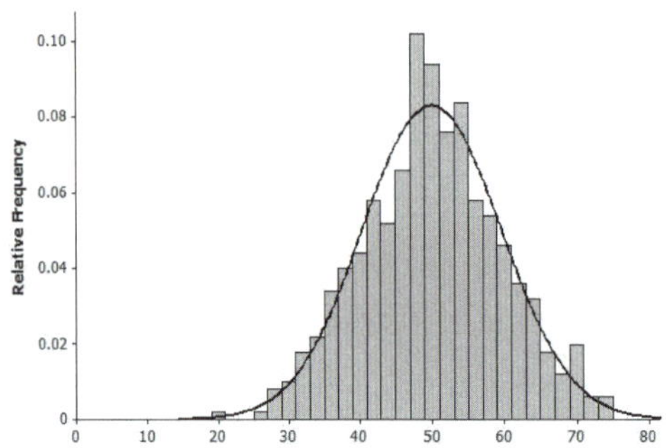

- **When a distribution is skewed, it is not appropriate to model the data distribution with a normal curve.**

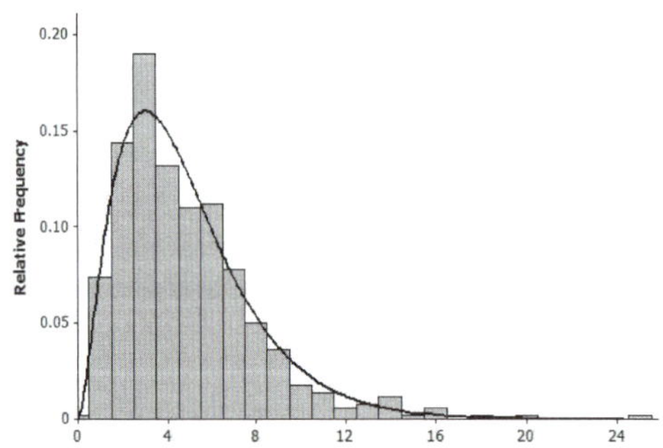

Exit Ticket (8 minutes)

Name ______________________________ Date ____________

Lesson 9: Using a Curve to Model a Data Distribution

Exit Ticket

The histogram below shows the distribution of heights (to the nearest inch) of $1{,}000$ young women.

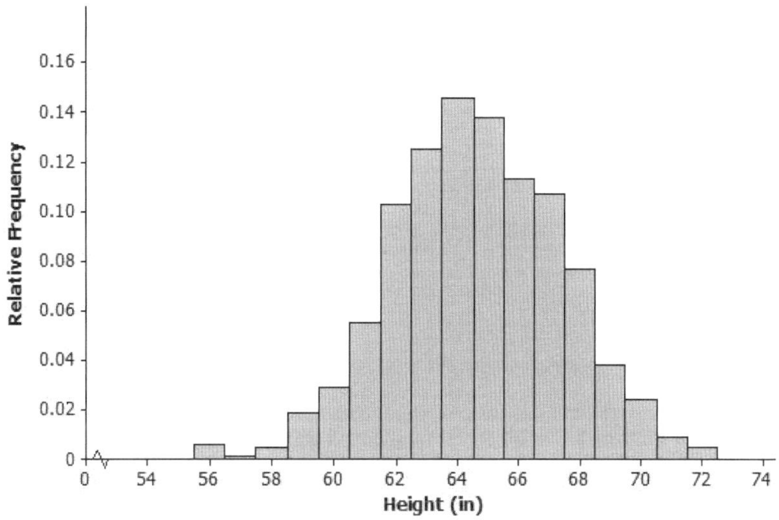

1. What does the width of each bar represent? What does the height of each bar represent?

2. The mean of the distribution of women's heights is 64.6 in., and the standard deviation is 2.75 in. Interpret the mean and standard deviation in this context.

3. Mark the mean on the graph, and mark *one* deviation above and below the mean. Approximately what proportion of the values in this data set are within one standard deviation of the mean?

4. Draw a smooth curve that comes reasonably close to passing through the midpoints of the tops of the bars in the histogram. Describe the shape of the distribution.

5. Shade the area of the histogram that represents the proportion of heights that are within one standard deviation of the mean.

Exit Ticket Sample Solutions

The histogram below shows the distribution of heights (to the nearest inch) of $1,000$ young women.

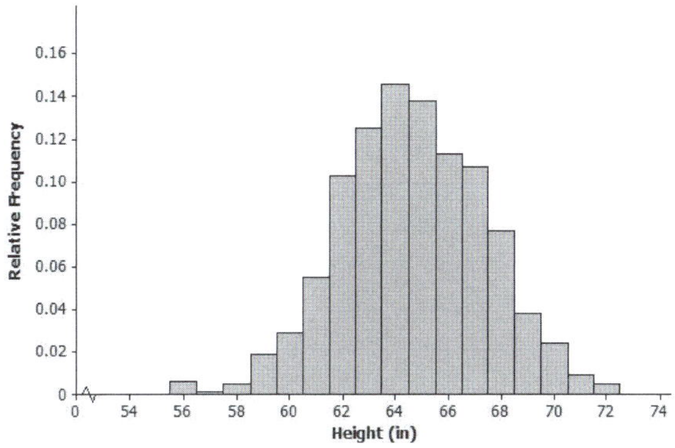

1. **What does the width of each bar represent? What does the height of each bar represent?**

   Each bar represents a 1-inch range of heights. For example, the bar above 56 represents heights between 55.50 and 56.49 inches. The height of each bar represents the proportion of the $1,000$ women in that 1-inch height range.

2. **The mean of the distribution of women's heights is 64.6 in., and the standard deviation is 2.75 in. Interpret the mean and standard deviation in this context.**

   The mean is the average height of the $1,000$ women, and it can be interpreted as a typical height value.

   The standard deviation is the typical number of inches that a woman's height is from the mean.

3. **Mark the mean on the graph, and mark *one* deviation above and below the mean. Approximately what proportion of the values in this data set are within one standard deviation of the mean?**

   $0.10 + 0.12 + 0.14 + 0.13 + 0.11 + 0.105 = 0.705 \approx 0.71$

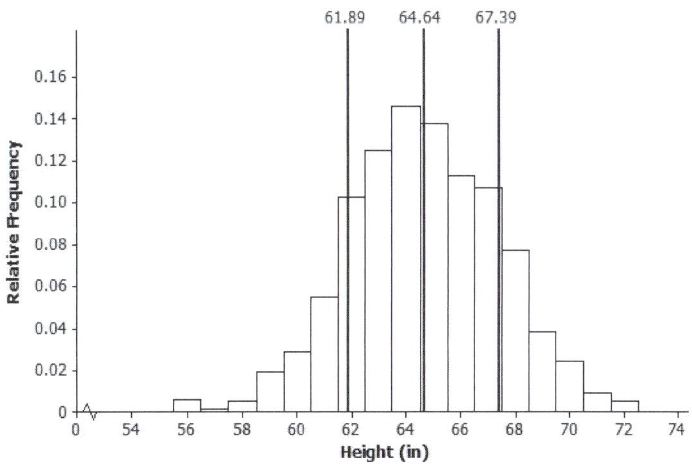

4. **Draw a smooth curve that comes reasonably close to passing through the midpoints of the tops of the bars in the histogram. Describe the shape of the distribution.**

Approximately normal (bell shaped and approximately symmetric)

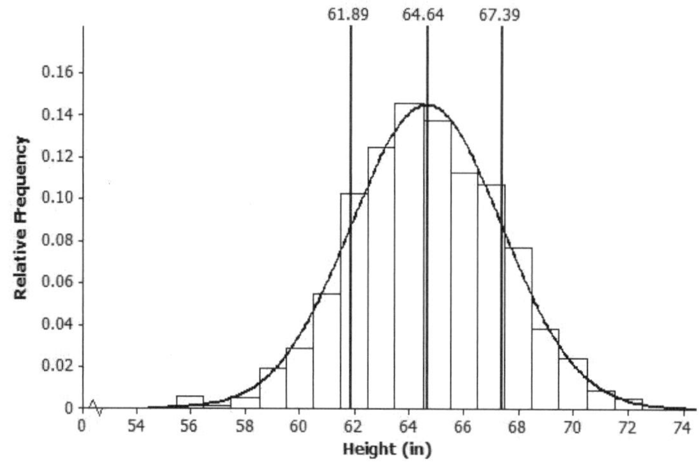

5. **Shade the area of the histogram that represents the proportion of heights that are within one standard deviation of the mean.**

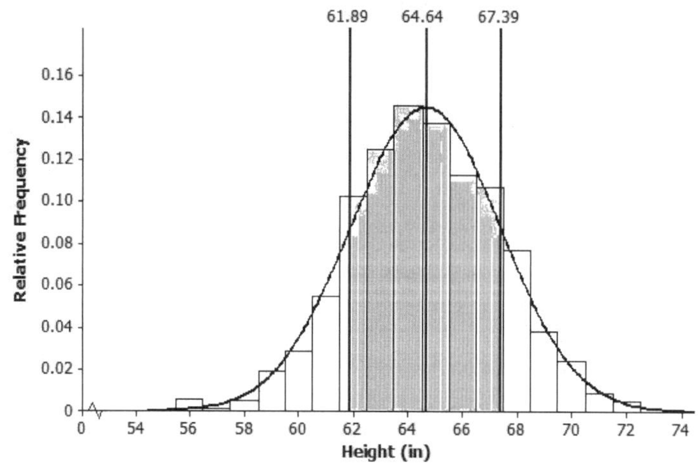

Problem Set Sample Solutions

1. **Periodically the U.S. Mint checks the weight of newly minted nickels. Below is a histogram of the weights (in grams) of a random sample of 100 new nickels.**

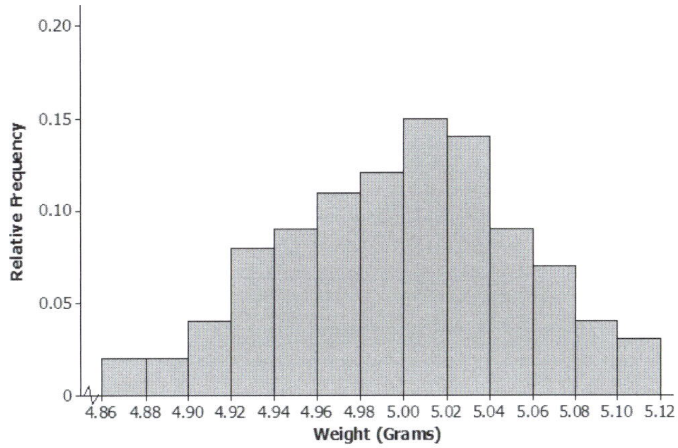

   a. **The mean and standard deviation of the distribution of nickel weights are 5.00 grams and 0.06 gram, respectively. Mark the mean on the histogram. Mark one standard deviation above the mean and one standard deviation below the mean.**

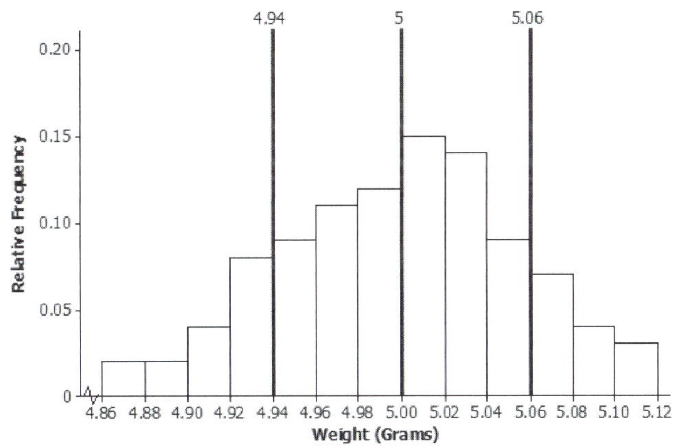

b. **Describe the shape of the distribution. Draw a smooth curve that comes reasonably close to passing through the midpoints of the tops of the bars in the histogram. Is this approximately a normal curve?**

The shape is approximately normal.

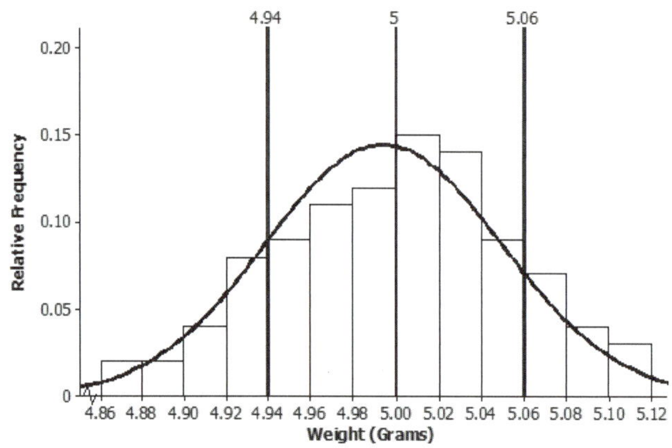

c. **Shade the area of the histogram that represents the proportion of weights within one standard deviation of the mean. Find the proportion of the data within one standard deviation of the mean.**

$0.09 + 0.11 + 0.12 + 0.15 + 0.14 + 0.09 = 0.70$

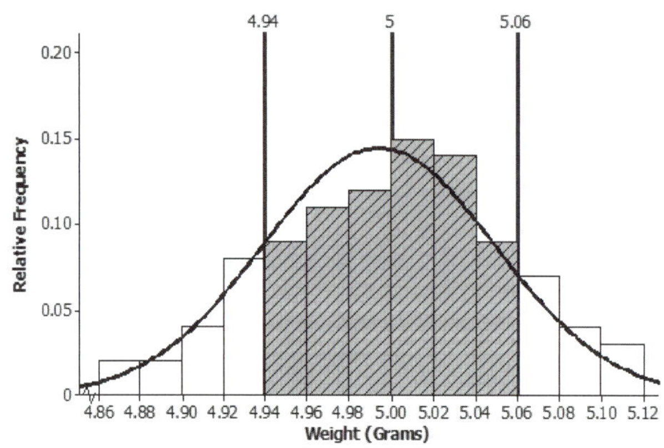

2. Below is a relative frequency histogram of the gross (in millions of dollars) for the all-time top-grossing American movies (as of the end of 2012). *Gross* is the total amount of money made before subtracting out expenses, like advertising costs and actors' salaries.

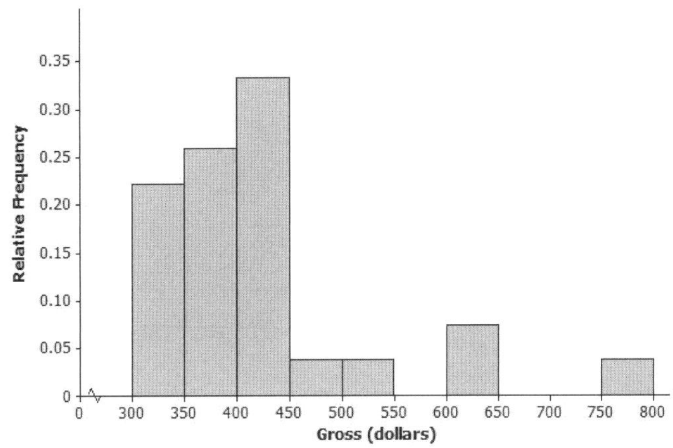

a. Describe the shape of the distribution of all-time top-grossing movies. Would a normal curve be the best curve to model this distribution? Explain your answer.

The shape is skewed to the right. A normal curve would not be the best curve to model the distribution.

b. Which of the following is a reasonable estimate for the mean of the distribution? Explain your choice.

i. 325 million

ii. 375 million

iii. 425 million

(iii) 425 million is a reasonable estimate since this is a skewed distribution, and the mean will be pulled toward the outliers.

c. Which of the following is a reasonable estimate for the sample standard deviation? Explain your choice.

i. 50 million

ii. 100 million

iii. 200 million

(ii) 100 million is a reasonable estimate because 50 million is too small, and 200 million is too large to be considered a typical deviation from the mean.

3. **Below is a histogram of the top speed of different types of animals.**

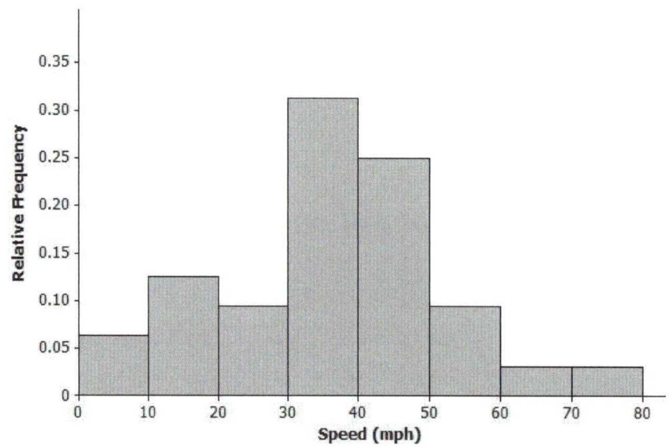

a. **Describe the shape of the top speed distribution.**

Approximately normal

b. **Estimate the mean and standard deviation of this distribution. Describe how you made your estimate.**

Mean is approximately 40 mph*, and standard deviation is about* 15 mph*. Answers will vary in terms of how the estimate was made.*

Note: Students could roughly estimate the mean by locating a balance point of the distribution. The standard deviation could be estimated by a typical deviation from the mean that was developed in Lesson 8.

c. **Draw a smooth curve that is approximately a normal curve. The actual mean and standard deviation of this data set are** 34.1 mph **and** 15.3 mph**, respectively. Shade the area of the histogram that represents the proportion of speeds that are within one standard deviation of the mean.**

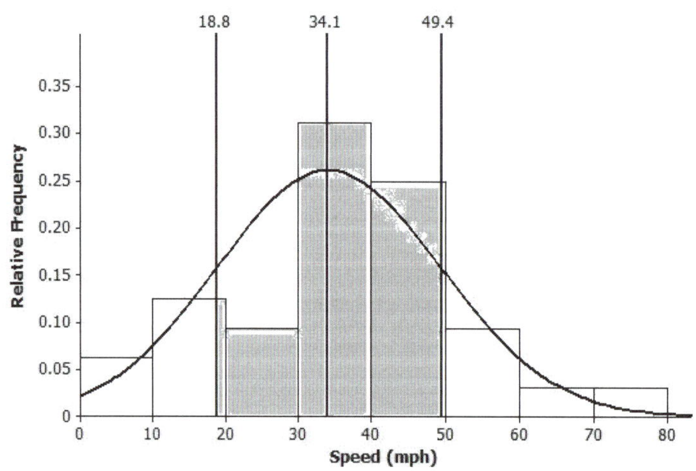

Lesson 10: Normal Distributions

Student Outcomes

- Students calculate z-scores.
- Students use technology and tables to estimate the area under a normal curve.
- Students interpret probabilities in context.

Lesson Notes

In this lesson, students calculate z-scores and use technology and tables to estimate the area under a normal curve. Depending on technology resources available to students, teachers may need to have students work with a partner or in small groups.

Classwork

Exercise 1 (3 minutes)

This first exercise is a review of what is meant by a normal distribution. Use this as an opportunity to informally assess students' understanding of different distribution types by having students attempt the exercises independently.

Exercise 1

Consider the following data distributions. In the previous lesson, you distinguished between distributions that were approximately normal and those that were not. For each of the following distributions, indicate if it is approximately normal, skewed, or neither, and explain your choice:

a.

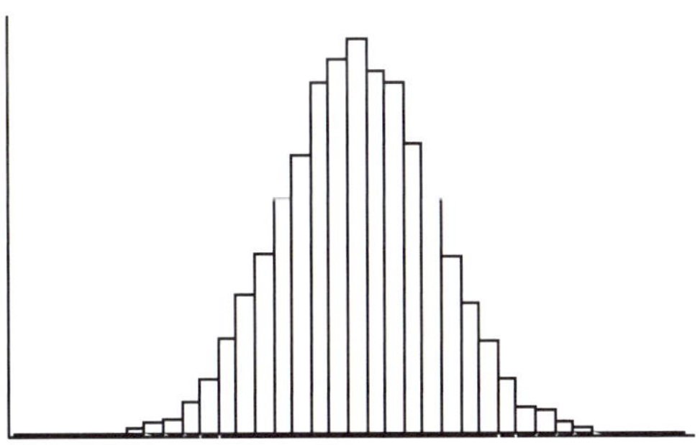

This distribution is approximately normal. It is approximately symmetric and mound shaped.

b.

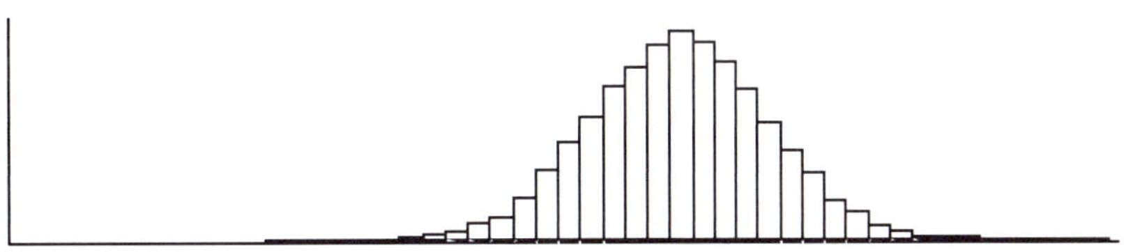

This distribution is approximately normal. It is approximately symmetric and mound shaped.

c.

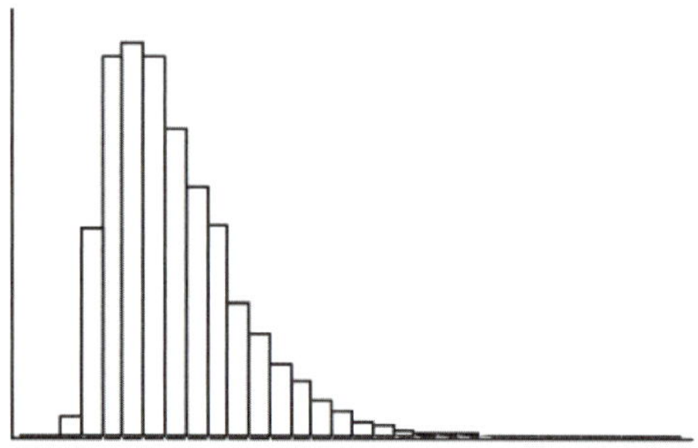

This distribution is not symmetric; therefore, it is not approximately normal. This distribution is skewed to the right as it has most of the data values in the beginning and then tapers off. It has a longer tail on the right.

d.

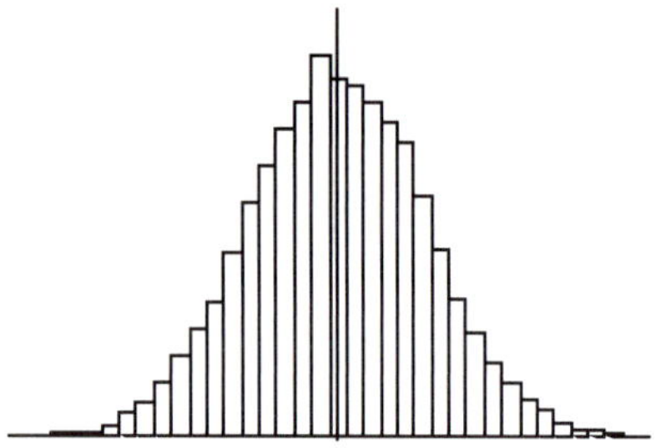

This distribution is approximately normal. It is symmetric and mound shaped.

e.

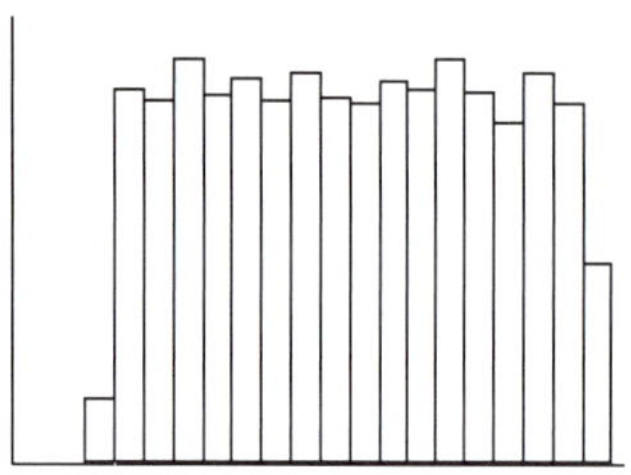

This distribution is not approximately normal. It is symmetric but not mound shaped. This distribution, however, is also not skewed. It does not have a longer tail on one side. It would be described as approximately a uniform distribution. Note that this distribution was discussed in earlier grades. It is important to emphasize that there are other types of data distributions and that not all are either approximately normal or skewed.

A normal distribution is a distribution that has a particular symmetric mound shape, as shown below.

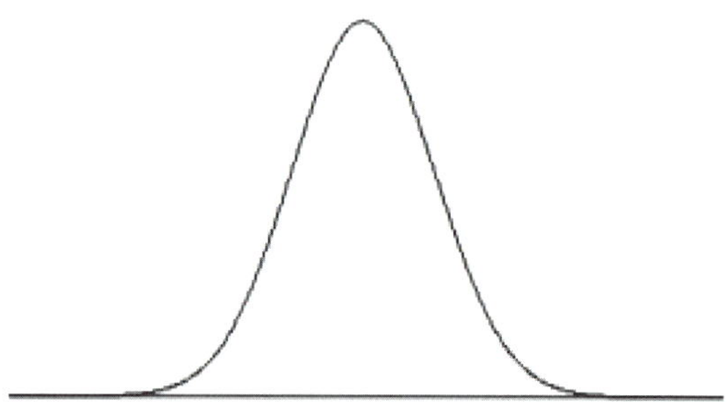

Exercise 2 (5 minutes)

Prior to having students tackle this introduction to z-scores (or after they have completed it), it might be helpful to familiarize them with the reason why z-scores are important. Open with the following discussion:

MP.3

- Suppose that you took a math test and a Spanish test. The mean score for both tests was 80. You got an 86 in math and a 90 in Spanish. Did you necessarily do better in Spanish relative to your fellow students?
  - *No. For example, suppose the standard deviation of the math scores was 4 and the standard deviation of the Spanish scores was 8. Then, my score of 86 in math is $1\frac{1}{2}$ standard deviations above the mean, and my score in Spanish is only $1\frac{1}{4}$ standard deviations above the mean. Relative to the other students, I did better in math.*

In the above example, the z-score for math is 1.5, and the z-score for Spanish is 1.25. Negative z-scores indicate values that are below the mean.

Ask students to state in their own words what a z-score is. Essentially, look for an early understanding that a z-score measures distance in units of the standard deviation. For example, a z-score of 1 represents an observation that is at a distance of 1 standard deviation above the mean.

Scaffolding:

- For students working below grade level, consider framing the question this way: "Suppose that you took a math test and a Spanish test. The mean for both tests was 80; the standard deviation of the math scores was 4, and the standard deviation of the Spanish scores was 8. You got an 86 in math and a 90 in Spanish. On which test did you do better than most of your classmates?" Consider showing a visual representation of sample data distributions for each to aid understanding.
- For students working above grade level, consider posing the question this way: "Suppose that you took a math test and a Spanish test. The mean score for both was 80. You got an 86 in math and a 90 in Spanish. On which test did you do better than most of your classmates? Explain your reasoning."

Exercise 2

When calculating probabilities associated with normal distributions, z-scores are used. A z-score for a particular value measures the number of standard deviations away from the mean. A positive z-score corresponds to a value that is above the mean, and a negative z-score corresponds to a value that is below the mean. The letter z is used to represent a variable that has a standard normal distribution where the mean is 0 and standard deviation is 1. This distribution was used to define a z-score. A z-score is calculated by

$$z = \frac{\text{value} - \text{mean}}{\text{standard deviation}}.$$

a. **The prices of the printers in a store have a mean of $\$240$ and a standard deviation of $\$50$. The printer that you eventually choose costs $\$340$.**

i. **What is the z-score for the price of your printer?**

$$z = \frac{340 - 240}{50} = 2$$

ii. **How many standard deviations above the mean was the price of your printer?**

The price of my printer was 2 standard deviations above the mean price.

b. **Ashish's height is 63 inches. The mean height for boys at his school is 68.1 inches, and the standard deviation of the boys' heights is 2.8 inches.**

i. **What is the z-score for Ashish's height? (Round your answer to the nearest hundredth.)**

$$z = \frac{63 - 68.1}{2.8} \approx -1.82$$

ii. **What is the meaning of this value?**

Ashish's height is 1.82 standard deviations below the mean height for boys at his school.

c. **Explain how a z-score is useful in describing data.**

A z-score is useful in describing how far a particular point is from the mean.

Example 1 (10 minutes): Use of z-Scores and a Graphing Calculator to Find Normal Probabilities

In this example, students are introduced to the process of calculating normal probabilities, and in this example and the two exercises that follow, z-scores are used along with a graphing calculator. (The use of tables of normal probabilities is introduced later in this lesson, and the use of spreadsheets is introduced in the next lesson.) Encourage students to always draw normal distribution curves and to show their work on the graph when working problems that involve a normal distribution.

Work through Example 3 with the class showing students how to calculate the relevant z-scores, and use a graphing calculator* to find the probability of interest. Students new to the curriculum may need additional support with the graphing calculator.

*Calculator note: The general form of this is *Normalcdf([left z bound],[right z bound])*. The *Normalcdf* function is accessed using *2nd, DISTR*. On selecting *2nd, DISTR*, some students using the more recent TI-84 operating systems might be presented with a menu asking for left bound, right bound, mean, and standard deviation. This can be avoided by having these students do the following: Press *2nd, QUIT* (to return to the home screen); press *MODE*; scroll down to the *NEXT* screen; set *STAT WIZARDS* to *OFF*.

Example 1: Use of z-Scores and a Graphing Calculator to Find Normal Probabilities

A swimmer named Amy specializes in the 50-meter backstroke. In competition, her mean time for the event is 39.7 seconds, and the standard deviation of her times is 2.3 seconds. Assume that Amy's times are approximately normally distributed.

a. **Estimate the probability that Amy's time is between 37 and 44 seconds.**

The first time is a little less than 2 standard deviations from her mean time of 39.7 seconds. The second time is nearly 2 standard deviations above her mean time. As a result, the probability of a time between the two values covers nearly 4 standard deviations and would be rather large. I estimate 0.9, or 90%.

b. Using z-scores and a graphing calculator and rounding your answers to the nearest thousandth, find the probability that Amy's time in her next race is between 37 and 44 seconds.

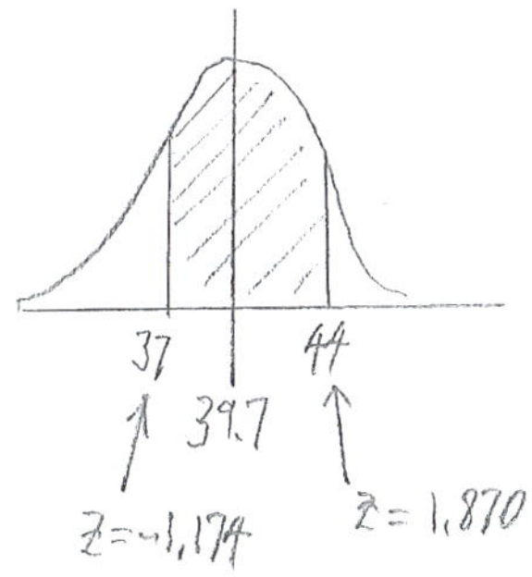

The z-score for 44 is $z = \frac{44 - 39.7}{2.3} \approx 1.870$, *and the z-score for 37 is* $z = \frac{37 - 39.7}{2.3} \approx -1.174$.

The probability that Amy's time is between 37 and 44 seconds is then found to be 0.849.

Note: If students are using TI-83 or TI-84 calculators, this result is found by entering Normalcdf$(-1.174, 1.870)$.

c. Estimate the probability that Amy's time is more than 45 seconds.

Amy's time of 45 seconds is more than 2 standard deviations from the mean of 39.7 *seconds. There is a small probability that her time will be greater than 45 seconds. I estimate* 0.03*, or* 3%.

d. Using z-scores and a graphing calculator and rounding your answers to the nearest thousandth, find the probability that Amy's time in her next race is more than 45 seconds.

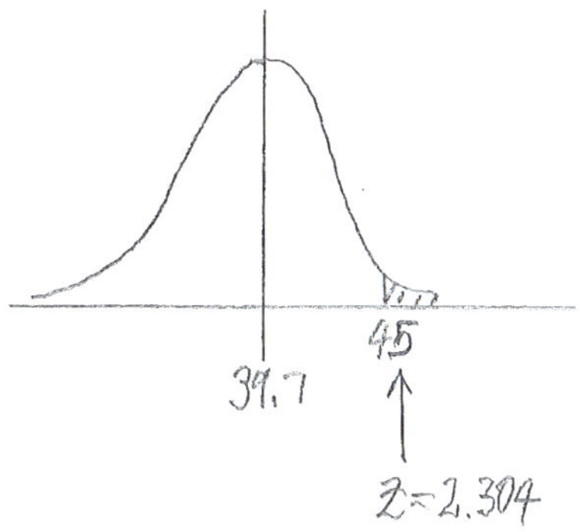

The z-score for 45 is $z = \frac{45 - 39.7}{2.3} \approx 2.304$.

The probability that Amy's time is more than 45 seconds is then found to be 0.011.

Note: This result is found by entering Normalcdf$(2.304, 999)$ *or the equivalent of this for other brands of calculators. The keystroke for positive infinity on some calculators is 1EE99. The number 999 can be replaced by any large positive number. Strictly speaking, the aim here is to find the area under the normal curve between* $z = 2.304$ *and positive infinity. However, it is impossible to enter positive infinity into the calculator, so any large positive number can be used in its place.*

e. What is the probability that Amy's time would be at least 45 seconds?

Since Amy's times have a continuous distribution, the probability of "more than 45 seconds" and the probability of "at least 45 seconds" are the same. So, the answer is the same as part (d), 0.011.
It is worth pointing out to students that this applies to any question about the normal distribution.

f. Using z-scores and a graphing calculator and rounding your answers to the nearest thousandth, find the probability that Amy's time in her next race is less than 36 seconds.

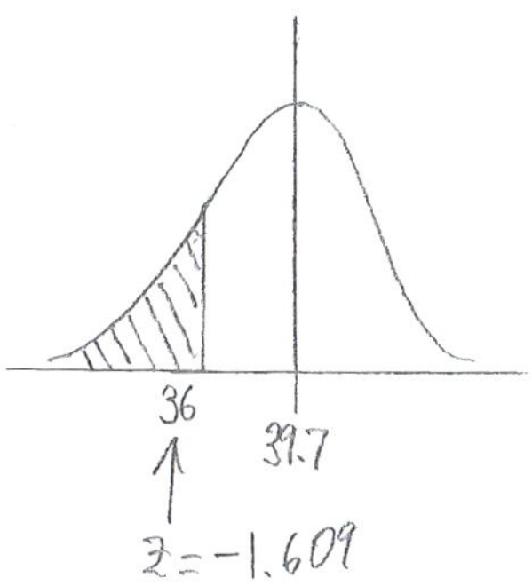

The z-score for 36 is $z = \frac{36 - 39.7}{2.3} \approx -1.609$.

The probability that Amy's time is less than 36 seconds is then found to be 0.054.

Note: This result is found by entering Normalcdf$(-999, -1.609)$ *or the equivalent of this command for other brands of calculators. Here,* -999 *is being used in place of negative infinity; any large negative number can be used in its place. The keystroke on some calculators for negative infinity is –1EE99.*

Exercise 3 (8 minutes)

MP.4 This exercise provides practice with the skills students have learned in Example 2. Again, encourage students to draw a normal distribution curve for each part of the exercise showing work on their graphs in order to answer questions about the distribution. Let students work independently (if technology resources allow) and confirm answers with a neighbor.

Exercise 3

The distribution of lifetimes of a particular brand of car tires has a mean of $51,200$ miles and a standard deviation of $8,200$ miles.

a. **Assuming that the distribution of lifetimes is approximately normally distributed and rounding your answers to the nearest thousandth, find the probability of each event.**

i. **A randomly selected tire lasts between $55,000$ and $65,000$ miles.**

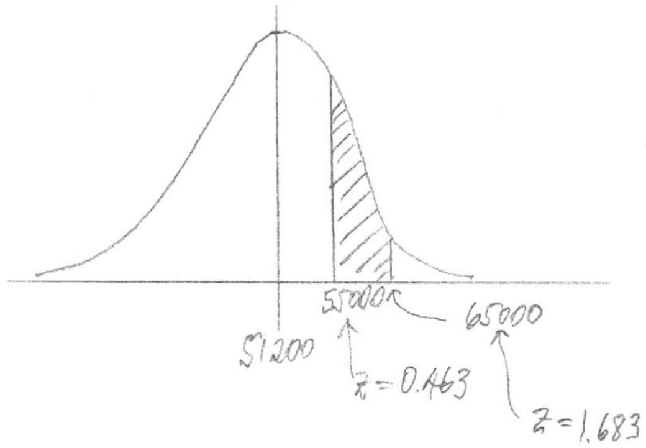

$\boldsymbol{P(\text{between } 55000 \text{ and } 65000) = 0.275}$

ii. **A randomly selected tire lasts less than $48,000$ miles.**

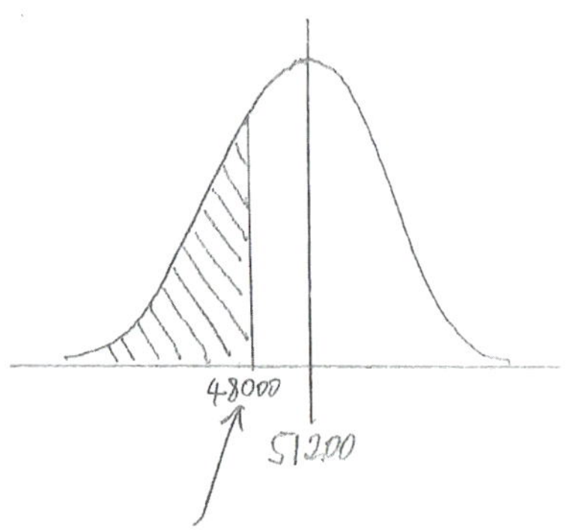

$\boldsymbol{P(\text{less than } 48000) = 0.348}$

iii. A randomly selected tire lasts at least $41,000$ miles.

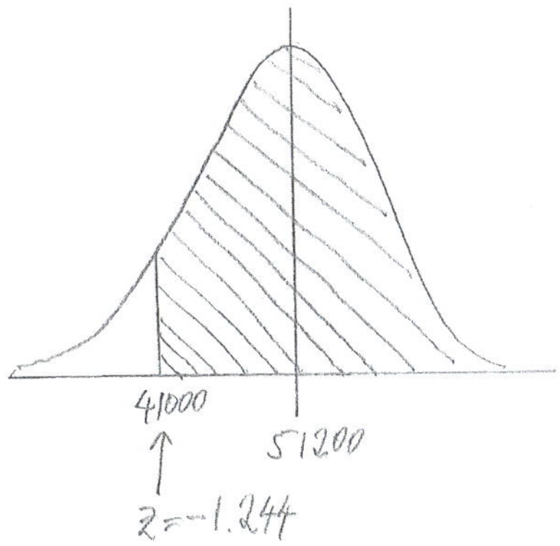

$P(\text{greater than } 41000) = 0.893$

b. Explain the meaning of the probability that you found in part (a)(iii).

If a large number of tires of this brand were to be randomly selected, then you would expect about 89.3% of them to last more than $41,000$ miles.

Exercise 4 (5 minutes)

MP.3

Here, students have to understand the idea of the lifetime of a tire being "within 10,000 miles of the mean." Some discussion of this idea might be necessary. Ask students what they think the statement is indicating about the lifetime of a tire. As they share their interpretations, record their summaries. Encourage students to use pictures, possibly involving a normal distribution, in their explanations. In this exercise, students can practice constructing arguments and critiquing the reasoning of others based on their understanding of a normal distribution.

Exercise 4

Think again about the brand of tires described in Exercise 3. What is the probability that the lifetime of a randomly selected tire is within $10,000$ miles of the mean lifetime for tires of this brand?

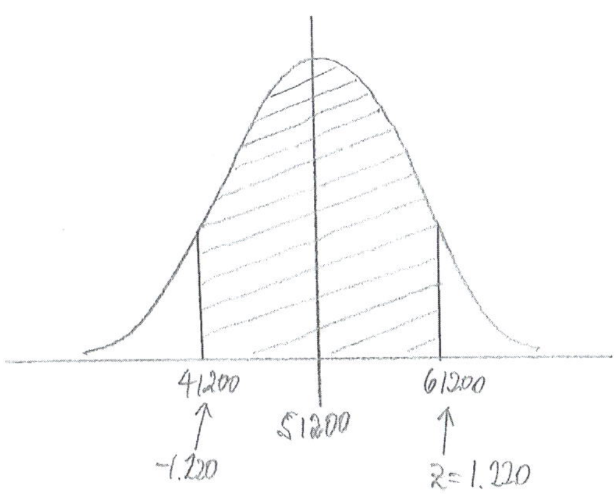

$P(\text{within } 10000 \text{ of mean}) = 0.778$

Example 2 (7 minutes): Using Table of Standard Normal Curve Areas

The last part of this lesson is devoted to the process of using tables of normal probabilities in place of the *Normalcdf* (or equivalent) function on the calculator. Completion of this example enables students who do not have access to a graphing calculator at home to complete the Problem Set. (However, if time is short, then this example could be reserved for Lesson 11, in which case Problem Set Problem 4 should be omitted from the assignment, and Lesson 11 should begin with this example.) Work through this example as a class showing students how to use the table to estimate the area under a normal curve.

Example 2: Using Table of Standard Normal Curve Areas

The standard normal distribution is the normal distribution with a mean of 0 and a standard deviation of 1. The diagrams below show standard normal distribution curves. Use a table of standard normal curve areas to determine the shaded areas.

a.

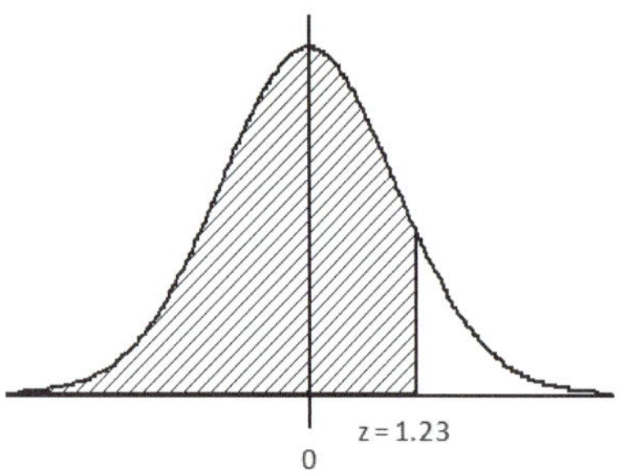

The provided table of normal areas gives the area to the left of the selected z-score. So, here, you find 1.2 in the column on the very left of the table, and then move horizontally to the column labeled 0.03. The table gives the required area (probability) to be 0.8907.

b.

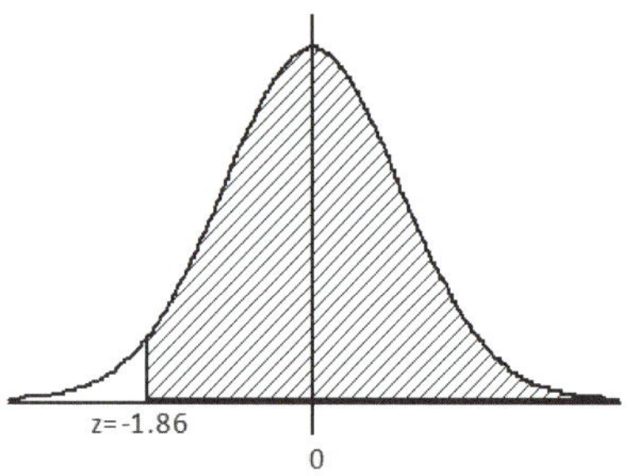

In the left-hand column of the table, find -1.8, and move horizontally to the column labeled 0.06. The table gives a probability of 0.0314. Note that the table always supplies the area to the left of the chosen z-score. Also, note that the total area under any normal curve is 1. (This is the case for any probability distribution curve.) So, the required area, which is the area to the right of -1.86, is $1 - 0.0314 = 0.9686$.

c.

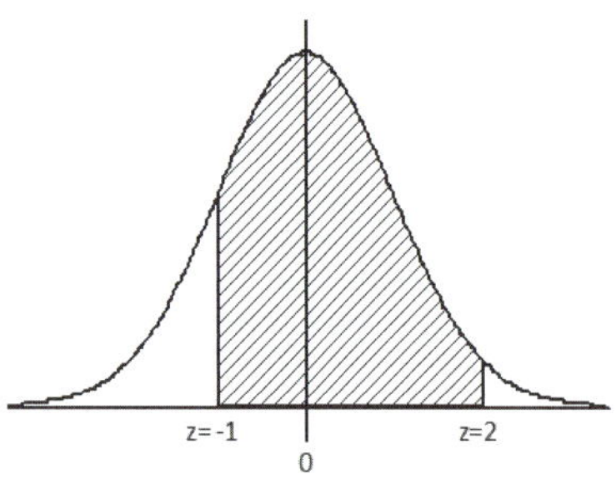

The approach here is to find the area to the left of $z = 2$ and to subtract the area to the left of $z = -1$. The table gives the area to the left of $z = 2.00$ to be 0.9772 and the area to the left of $z = -1.00$ to be 0.1587. So, the required area is $0.9772 - 0.1587 = 0.8185$.

Closing (2 minutes)

Remind students to be aware of order when calculating z-scores. Refer back to Exercise 2.

- The prices of printers in a store have a mean of $\$240$ and a standard deviation of $\$50$. The printer that you eventually choose costs $\$340$. What is wrong with the following z-score? How do you know?

$$z = \frac{240 - 340}{50} = \frac{-100}{50} = -2$$

  - *The z-score is negative, when it should be positive. The printer I chose is greater than the mean, which should result in a positive z-score.*

Have students interpret a probability in their own words. For example, refer back to Example 1.

- A swimmer named Amy specializes in the 50-meter backstroke. You found that the probability that Amy's time is between 37 and 44 seconds is 0.849. How would you interpret this?
  - *Approximately 84.9% of Amy's finish times are between 37 and 44 seconds.*

Ask students to summarize the main ideas of the lesson in writing or with a neighbor. Use this as an opportunity to informally assess comprehension of the lesson. The Lesson Summary below offers some important ideas that should be included.

> **Lesson Summary**
>
> **A *normal distribution* is a continuous distribution that has the particular symmetric mound-shaped curve that is shown at the beginning of the lesson.**
>
> **Probabilities associated with normal distributions are determined using z-scores and can be found using a graphing calculator or tables of standard normal curve areas.**

Exit Ticket (5 minutes)

Name ______________________________ Date ______________

Lesson 10: Normal Distributions

Exit Ticket

The weights of cars passing over a bridge have a mean of $3,550$ pounds and standard deviation of 870 pounds. Assume that the weights of the cars passing over the bridge are normally distributed. Determine the probability of each instance, and explain how you found each answer.

a. The weight of a randomly selected car is more than $4,000$ pounds.

b. The weight of a randomly selected car is less than $3,000$ pounds.

c. The weight of a randomly selected car is between $2,800$ and $4,500$ pounds.

Exit Ticket Sample Solutions

Student answers may vary if using a table versus a graphing calculator to determine the area under the normal curve.

The weights of cars passing over a bridge have a mean of $3,550$ pounds and standard deviation of 870 pounds. Assume that the weights of the cars passing over the bridge are normally distributed. Determine the probability of each instance, and explain how you found each answer.

a. **The weight of a randomly selected car is more than $4,000$ pounds.**

$P(\text{greater than } 4000) = 0.303$

I calculated a z-score and then used my graphing calculator to find the area under the normal curve (above $z = 0.517$).

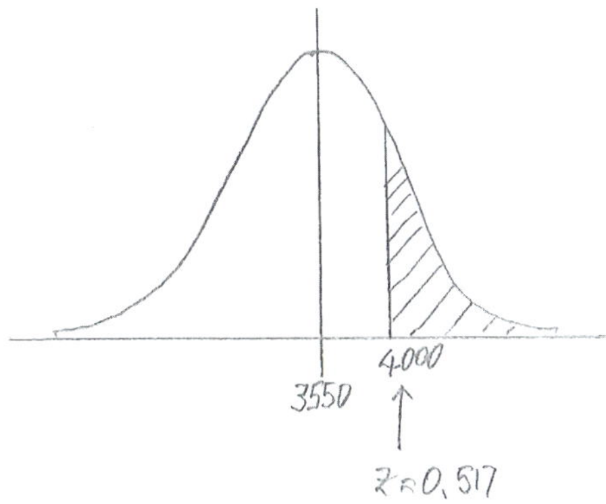

b. **The weight of a randomly selected car is less than $3,000$ pounds.**

$P(\text{less than } 3000) = 0.264$

I calculated a z-score and then used my graphing calculator to find the area under the normal curve (below $z = -0.632$).

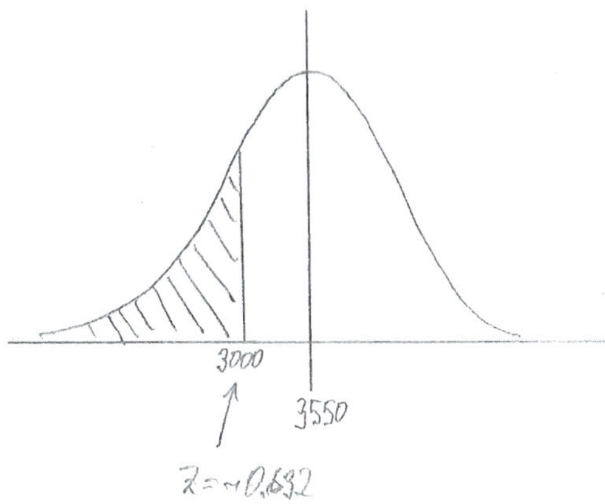

c. **The weight of a randomly selected car is between $2,800$ and $4,500$ pounds.**

$P(\text{between } 2800 \text{ and } 4500) = 0.668$

I calculated two z-scores and then used my graphing calculator to find the area under the normal curve (between $z = -0.862$ and $z = 1.092$).

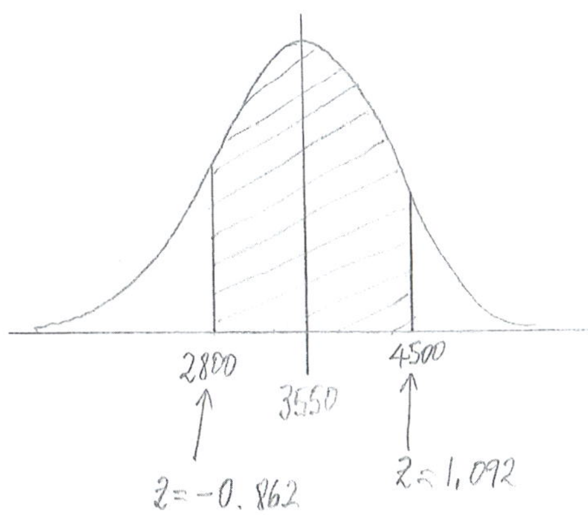

Problem Set Sample Solutions

1. Which of the following histograms show distributions that are approximately normal?

   a.

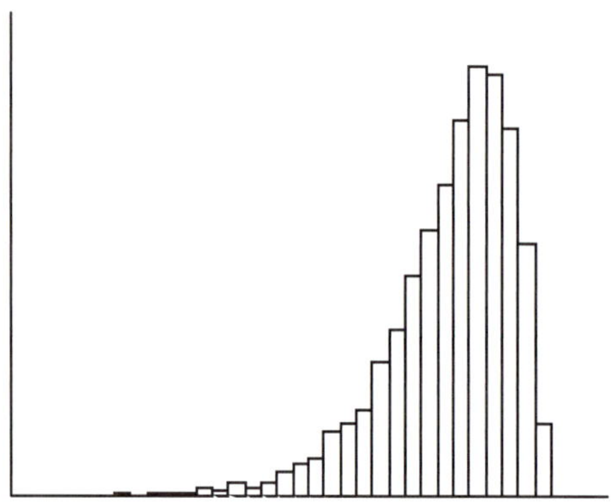

No, this distribution is not approximately normal.

   b.

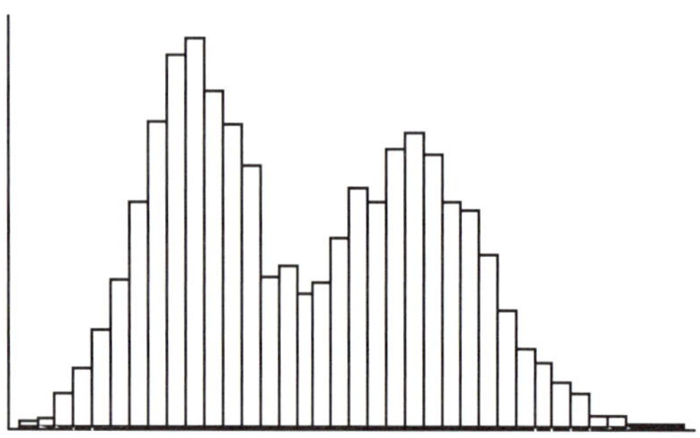

No, this distribution is not approximately normal.

   c.

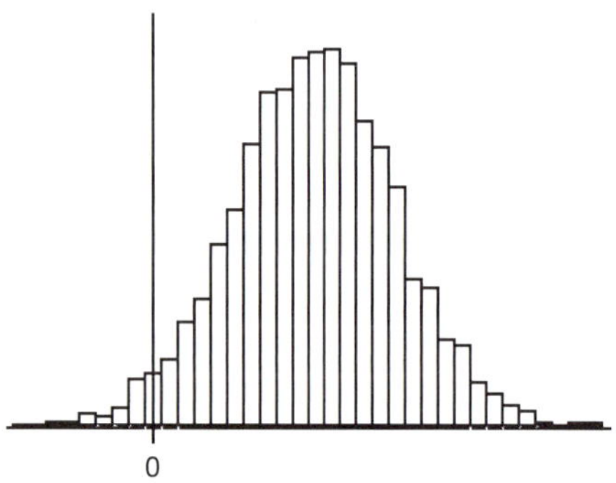

Yes, this distribution is approximately normal.

2. Suppose that a particular medical procedure has a cost that is approximately normally distributed with a mean of $\$19,800$ and a standard deviation of $\$2,900$. For a randomly selected patient, find the probabilities of the following events. (Round your answers to the nearest thousandth.)

   a. The procedure costs between $\$18,000$ and $\$22,000$.

$P(\text{between } 18000 \text{ and } 22000) = 0.509$

   b. The procedure costs less than $\$15,000$.

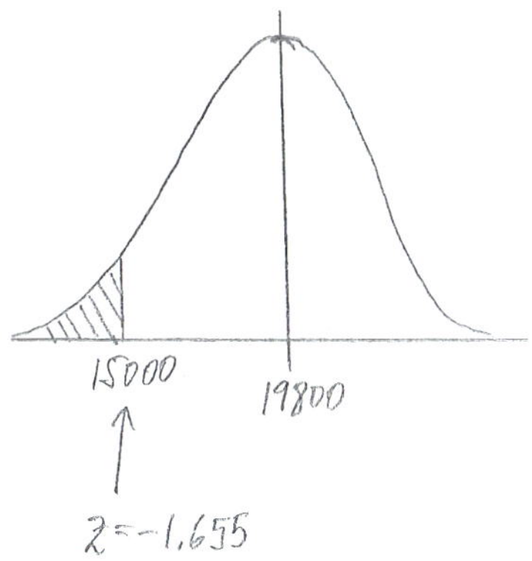

$P(\text{less than } 15000) = 0.049$

   c. The procedure costs more than $\$17,250$.

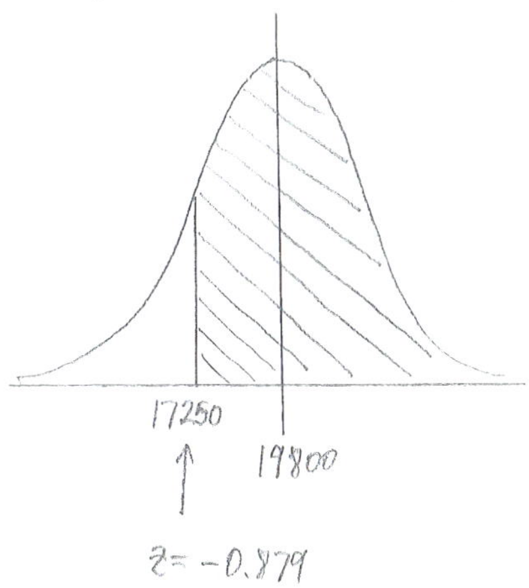

$P(\text{greater than } 17250) = 0.810$

3. **Consider the medical procedure described in the previous question, and suppose a patient is charged $\$24,900$ for the procedure. The patient is reported as saying, "I've been charged an outrageous amount!" How justified is this comment? Use probability to support your answer.**

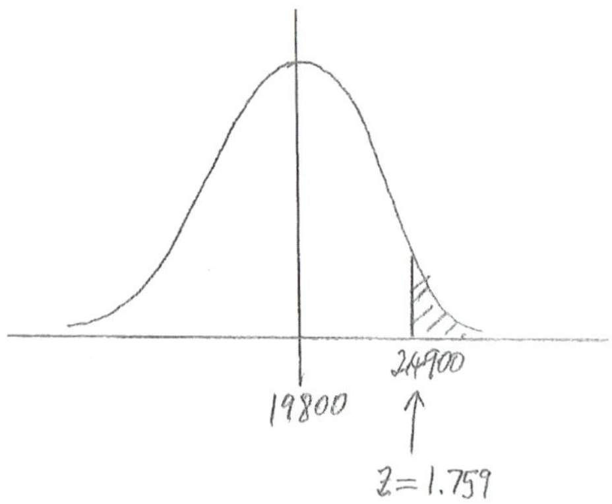

The probability that the procedure will cost at least $\$24,900$ is 0.039. So, this charge places the patient's bill in the top 4% of bills for this procedure. While the procedure turned out to be very expensive for this patient, use of the word "outrageous" could be considered a little extreme.

4. **Think again about the medical procedure described in Problem 2.**
   a. **Rounding your answers to the nearest thousandth, find the probability of each instance for a randomly selected patient.**
      i. **The cost of the procedure is within two standard deviations of the mean cost.**

The cost two standard deviations above the mean gives $z = 2$, and the price two standard deviations below the mean gives $z = -2$.

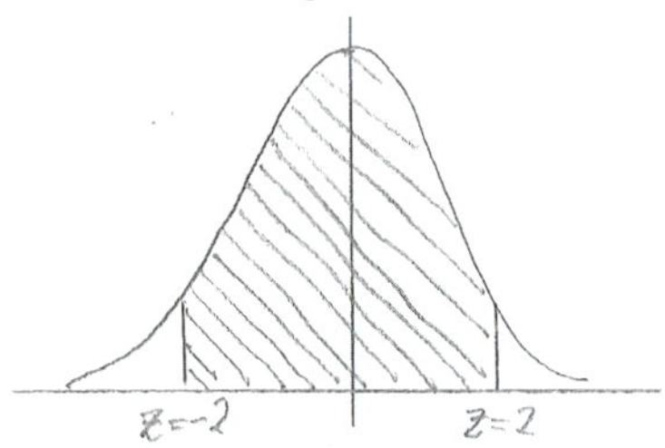

The probability that the price is within 2 standard deviations of the mean is 0.954.

      ii. **The cost of the procedure is more than one standard deviation from the mean cost.**

The cost one standard deviation above the mean gives $z = 1$, and the cost one standard deviation below the mean gives $z = -1$.

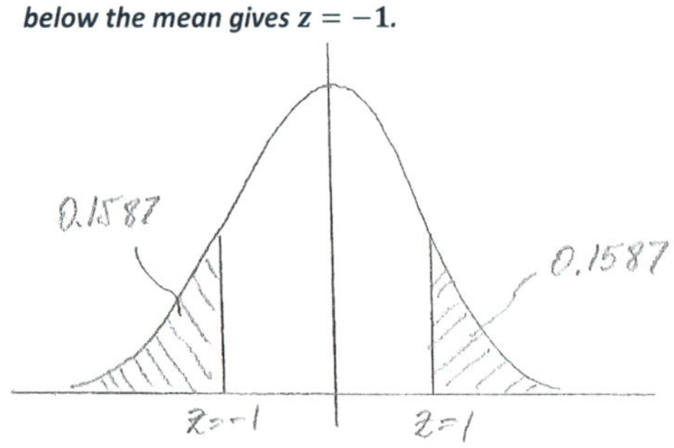

$2(0.1587) = 0.317$

The probability that the cost of the procedure is more than one standard deviation from the mean is 0.317.

b. If the mean or the standard deviation were to be changed, would your answers to part (a) be affected? Explain.

No. For example, looking at part (a)(i), the value two standard deviations above the mean will always have a z-score of 2, and the value two standard deviations below the mean will always have a z-score of -2. So, the answer will always be the same whatever the mean and the standard deviation. Similarly, in part (a)(ii), the z-scores will always be 1 and -1; therefore, the answer will always be the same, regardless of the mean and the standard deviation.

5. Use a table of standard normal curve areas to find the following:

a. The area to the left of $z = 0.56$

0.7123

b. The area to the right of $z = 1.20$

$1 - 0.8849 = 0.1151$

c. The area to the left of $z = -1.47$

0.0708

d. The area to the right of $z = -0.35$

$1 - 0.3632 = 0.6368$

e. The area between $z = -1.39$ and $z = 0.80$

$0.7881 - 0.0823 = 0.7058$

f. Choose a response from parts (a) through (f), and explain how you determined your answer.

Answers will vary.

Sample response: To find the area to the right of $z = -0.35$ in part (d), I used the table to find the cumulative area up to (which is to the left of) $z = -0.35$. Then, to find the area to the right of the z-score, I had to subtract the cumulative area from 1.

Standard Normal Curve Areas

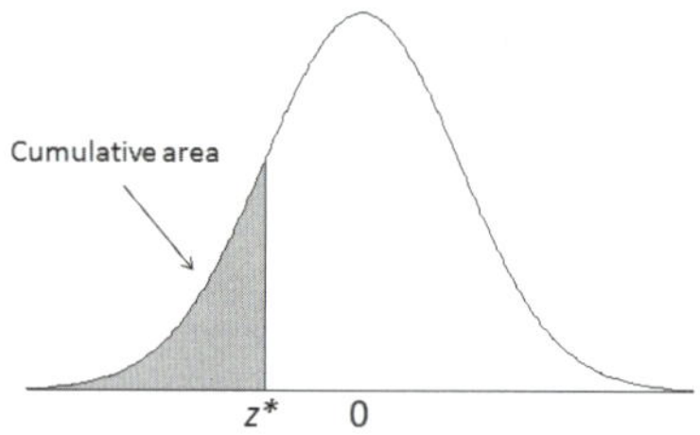

z	0.00	0.01	0.02	0.03	0.04	0.05	0.06	0.07	0.08	0.09
−3.8	0.0001	0.0001	0.0001	0.0001	0.0001	0.0001	0.0001	0.0001	0.0001	0.0001
−3.7	0.0001	0.0001	0.0001	0.0001	0.0001	0.0001	0.0001	0.0001	0.0001	0.0001
−3.6	0.0002	0.0002	0.0001	0.0001	0.0001	0.0001	0.0001	0.0001	0.0001	0.0001
−3.5	0.0002	0.0002	0.0002	0.0002	0.0002	0.0002	0.0002	0.0002	0.0002	0.0002
−3.4	0.0003	0.0003	0.0003	0.0003	0.0003	0.0003	0.0003	0.0003	0.0003	0.0002
−3.3	0.0005	0.0005	0.0005	0.0004	0.0004	0.0004	0.0004	0.0004	0.0004	0.0003
−3.2	0.0007	0.0007	0.0006	0.0006	0.0006	0.0006	0.0006	0.0005	0.0005	0.0005
−3.1	0.0010	0.0009	0.0009	0.0009	0.0008	0.0008	0.0008	0.0008	0.0007	0.0007
−3.0	0.0013	0.0013	0.0013	0.0012	0.0012	0.0011	0.0011	0.0011	0.0010	0.0010
−2.9	0.0019	0.0018	0.0018	0.0017	0.0016	0.0016	0.0015	0.0015	0.0014	0.0014
−2.8	0.0026	0.0025	0.0024	0.0023	0.0023	0.0022	0.0021	0.0021	0.0020	0.0019
−2.7	0.0035	0.0034	0.0033	0.0032	0.0031	0.0030	0.0029	0.0028	0.0027	0.0026
−2.6	0.0047	0.0045	0.0044	0.0043	0.0041	0.0040	0.0039	0.0038	0.0037	0.0036
−2.5	0.0062	0.0060	0.0059	0.0057	0.0055	0.0054	0.0052	0.0051	0.0049	0.0048
−2.4	0.0082	0.0080	0.0078	0.0075	0.0073	0.0071	0.0069	0.0068	0.0066	0.0064
−2.3	0.0107	0.0104	0.0102	0.0099	0.0096	0.0094	0.0091	0.0089	0.0087	0.0084
−2.2	0.0139	0.0136	0.0132	0.0129	0.0125	0.0122	0.0119	0.0116	0.0113	0.0110
−2.1	0.0179	0.0174	0.0160	0.0166	0.0162	0.0158	0.0154	0.0150	0.0146	0.0143
−2.0	0.0228	0.0222	0.0217	0.0212	0.0207	0.0202	0.0197	0.0192	0.0188	0.0183
−1.9	0.0287	0.0281	0.0274	0.0268	0.0262	0.0256	0.0250	0.0244	0.0239	0.0233
−1.8	0.0359	0.0351	0.0344	0.0336	0.0329	0.0322	0.0314	0.0307	0.0301	0.0294
−1.7	0.0446	0.0436	0.0427	0.0418	0.0409	0.0401	0.0392	0.0384	0.0375	0.0367
−1.6	0.0548	0.0537	0.0526	0.0516	0.0505	0.0495	0.0485	0.0475	0.0465	0.0455
−1.5	0.0668	0.0655	0.0643	0.0630	0.0618	0.0606	0.0594	0.0582	0.0571	0.0599
−1.4	0.0808	0.0793	0.0778	0.0764	0.0749	0.0735	0.0721	0.0708	0.0694	0.0681
−1.3	0.0968	0.0951	0.0934	0.0918	0.0901	0.0885	0.0869	0.0853	0.0838	0.0823
−1.2	0.1151	0.1131	0.1112	0.1093	0.1075	0.1056	0.1038	0.1020	0.1003	0.0985
−1.1	0.1357	0.1335	0.1314	0.1292	0.1271	0.1251	0.1230	0.1210	0.1190	0.1170
−1.0	0.1587	0.1562	0.1539	0.1515	0.1492	0.1469	0.1446	0.1423	0.1401	0.1379
−0.9	0.1841	0.1814	0.1788	0.1762	0.1736	0.1711	0.1685	0.1660	0.1635	0.1611
−0.8	0.2119	0.2090	0.2061	0.2033	0.2005	0.1977	0.1949	0.1922	0.1894	0.1867
−0.7	0.2420	0.2389	0.2358	0.2327	0.2296	0.2266	0.2236	0.2206	0.2177	0.2148
−0.6	0.2743	0.2709	0.2676	0.2643	0.2611	0.2578	0.2546	0.2514	0.2483	0.2451
−0.5	0.3085	0.3050	0.3015	0.2981	0.2946	0.2912	0.2877	0.2843	0.2810	0.2776
−0.4	0.3446	0.3409	0.3372	0.3336	0.3300	0.3264	0.3228	0.3192	0.3156	0.3121
−0.3	0.3821	0.3783	0.3745	0.3707	0.3669	0.3632	0.3594	0.3557	0.3520	0.3483
−0.2	0.4207	0.4168	0.4129	0.4090	0.4052	0.4013	0.3974	0.3936	0.3897	0.3859
−0.1	0.4602	0.4562	0.4522	0.4483	0.4443	0.4404	0.4364	0.4325	0.4286	0.4247
−0.0	0.5000	0.4960	0.4920	0.4880	0.4840	0.4801	0.4761	0.4721	0.4681	0.4641

EUREKA MATH™

z	0.00	0.01	0.02	0.03	0.04	0.05	0.06	0.07	0.08	0.09
0.0	0.5000	0.5040	0.5080	0.5120	0.5160	0.5199	0.5239	0.5279	0.5319	0.5359
0.1	0.5398	0.5438	0.5478	0.5517	0.5557	0.5596	0.5636	0.5675	0.5714	0.5753
0.2	0.5793	0.5832	0.5871	0.5910	0.5948	0.5987	0.6026	0.6064	0.6103	0.6141
0.3	0.6179	0.6217	0.6255	0.6293	0.6331	0.6368	0.6406	0.6443	0.6480	0.6517
0.4	0.6554	0.6591	0.6628	0.6664	0.6700	0.6736	0.6772	0.6808	0.6844	0.6879
0.5	0.6915	0.6950	0.6985	0.7019	0.7054	0.7088	0.7123	0.7157	0.7190	0.7224
0.6	0.7257	0.7291	0.7324	0.7357	0.7389	0.7422	0.7454	0.7486	0.7517	0.7549
0.7	0.7580	0.7611	0.7642	0.7673	0.7704	0.7734	0.7764	0.7794	0.7823	0.7852
0.8	0.7881	0.7910	0.7939	0.7967	0.7995	0.8023	0.8051	0.8078	0.8106	0.8133
0.9	0.8159	0.8186	0.8212	0.8238	0.8264	0.8289	0.8315	0.8340	0.8365	0.8389
1.0	0.8413	0.8438	0.8461	0.8485	0.8508	0.8531	0.8554	0.8577	0.8599	0.8621
1.1	0.8643	0.8665	0.8686	0.8708	0.8729	0.8749	0.8770	0.8790	0.8810	0.8830
1.2	0.8849	0.8869	0.8888	0.8907	0.8925	0.8944	0.8962	0.8980	0.8997	0.9015
1.3	0.9032	0.9049	0.9066	0.9082	0.9099	0.9115	0.9131	0.9147	0.9162	0.9177
1.4	0.9192	0.9207	0.9222	0.9236	0.9251	0.9265	0.9279	0.9292	0.9306	0.9319
1.5	0.9332	0.9345	0.9357	0.9370	0.9382	0.9394	0.9406	0.9418	0.9429	0.9441
1.6	0.9452	0.9463	0.9474	0.9484	0.9495	0.9505	0.9515	0.9525	0.9535	0.9545
1.7	0.9554	0.9564	0.9573	0.9582	0.9591	0.9599	0.9608	0.9616	0.9625	0.9633
1.8	0.9641	0.9649	0.9656	0.9664	0.9671	0.9678	0.9686	0.9693	0.9699	0.9706
1.9	0.9713	0.9719	0.9726	0.9732	0.9738	0.9744	0.9750	0.9756	0.9761	0.9767
2.0	0.9772	0.9778	0.9783	0.9788	0.9793	0.9798	0.9803	0.9808	0.9812	0.9817
2.1	0.9821	0.9826	0.9830	0.9834	0.9838	0.9842	0.9846	0.9850	0.9854	0.9857
2.2	0.9861	0.9864	0.9868	0.9871	0.9875	0.9878	0.9881	0.9884	0.9887	0.9890
2.3	0.9893	0.9896	0.9898	0.9901	0.9904	0.9906	0.9909	0.9911	0.9913	0.9916
2.4	0.9918	0.9920	0.9922	0.9925	0.9927	0.9929	0.9931	0.9932	0.9934	0.9936
2.5	0.9938	0.9940	0.9941	0.9943	0.9945	0.9946	0.9948	0.9949	0.9951	0.9952
2.6	0.9953	0.9955	0.9956	0.9957	0.9959	0.9960	0.9961	0.9962	0.9963	0.9964
2.7	0.9965	0.9966	0.9967	0.9968	0.9969	0.9970	0.9971	0.9972	0.9973	0.9974
2.8	0.9974	0.9975	0.9976	0.9977	0.9977	0.9978	0.9979	0.9979	0.9980	0.9981
2.9	0.9981	0.9982	0.9982	0.9983	0.9984	0.9984	0.9985	0.9985	0.9986	0.9986
3.0	0.9987	0.9987	0.9987	0.9988	0.9988	0.9989	0.9989	0.9989	0.9990	0.9990
3.1	0.9990	0.9991	0.9991	0.9991	0.9992	0.9992	0.9992	0.9992	0.9993	0.9993
3.2	0.9993	0.9993	0.9994	0.9994	0.9994	0.9994	0.9994	0.9995	0.9995	0.9995
3.3	0.9995	0.9995	0.9995	0.9996	0.9996	0.9996	0.9996	0.9996	0.9996	0.9997
3.4	0.9997	0.9997	0.9997	0.9997	0.9997	0.9997	0.9997	0.9997	0.9997	0.9998
3.5	0.9998	0.9998	0.9998	0.9998	0.9998	0.9998	0.9998	0.9998	0.9998	0.9998
3.6	0.9998	0.9998	0.9999	0.9999	0.9999	0.9999	0.9999	0.9999	0.9999	0.9999
3.7	0.9999	0.9999	0.9999	0.9999	0.9999	0.9999	0.9999	0.9999	0.9999	0.9999
3.8	0.9999	0.9999	0.9999	0.9999	0.9999	0.9999	0.9999	0.9999	0.9999	0.9999

Lesson 11: Normal Distributions

Student Outcomes

- Students use tables and technology to estimate the area under a normal curve.
- Students interpret probabilities in context.
- When appropriate, students select an appropriate normal distribution to serve as a model for a given data distribution.

Lesson Notes

In Lesson 10, students first learn how to calculate z-scores and are then shown how to use z-scores and a graphing calculator to find normal probabilities. Students are then introduced to the process of calculating normal probabilities using tables of standard normal curve areas. In this lesson, students calculate normal probabilities using tables and spreadsheets. They also learn how to use a graphing calculator to find normal probabilities directly (without using z-scores) and are introduced to the idea of fitting a normal curve to a data distribution that seems to be approximately normal.

Classwork

Example 1 (7 minutes): Calculation of Normal Probabilities Using z-Scores and Tables of Standard Normal Areas

In this example, two of the techniques learned in Lesson 10—evaluation of z-scores and use of tables of standard normal areas—are combined to find normal probabilities. Consider asking students to work independently or with a partner, and use this as an opportunity to informally assess student progress.

Example 1: Calculation of Normal Probabilities Using z-Scores and Tables of Standard Normal Areas

The U.S. Department of Agriculture (USDA), in its Official Food Plans (www.cnpp.usda.gov), states that the average cost of food for a 14- to 18-year-old male (on the Moderate-Cost Plan) is $\$261.50$ per month. Assume that the monthly food cost for a 14- to 18-year-old male is approximately normally distributed with a mean of $\$261.50$ and a standard deviation of $\$16.25$.

a. **Use a table of standard normal curve areas to find the probability that the monthly food cost for a randomly selected 14- to 18-year-old male is**

   i. **Less than $\$280$.**

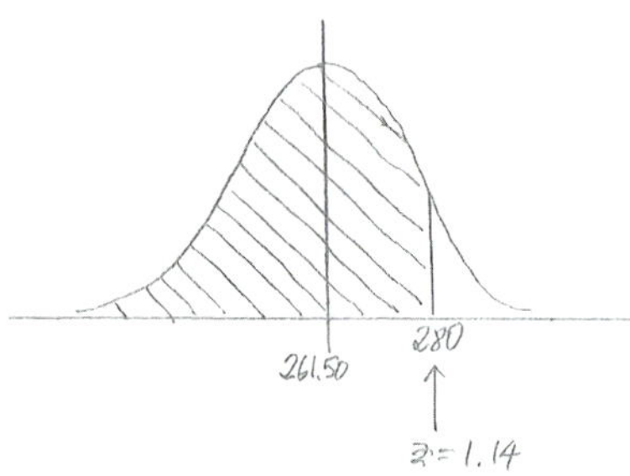

$P(\text{less than } 280) = 0.8729$

ii. More than $270.

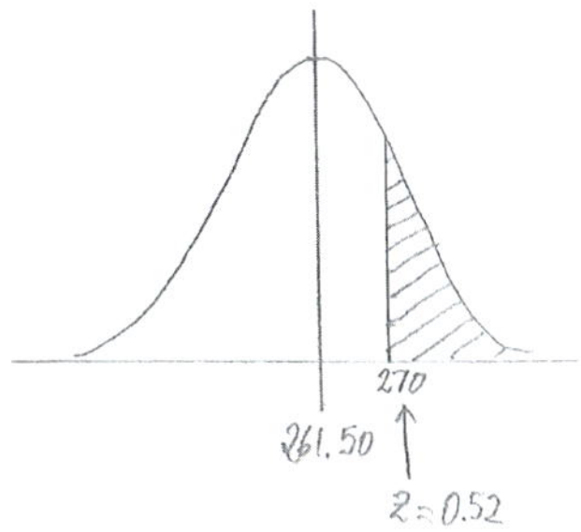

$P(\text{greater than } 270)$
$= 1 - 0.6985$
$= 0.3015$

iii. More than $250.

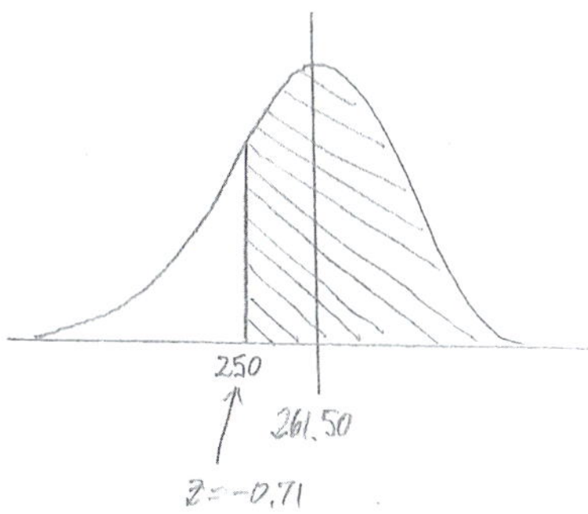

$P(\text{greater than } 250)$
$= 1 - 0.2389$
$= 0.7611$

iv. Between $240 and $275.

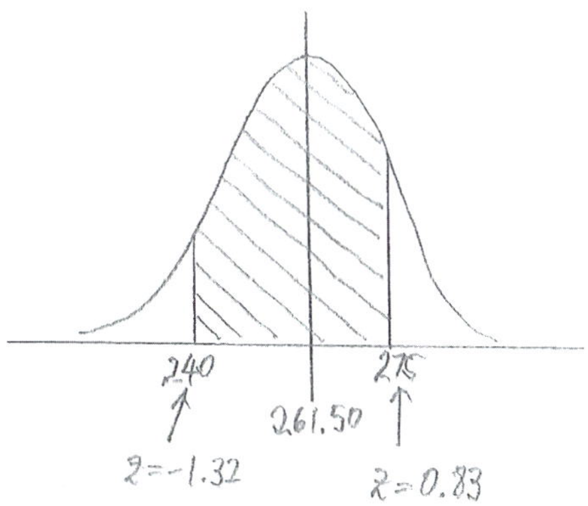

$P(\text{between } 240 \text{ and } 275)$
$= 0.7967 - 0.0934$
$= 0.7033$

b. **Explain the meaning of the probability that you found in part (a)(iv).**

If a very large number of 14- to 18-year-old males were to be selected at random, then you would expect about 70.33% *of them to have monthly food costs between* $240 *and* $275.

Exercise 1 (5 minutes)

MP.4 This exercise provides practice with the approach demonstrated in Example 1. Continue to encourage students to include a normal distribution curve with work shown for each part of the exercise.

Exercise 1

The USDA document described in Example 1 also states that the average cost of food for a 14- to 18-year-old female (again, on the Moderate-Cost Plan) is $\$215.20$ per month. Assume that the monthly food cost for a 14- to 18-year-old female is approximately normally distributed with a mean of $\$215.20$ and a standard deviation of $\$14.85$.

a. **Use a table of standard normal curve areas to find the probability that the monthly food cost for a randomly selected 14- to 18-year-old female is**

i. **Less than $\$225$.**

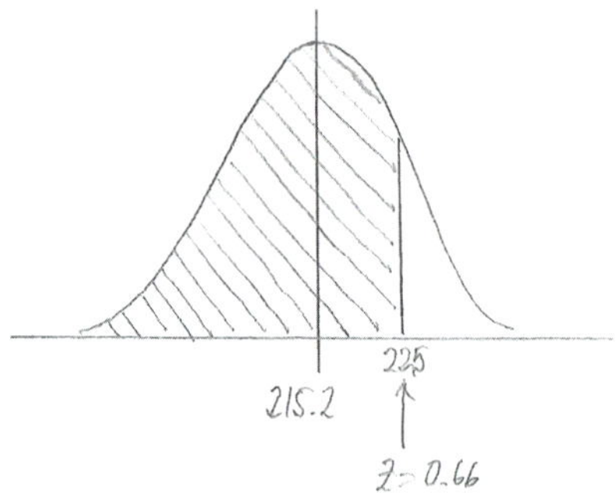

$P(\text{less than } 225) = 0.7454$

ii. **Less than $\$200$.**

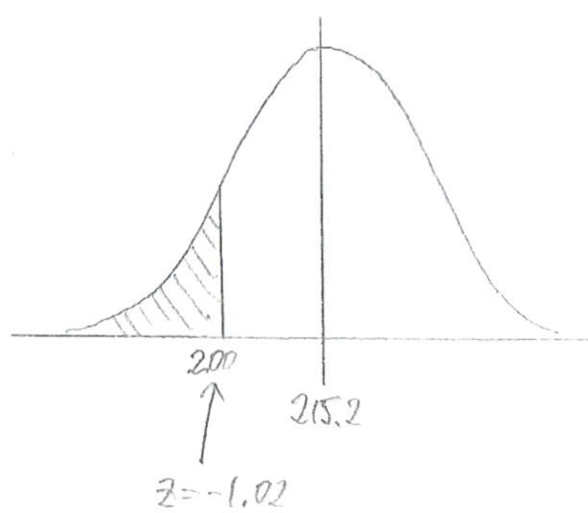

$P(\text{less than } 200) = 0.1539$

iii. **More than $\$250$.**

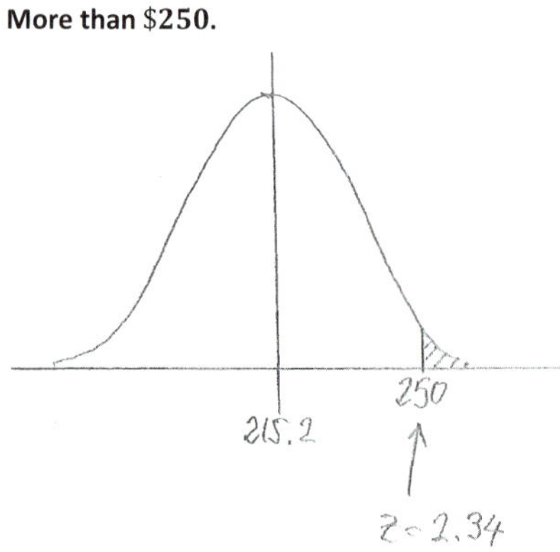

$P(\text{more than } 250) = 0.0096$

iv. Between $\$190$ and $\$220$.

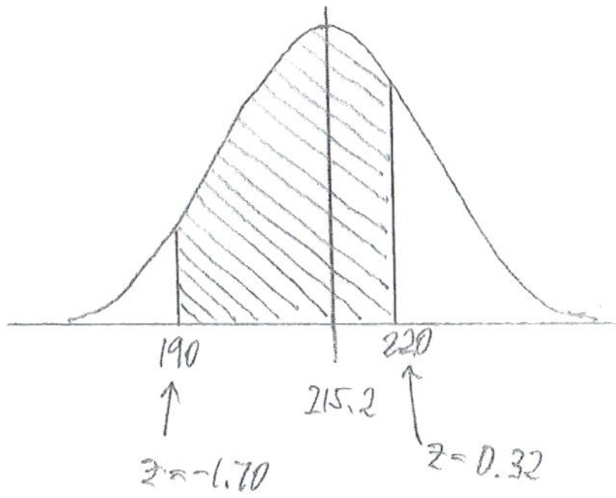

$P(\text{between 190 and 220}) = 0.6255 - 0.0446$
$= 0.5809$

b. Explain the meaning of the probability that you found in part (a)(iv).

If a very large number of 14- to 18-year-old females were to be selected at random, then you would expect about 58.09% of them to have monthly food costs between $\$190$ and $\$220$.

Example 2 (5 minutes): Use of a Graphing Calculator to Find Normal Probabilities Directly

In this example, students learn how to calculate normal probabilities using a graphing calculator without using z-scores. Use this example to show the class how to do this using a graphing calculator*.

*Calculator note: The general form of this is *Normalcdf([left bound],[right bound],[mean],[standard deviation])*. The *Normalcdf* function is accessed using *2nd, DISTR*.

Example 2: Use of a Graphing Calculator to Find Normal Probabilities Directly

Return to the information given in Example 1. Using a graphing calculator, and *without* using z-scores, find the probability (rounded to the nearest thousandth) that the monthly food cost for a randomly selected 14- to 18-year-old male is

a. **Between $\$260$ and $\$265$.**

$P(\text{between } \$260 \text{ and } \$265) = 0.122$

If students are using TI-83 or TI-84 calculators, this result is found by entering Normalcdf$(260, 265, 261.5, 16.25)$.

b. **At least $\$252$.**

$P(\geq 252) = 0.721$

This result is found by entering Normalcdf$(252, 999, 261.5, 16.25)$ or the equivalent for other brands of calculators. Any large positive number or 1EE99 can be used in place of 999, as long as the number is at least four standard deviations above the mean.

c. **At most $\$248$.**

$P(\leq 248) = 0.203$

This result is found by entering Normalcdf$(-999, 248, 261.5, 16.25)$ or the equivalent for other brands of calculators. Any large negative number or −1EE99 can be used in place of -999, as long as the number is at least four standard deviations below the mean.

Exercise 2 (5 minutes)

MP.5 Here, students have an opportunity to practice using a graphing calculator to find normal probabilities directly and reflect on the use of technology compared to the use of tables of normal curve areas.

Exercise 2

Return to the information given in Exercise 1.

a. **In Exercise 1, you calculated the probability that the monthly food cost for a randomly selected 14- to 18-year-old female is between $\$190$ and $\$220$. Would the probability that the monthly food cost for a randomly selected 14- to 18-year-old female is between $\$195$ and $\$230$ be greater than or smaller than the probability for between $\$190$ and $\$220$? Explain your thinking.**

The probability would be greater between $\$195$ and $\$230$. If you look at a sketch of a normal curve with mean $\$215.20$ and standard deviation $\$14.85$, there is more area under the curve for the wider interval of $\$195$ to $\$230$.

b. **Do you think that the probability that the monthly food cost for a randomly selected 14- to 18-year-old female is between $\$195$ and $\$230$ is closer to 0.50, 0.75, or 0.90? Explain your thinking.**

Closer to 0.75. Based on my answer to part (a), I expect the probability to be greater than the probability for between $\$190$ and $\$220$, which was 0.5809, but I do not think it would be as great as 0.90.

c. **Using a graphing calculator, and without using z-scores, find the probability (rounded to the nearest thousandth) that the monthly food cost for a randomly selected 14- to 18-year-old female is between $\$195$ and $\$230$. Is this probability consistent with your answer to part (b)?**

$P(\text{between } 195 \text{ and } 230) = 0.754$

This probability is close to 0.75, which was my answer in part (b).

d. **How does the probability you calculated in part (c) compare to the probability that would have been obtained using the table of normal curve areas?**

The z-score for $\$195$ is $z = \frac{195-215.20}{14.85} = -1.36$, and the z-score for $\$230$ is $z = \frac{230-215.20}{14.85} = 1.00$. Using the table of normal curve areas, $P(\text{between } 195 \text{ and } 230) = 0.8413 - 0.0853 = 0.756$. This is very close to the answer that I got using the graphing calculator.

e. **What is one advantage to using a graphing calculator to calculate this probability?**

It is a lot faster to use the graphing calculator because I did not have to calculate z-scores in order to get the probability.

f. **In Exercise 1, you calculated the probability that the monthly food cost for a randomly selected 14- to 18-year-old female is at most $\$200$. Would the probability that the monthly food cost for a randomly selected 14- to 18-year-old female is at most $\$210$ be greater than or less than the probability for at most $\$200$? Explain your thinking.**

The probability would be greater for at most $\$210$. There is more area under the normal curve to the left of $\$210$ than to the left of $\$200$.

g. **Do you think that the probability that the monthly food cost for a randomly selected 14- to 18-year-old female is at most $\$210$ is closer to 0.10, 0.30, or 0.50? Explain your thinking.**

Closer to 0.30. Based on my answer to part (f), I expect the probability to be greater than the probability for at most $\$200$, which was 0.1539, but I do not think it would be as great as 0.50 because $\$210$ is less than the mean of $\$215.20$, and the area to the left of $\$215.20$ is 0.50.

h. Using a graphing calculator, and without using z-scores, find the probability (rounded to the nearest thousandth) that the monthly food cost for a randomly selected 14- to 18-year-old female is at most $\$210$.

$P(\leq 210) = 0.363$

i. Using a graphing calculator, and without using z-scores, find the probability (rounded to the nearest thousandth) that the monthly food cost for a randomly selected 14- to 18-year-old female is at least $\$235$.

$P(\geq 235) = 0.091$

Example 3 (5 minutes): Using a Spreadsheet to Find Normal Probabilities

In this example, students learn how to calculate normal probabilities using a spreadsheet. This example and the exercise that follows revisits Example 1 and Exercise 1 but has students use a spreadsheet rather than a table of normal curve areas. Use this example to show the class how to do this using a spreadsheet.*

*Spreadsheet note: Many spreadsheet programs, such as Excel, have a built-in function to calculate normal probabilities. In Excel, this can be done by finding the area to the left of any particular cutoff value by typing the following into a cell of the spreadsheet and then hitting the return key:

= *NORMDIST*(cutoff,mean,stddev,true).

The *true* needs to be included at the end in order to get the area to the left of the cutoff. For example,

= *NORMDIST*(50,40,10,true) gives 0.84124, which is the area to the left of 50 under the normal curve with mean 40 and standard deviation 10.

Example 3: Using a Spreadsheet to Find Normal Probabilities

Return to the information given in Example 1. The USDA, in its Official Food Plans (www.cnpp.usda.gov), states that the average cost of food for a 14- to 18-year-old male (on the Moderate-Cost Plan) is $\$261.50$ per month. Assume that the monthly food cost for a 14- to 18-year-old male is approximately normally distributed with a mean of $\$261.50$ and a standard deviation of $\$16.25$. Round your answers to four decimal places.

Use a spreadsheet to find the probability that the monthly food cost for a randomly selected 14- to 18-year-old male is

a. **Less than $\$280$.**

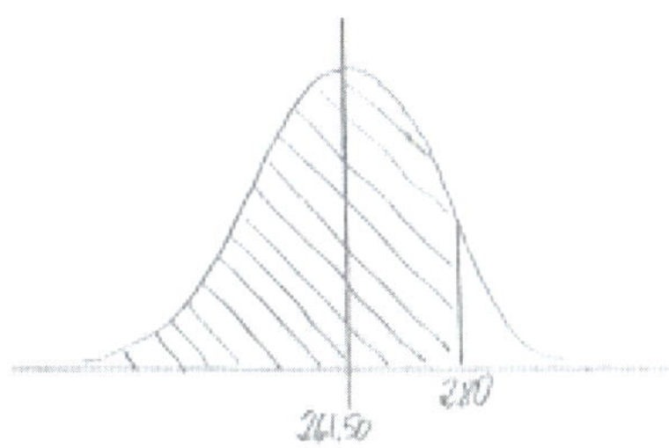

If students are using Excel, this would be found by using = *NORMDIST*($280, 261.50, 16.25$, true). *The difference between this answer and the answer using the table of normal curve areas is due to rounding in calculating the z-score needed in order to use the table.*

$P(\text{less than } 280) = 0.8725$

b. More than $\$270$.

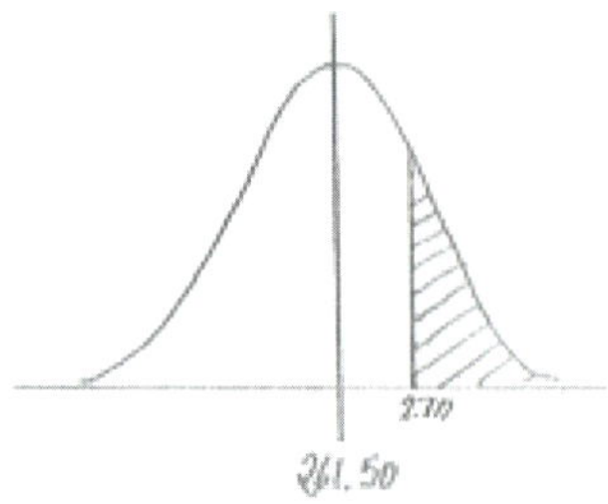

$P(\text{greater than } 270) = 1 - 0.6995$
$= 0.3005$

c. More than $\$250$.

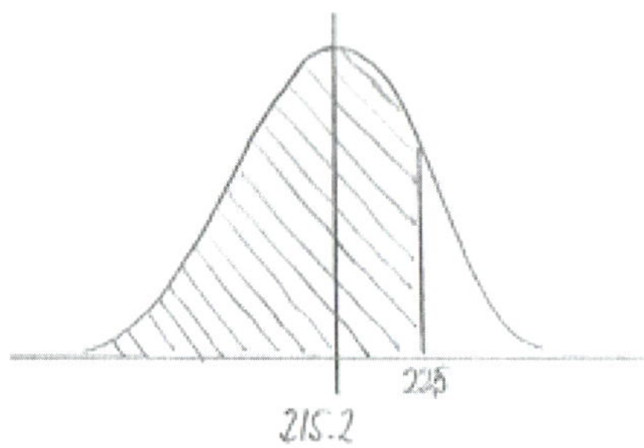

$P(greater\ than\ 250) = 1 - 0.2396$
$= 0.7604$

d. Between $\$240$ and $\$275$.

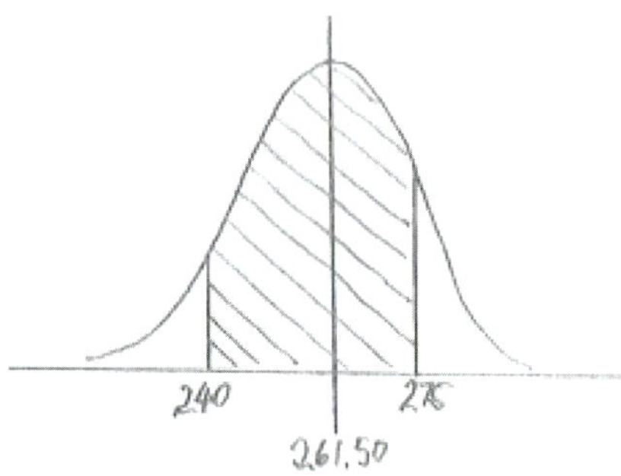

$P(\text{between } 240 \text{ and } 275) = 0.7969 - 0.0929$
$= 0.7040$

Exercise 3 (5 minutes)

Exercise 3

The USDA document described in Example 1 also states that the average cost of food for a 14- to 18-year-old female (again, on the Moderate-Cost Plan) is $\$215.20$ per month. Assume that the monthly food cost for a 14- to 18-year-old female is approximately normally distributed with a mean of $\$215.20$ and a standard deviation of $\$14.85$. Round your answers to 4 decimal places.

Use a spreadsheet to find the probability that the monthly food cost for a randomly selected 14- to 18-year-old female is

a. Less than $\$225$.

$P(\text{less than } 225) = 0.7454$

b. Less than $200.

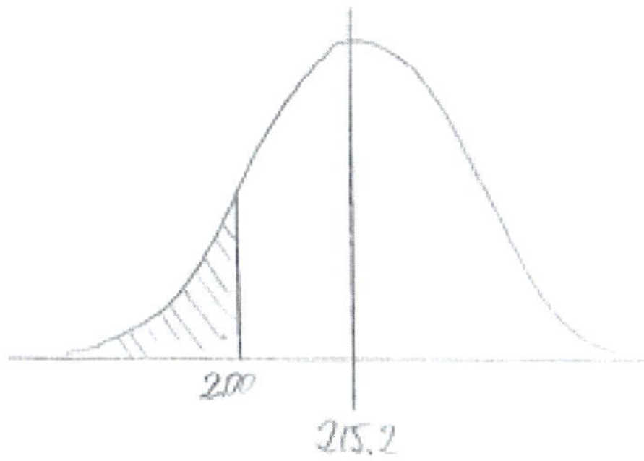

$P(\text{less than } 200) = 0.1530$

c. More than $250.

$P(\text{more than } 250) = 1 - 0.9904$
$= 0.0096$

d. Between $190 and $220.

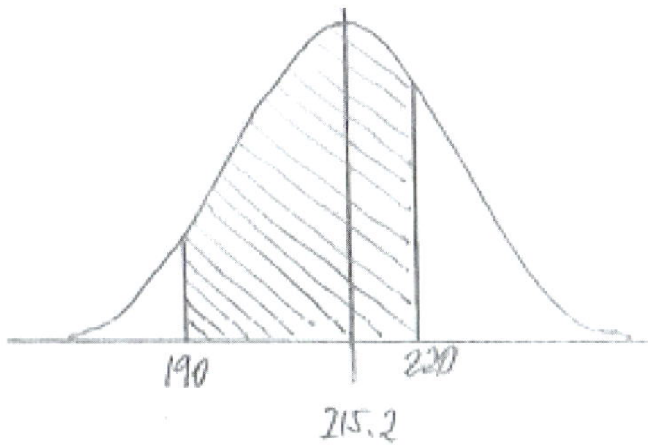

$P(\text{between } 190 \text{ and } 220) = 0.6267 - 0.0449$
$= 0.5818$

Exercise 4 (6 minutes)

Here, students are led through the process of choosing a normal distribution to model a given data set. Students might need some assistance prior to tackling this exercise. Teachers may wish to provide a quick review of how to draw a histogram. Students may need to be reminded how to calculate the mean and the standard deviation for data given in the form of a frequency distribution. In fact, this example provides the additional complication that the data are given in the form of a *grouped* frequency distribution, so the midpoints of the intervals have to be used as the data values.

If time is short, simply provide students with reminders of these techniques mentioned above, and the exercise can then be finished as part of the homework assignment.

Exercise 4

The reaction times of 490 people were measured. The results are shown in the frequency distribution below.

Reaction Time (seconds)	0.1 to < 0.15	0.15 to < 0.2	0.2 to < 0.25	0.25 to < 0.3	0.3 to < 0.35	0.35 to < 0.4
Frequency	9	82	220	138	37	4

a. **Construct a histogram that displays these results.**

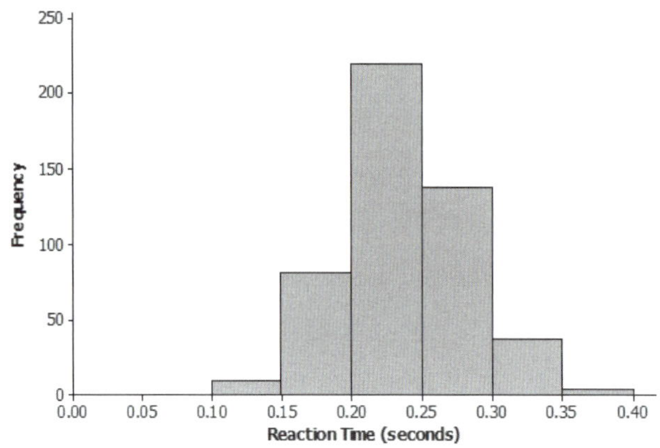

b. **Looking at the histogram, do you think a normal distribution would be an appropriate model for this distribution?**

Yes, the histogram is approximately symmetric and mound shaped.

c. **The mean of the reaction times for these 490 people is 0.2377, and the standard deviation of the reaction times is 0.0457. For a normal distribution with this mean and standard deviation, what is the probability that a randomly selected reaction time is at least 0.25?**

Using Normalcdf$(0.25, 999, 0.2377, 0.0457)$*, you get* $P(\geq 0.25) = 0.394$.

d. **The actual proportion of these 490 people who had a reaction time that was at least 0.25 is 0.365 (this can be calculated from the frequency distribution). How does this proportion compare to the probability that you calculated in part (c)? Does this confirm that the normal distribution is an appropriate model for the reaction time distribution?**

0.365 *is reasonably close to the probability based on the normal distribution, which was* 0.394*. I think that the normal model was appropriate.*

Closing (2 minutes)

Refer to Exercise 3.

- How would you interpret the probability that you found using a calculator in part (c)?
  - *Approximately* 39.4% *of the people had a reaction time of* 0.25 *second or higher.*

Ask students to summarize the main ideas of the lesson in writing or with a neighbor. Use this as an opportunity to informally assess comprehension of the lesson. The Lesson Summary below offers some important ideas that should be included.

Lesson Summary

Probabilities associated with normal distributions can be found using z-scores and tables of standard normal curve areas.

Probabilities associated with normal distributions can be found directly (without using z-scores) using a graphing calculator.

When a data distribution has a shape that is approximately normal, a normal distribution can be used as a model for the data distribution. The normal distribution with the same mean and the standard deviation as the data distribution is used.

Exit Ticket (5 minutes)

Name ______________________________ Date ______________

Lesson 11: Normal Distributions

Exit Ticket

1. SAT scores were originally scaled so that the scores for each section were approximately normally distributed with a mean of 500 and a standard deviation of 100. Assuming that this scaling still applies, use a table of standard normal curve areas to find the probability that a randomly selected SAT student scores
   a. More than 700.

   b. Between 440 and 560.

2. In 2012, the mean SAT math score was 514, and the standard deviation was 117. For the purposes of this question, assume that the scores were normally distributed. Using a graphing calculator, and without using z-scores, find the probability (rounded to the nearest thousandth), and explain how the answer was determined that a randomly selected SAT math student in 2012 scored
   a. Between 400 and 480.

   b. Less than 350.

EUREKA MATH™

Exit Ticket Sample Solutions

1. SAT scores were originally scaled so that the scores for each section were approximately normally distributed with a mean of 500 and a standard deviation of 100. Assuming that this scaling still applies, use a table of standard normal curve areas to find the probability that a randomly selected SAT student scores

   a. More than 700.

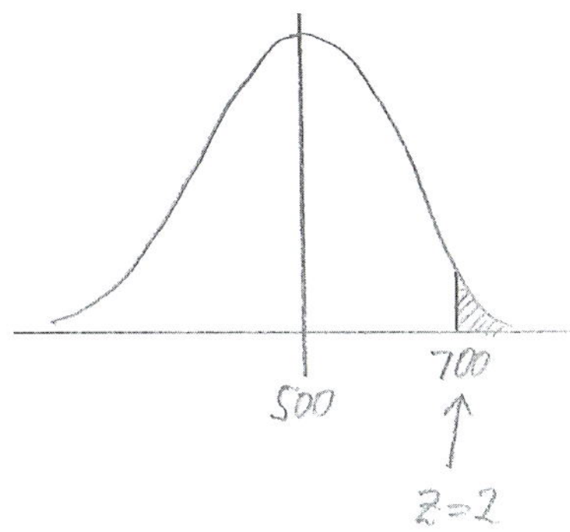

$$P(\text{greater than } 700) = 1 - 0.9772$$
$$= 0.0228$$

   b. Between 440 and 560.

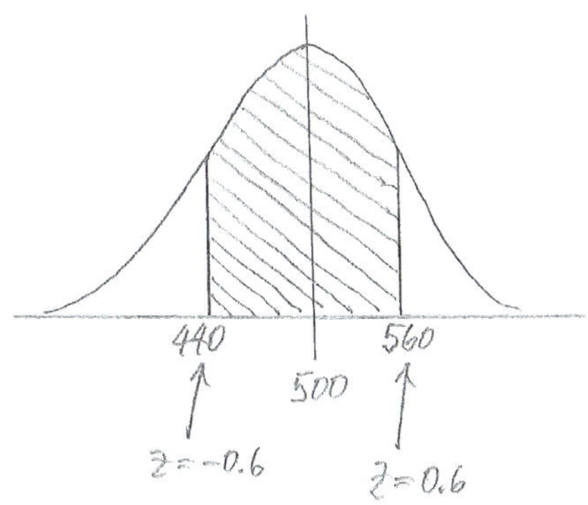

$$P(\text{between } 440 \text{ and } 560) = 0.7257 - 0.2743$$
$$= 0.4514$$

2. In 2012, the mean SAT math score was 514, and the standard deviation was 117. For the purposes of this question, assume that the scores were normally distributed. Using a graphing calculator, and without using z-scores, find the probability (rounded to the nearest thousandth), and explain how the answer was determined that a randomly selected SAT math student in 2012 scored

   a. Between 400 and 480.

   $P(\text{between } 400 \text{ and } 480) = 0.221$

   I used a TI-84 graphing calculator: Normalcdf$(400, 480, 514, 117)$.

   b. Less than 350.

   $P(\text{less than } 350) = 0.081$

   I used a TI-84 graphing calculator: Normalcdf$(0, 350, 514, 117)$.

Problem Set Sample Solutions

1. Use a table of standard normal curve areas to find the following:

   a. The area to the left of $z = 1.88$

      $\mathbf{0.9699}$

   b. The area to the right of $z = 1.42$

      $\mathbf{1 - 0.9222 = 0.0778}$

   c. The area to the left of $z = -0.39$

      $\mathbf{0.3483}$

   d. The area to the right of $z = -0.46$

      $\mathbf{1 - 0.3228 = 0.6772}$

   e. The area between $z = -1.22$ and $z = -0.5$

      $\mathbf{0.3085 - 0.1112 = 0.1973}$

2. Suppose that the durations of high school baseball games are approximately normally distributed with mean 105 minutes and standard deviation 11 minutes. Use a table of standard normal curve areas to find the probability that a randomly selected high school baseball game lasts

   a. Less than 115 minutes.

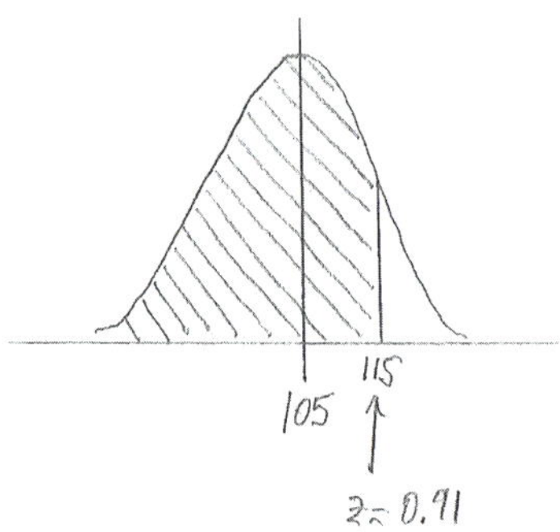

$\mathbf{P(\text{less than } 115) = 0.8186}$

   b. More than 100 minutes.

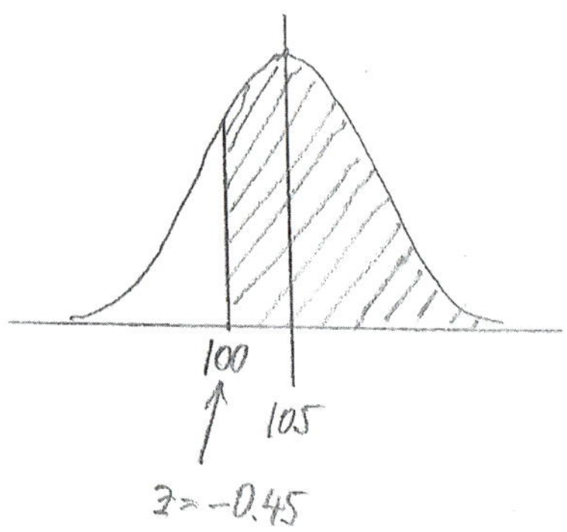

$\mathbf{P(\text{greater than } 100) = 1 - 0.3264}$
$\mathbf{= 0.6736}$

EUREKA MATH™

c. Between 90 and 110 minutes.

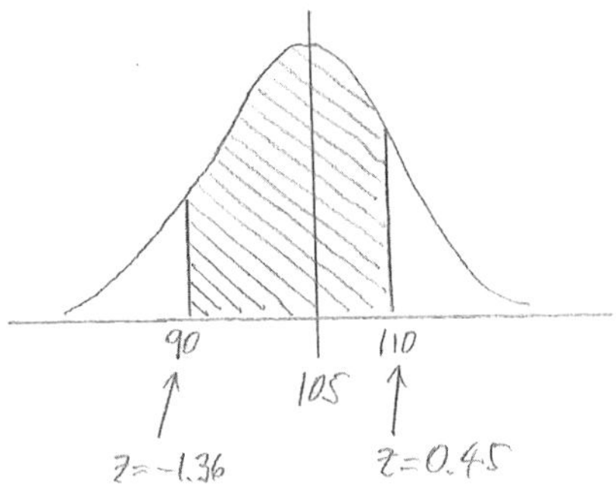

$P(\text{between 90 and 110}) = 0.6736 - 0.0869$
$= 0.5867$

3. Using a graphing calculator, and without using z-scores, check your answers to Problem 2. (Round your answers to the nearest thousandth.)

   a. $P(\text{less than } 115) = 0.818$

   b. $P(\text{greater than} 100) = 0.675$

   c. $P(\text{between 90 and 110}) = 0.589$

4. In Problem 2, you were told that the durations of high school baseball games are approximately normally distributed with a mean of 105 minutes and a standard deviation of 11 minutes. Suppose also that the durations of high school softball games are approximately normally distributed with a mean of 95 minutes and the same standard deviation, 11 minutes. Is it more likely that a high school baseball game will last between 100 and 110 minutes or that a high school softball game will last between 100 and 110 minutes? Answer this question without doing any calculations.

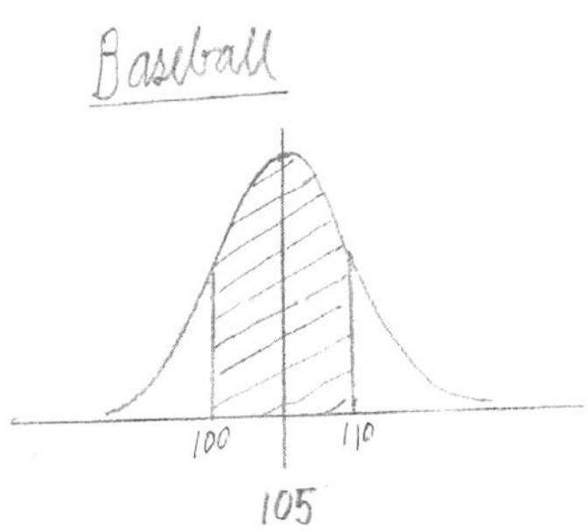

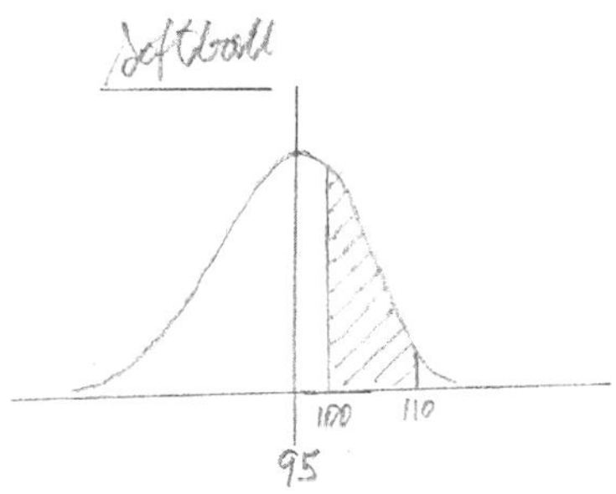

The heights of the two normal distribution graphs are the same; the only difference between the graphs is that the softball graph has a smaller mean. So, when the region under the graph between 100 and 110 is shaded, you get a larger area for the baseball graph than for the softball graph. Therefore, it is more likely that the baseball game will last between 100 and 110 minutes.

5. A farmer has 625 female adult sheep. The sheep have recently been weighed, and the results are shown in the table below.

Weight (pounds)	140 to $<$ 150	150 to $<$ 160	160 to $<$ 170	170 to $<$ 180	180 to $<$ 190	190 to $<$ 200	200 to $<$ 210
Frequency	8	36	173	221	149	33	5

a. Construct a histogram that displays these results.

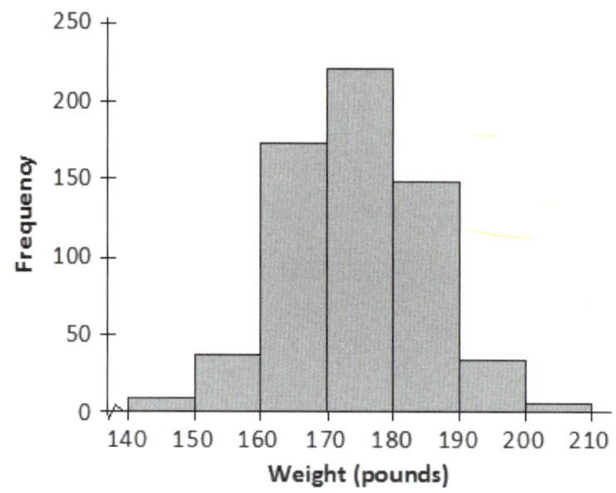

b. Looking at the histogram, do you think a normal distribution would be an appropriate model for this distribution?

Yes, the histogram is approximately symmetric and mound shaped.

c. The weights of the 625 sheep have mean 174.21 pounds and standard deviation 10.11 pounds. For a normal distribution with this mean and standard deviation, what is the probability that a randomly selected sheep has a weight of at least 190 pounds? (Round your answer to the nearest thousandth.)

Using Normalcdf$(190, 999, 174.21, 10.11)$, *you get* $P(\text{greater than } 190) = 0.059$.

Standard Normal Curve Areas

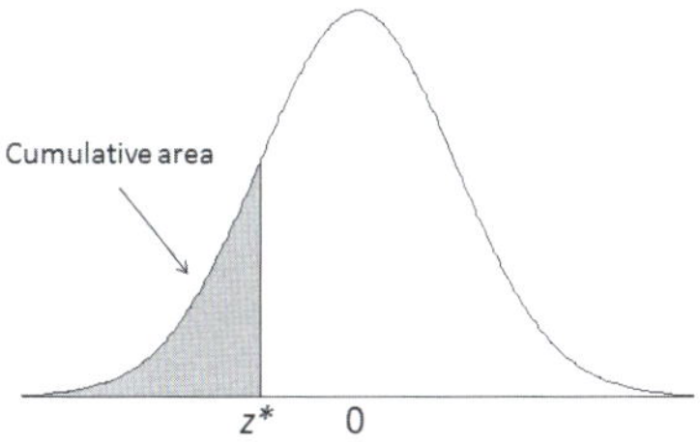

z	0.00	0.01	0.02	0.03	0.04	0.05	0.06	0.07	0.08	0.09
−3.8	0.0001	0.0001	0.0001	0.0001	0.0001	0.0001	0.0001	0.0001	0.0001	0.0001
−3.7	0.0001	0.0001	0.0001	0.0001	0.0001	0.0001	0.0001	0.0001	0.0001	0.0001
−3.6	0.0002	0.0002	0.0001	0.0001	0.0001	0.0001	0.0001	0.0001	0.0001	0.0001
−3.5	0.0002	0.0002	0.0002	0.0002	0.0002	0.0002	0.0002	0.0002	0.0002	0.0002
−3.4	0.0003	0.0003	0.0003	0.0003	0.0003	0.0003	0.0003	0.0003	0.0003	0.0002
−3.3	0.0005	0.0005	0.0005	0.0004	0.0004	0.0004	0.0004	0.0004	0.0004	0.0003
−3.2	0.0007	0.0007	0.0006	0.0006	0.0006	0.0006	0.0006	0.0005	0.0005	0.0005
−3.1	0.0010	0.0009	0.0009	0.0009	0.0008	0.0008	0.0008	0.0008	0.0007	0.0007
−3.0	0.0013	0.0013	0.0013	0.0012	0.0012	0.0011	0.0011	0.0011	0.0010	0.0010
−2.9	0.0019	0.0018	0.0018	0.0017	0.0016	0.0016	0.0015	0.0015	0.0014	0.0014
−2.8	0.0026	0.0025	0.0024	0.0023	0.0023	0.0022	0.0021	0.0021	0.0020	0.0019
−2.7	0.0035	0.0034	0.0033	0.0032	0.0031	0.0030	0.0029	0.0028	0.0027	0.0026
−2.6	0.0047	0.0045	0.0044	0.0043	0.0041	0.0040	0.0039	0.0038	0.0037	0.0036
−2.5	0.0062	0.0060	0.0059	0.0057	0.0055	0.0054	0.0052	0.0051	0.0049	0.0048
−2.4	0.0082	0.0080	0.0078	0.0075	0.0073	0.0071	0.0069	0.0068	0.0066	0.0064
−2.3	0.0107	0.0104	0.0102	0.0099	0.0096	0.0094	0.0091	0.0089	0.0087	0.0084
−2.2	0.0139	0.0136	0.0132	0.0129	0.0125	0.0122	0.0119	0.0116	0.0113	0.0110
−2.1	0.0179	0.0174	0.0160	0.0166	0.0162	0.0158	0.0154	0.0150	0.0146	0.0143
−2.0	0.0228	0.0222	0.0217	0.0212	0.0207	0.0202	0.0197	0.0192	0.0188	0.0183
−1.9	0.0287	0.0281	0.0274	0.0268	0.0262	0.0256	0.0250	0.0244	0.0239	0.0233
−1.8	0.0359	0.0351	0.0344	0.0336	0.0329	0.0322	0.0314	0.0307	0.0301	0.0294
−1.7	0.0446	0.0436	0.0427	0.0418	0.0409	0.0401	0.0392	0.0384	0.0375	0.0367
−1.6	0.0548	0.0537	0.0526	0.0516	0.0505	0.0495	0.0485	0.0475	0.0465	0.0455
−1.5	0.0668	0.0655	0.0643	0.0630	0.0618	0.0606	0.0594	0.0582	0.0571	0.0599
−1.4	0.0808	0.0793	0.0778	0.0764	0.0749	0.0735	0.0721	0.0708	0.0694	0.0681
−1.3	0.0968	0.0951	0.0934	0.0918	0.0901	0.0885	0.0869	0.0853	0.0838	0.0823
−1.2	0.1151	0.1131	0.1112	0.1093	0.1075	0.1056	0.1038	0.1020	0.1003	0.0985
−1.1	0.1357	0.1335	0.1314	0.1292	0.1271	0.1251	0.1230	0.1210	0.1190	0.1170
−1.0	0.1587	0.1562	0.1539	0.1515	0.1492	0.1469	0.1446	0.1423	0.1401	0.1379
−0.9	0.1841	0.1814	0.1788	0.1762	0.1736	0.1711	0.1685	0.1660	0.1635	0.1611
−0.8	0.2119	0.2090	0.2061	0.2033	0.2005	0.1977	0.1949	0.1922	0.1894	0.1867
−0.7	0.2420	0.2389	0.2358	0.2327	0.2296	0.2266	0.2236	0.2206	0.2177	0.2148
−0.6	0.2743	0.2709	0.2676	0.2643	0.2611	0.2578	0.2546	0.2514	0.2483	0.2451
−0.5	0.3085	0.3050	0.3015	0.2981	0.2946	0.2912	0.2877	0.2843	0.2810	0.2776
−0.4	0.3446	0.3409	0.3372	0.3336	0.3300	0.3264	0.3228	0.3192	0.3156	0.3121
−0.3	0.3821	0.3783	0.3745	0.3707	0.3669	0.3632	0.3594	0.3557	0.3520	0.3483
−0.2	0.4207	0.4168	0.4129	0.4090	0.4052	0.4013	0.3974	0.3936	0.3897	0.3859
−0.1	0.4602	0.4562	0.4522	0.4483	0.4443	0.4404	0.4364	0.4325	0.4286	0.4247
−0.0	0.5000	0.4960	0.4920	0.4880	0.4840	0.4801	0.4761	0.4721	0.4681	0.4641

z	0.00	0.01	0.02	0.03	0.04	0.05	0.06	0.07	0.08	0.09
0.0	0.5000	0.5040	0.5080	0.5120	0.5160	0.5199	0.5239	0.5279	0.5319	0.5359
0.1	0.5398	0.5438	0.5478	0.5517	0.5557	0.5596	0.5636	0.5675	0.5714	0.5753
0.2	0.5793	0.5832	0.5871	0.5910	0.5948	0.5987	0.6026	0.6064	0.6103	0.6141
0.3	0.6179	0.6217	0.6255	0.6293	0.6331	0.6368	0.6406	0.6443	0.6480	0.6517
0.4	0.6554	0.6591	0.6628	0.6664	0.6700	0.6736	0.6772	0.6808	0.6844	0.6879
0.5	0.6915	0.6950	0.6985	0.7019	0.7054	0.7088	0.7123	0.7157	0.7190	0.7224
0.6	0.7257	0.7291	0.7324	0.7357	0.7389	0.7422	0.7454	0.7486	0.7517	0.7549
0.7	0.7580	0.7611	0.7642	0.7673	0.7704	0.7734	0.7764	0.7794	0.7823	0.7852
0.8	0.7881	0.7910	0.7939	0.7967	0.7995	0.8023	0.8051	0.8078	0.8106	0.8133
0.9	0.8159	0.8186	0.8212	0.8238	0.8264	0.8289	0.8315	0.8340	0.8365	0.8389
1.0	0.8413	0.8438	0.8461	0.8485	0.8508	0.8531	0.8554	0.8577	0.8599	0.8621
1.1	0.8643	0.8665	0.8686	0.8708	0.8729	0.8749	0.8770	0.8790	0.8810	0.8830
1.2	0.8849	0.8869	0.8888	0.8907	0.8925	0.8944	0.8962	0.8980	0.8997	0.9015
1.3	0.9032	0.9049	0.9066	0.9082	0.9099	0.9115	0.9131	0.9147	0.9162	0.9177
1.4	0.9192	0.9207	0.9222	0.9236	0.9251	0.9265	0.9279	0.9292	0.9306	0.9319
1.5	0.9332	0.9345	0.9357	0.9370	0.9382	0.9394	0.9406	0.9418	0.9429	0.9441
1.6	0.9452	0.9463	0.9474	0.9484	0.9495	0.9505	0.9515	0.9525	0.9535	0.9545
1.7	0.9554	0.9564	0.9573	0.9582	0.9591	0.9599	0.9608	0.9616	0.9625	0.9633
1.8	0.9641	0.9649	0.9656	0.9664	0.9671	0.9678	0.9686	0.9693	0.9699	0.9706
1.9	0.9713	0.9719	0.9726	0.9732	0.9738	0.9744	0.9750	0.9756	0.9761	0.9767
2.0	0.9772	0.9778	0.9783	0.9788	0.9793	0.9798	0.9803	0.9808	0.9812	0.9817
2.1	0.9821	0.9826	0.9830	0.9834	0.9838	0.9842	0.9846	0.9850	0.9854	0.9857
2.2	0.9861	0.9864	0.9868	0.9871	0.9875	0.9878	0.9881	0.9884	0.9887	0.9890
2.3	0.9893	0.9896	0.9898	0.9901	0.9904	0.9906	0.9909	0.9911	0.9913	0.9916
2.4	0.9918	0.9920	0.9922	0.9925	0.9927	0.9929	0.9931	0.9932	0.9934	0.9936
2.5	0.9938	0.9940	0.9941	0.9943	0.9945	0.9946	0.9948	0.9949	0.9951	0.9952
2.6	0.9953	0.9955	0.9956	0.9957	0.9959	0.9960	0.9961	0.9962	0.9963	0.9964
2.7	0.9965	0.9966	0.9967	0.9968	0.9969	0.9970	0.9971	0.9972	0.9973	0.9974
2.8	0.9974	0.9975	0.9976	0.9977	0.9977	0.9978	0.9979	0.9979	0.9980	0.9981
2.9	0.9981	0.9982	0.9982	0.9983	0.9984	0.9984	0.9985	0.9985	0.9986	0.9986
3.0	0.9987	0.9987	0.9987	0.9988	0.9988	0.9989	0.9989	0.9989	0.9990	0.9990
3.1	0.9990	0.9991	0.9991	0.9991	0.9992	0.9992	0.9992	0.9992	0.9993	0.9993
3.2	0.9993	0.9993	0.9994	0.9994	0.9994	0.9994	0.9994	0.9995	0.9995	0.9995
3.3	0.9995	0.9995	0.9995	0.9996	0.9996	0.9996	0.9996	0.9996	0.9996	0.9997
3.4	0.9997	0.9997	0.9997	0.9997	0.9997	0.9997	0.9997	0.9997	0.9997	0.9998
3.5	0.9998	0.9998	0.9998	0.9998	0.9998	0.9998	0.9998	0.9998	0.9998	0.9998
3.6	0.9998	0.9998	0.9999	0.9999	0.9999	0.9999	0.9999	0.9999	0.9999	0.9999
3.7	0.9999	0.9999	0.9999	0.9999	0.9999	0.9999	0.9999	0.9999	0.9999	0.9999
3.8	0.9999	0.9999	0.9999	0.9999	0.9999	0.9999	0.9999	0.9999	0.9999	0.9999

Name ______________________________ Date ______________

1. On his way to work every day, Frank passes through two intersections with traffic signals. Sometimes the lights are green when he arrives at the light; sometimes they are red, and he must stop. The probability that he must stop at the first signal is $P(\text{first}) = 0.4$. The probability that he must stop at the second signal is $P(\text{second}) = 0.5$. The probability that he must stop at both lights is 0.3. Suppose we randomly select one morning that he travels to work, and we look at the outcomes of the two lights.

   a. Describe the event "first, not second" in words.

   b. List all of the outcomes in the sample space.

   c. Calculate $P(\text{first or second})$, and interpret your result in context.

d. Calculate the probability of the event described in part (a).

e. Are the events "stopping at first light" and "stopping at second light" independent? Explain your answer.

f. Assuming the probability of stopping at the first signal does not change from day to day, how surprising would it be for Frank to have to stop at the first light five days in a row? If he does have to stop at the first light five days in a row, would you question the model that assigns a probability of 0.5 to the first light each day?

2. An online bookstore sells both print books and e-books (books in an electronic format). Customers can pay with either a gift card or a credit card.

   a. Suppose that the probability of the event "print book is purchased" is 0.6 and that the probability of the event "customer pays using gift card" is 0.2. If these two events are independent, what is the probability that a randomly selected book purchase is a print book paid for using a gift card?

   b. Suppose that the probability of the event "e-book is purchased" is 0.4; the probability of the event "customer pays using gift card" is 0.2; and the probability of the event "e-book is purchased and customer pays using a gift card" is 0.1. Are the two events "e-book is purchased" and "customer pays using a gift card" independent? Explain why or why not.

3. Airlines post the estimated arrival times for all of their flights. However, sometimes the flights arrive later than expected. The following data report the number of flights that were "on time" or "late" for two different airlines in November 2012 for all flights to Houston, Chicago, and Los Angeles:

	Houston		Chicago		Los Angeles	
	On Time	**Late**	**On Time**	**Late**	**On Time**	**Late**
Airline A	7,318	1,017	466	135	544	145
Airline B	598	70	8,330	1,755	2,707	566

   a. Use the data to estimate the probability that a randomly selected flight arriving in Houston will be on time.

Consider only the flights to Chicago and Los Angeles for these two airlines combined.

	Chicago	Los Angeles
On Time	8,796	3,251
Late	1,890	711

b. Explain what it means to say that the events "arriving on time" and "Chicago" are independent.

c. Do the events "arriving on time" and "Chicago" appear to be approximately independent? Explain your answer.

4. The average height of the 140 million U.S. males is 5 ft. and 10 in. Some males from the U.S. become professional basketball players. The average height of the 350–450 professional basketball players in the NBA (National Basketball Association) is about 6 ft. and 7 in. Which of the following probabilities should be larger? Or would they be similar?

- The probability that a U.S. male over 6 ft. tall is a professional basketball player
- The probability that a professional basketball player is over 6 ft. tall

Explain your reasoning.

5. A researcher gathers data on how long teenagers spend on individual cell phone calls (in number of minutes). Suppose the research determines that these calls have a mean 10 minutes and standard deviation 7 minutes.

   a. Suppose the researcher also claims that the distribution of the call lengths follows a normal distribution. Sketch a graph displaying this distribution. Be sure to add a scale and to label your horizontal axis.

b. Using your graph, shade the area that represents the probability that a randomly selected call lasts more than 12 minutes. Is this probability closer to 0.50 or to 0.05?

c. After looking at the above mean and standard deviation of the call length data, a second researcher indicates that she does not think that a normal distribution is an appropriate model for the call length distribution. Which researcher (the first or the second) do you think is correct? Justify your choice.

A Progression Toward Mastery					
Assessment Task Item		**STEP 1** Missing or incorrect answer and little evidence of reasoning or application of mathematics to solve the problem.	**STEP 2** Missing or incorrect answer but evidence of some reasoning or application of mathematics to solve the problem.	**STEP 3** A correct answer with some evidence of reasoning or application of mathematics to solve the problem. OR An incorrect answer with substantial evidence of solid reasoning or application of mathematics to solve the problem.	**STEP 4** A correct answer supported by substantial evidence of solid reasoning or application of mathematics to solve the problem.
1	a S-CP.A.1	Student provides an answer that does not relate to the provided notation.	Student appears to have misread the event or describes a probability rather than the event.	Student does not answer in context.	Student describes the combined outcome for both lights and in the context of the problem.
	b S-CP.A.1	Student provides an answer that does not relate to the idea of *sample space* (i.e., all possible outcomes of the probability experiment).	Student lists possible outcomes but not as paired outcomes (e.g., describes the lights individually rather than the joint outcomes of the two lights).	Student provides an answer that does not include all of the possible outcomes.	Student provides an answer that includes all of the outcomes and is presented in an organized fashion.
	c S-CP.B.7	Student misrepresents the event (e.g., the interaction rather than the union) or determines the probability incorrectly and obtains a value below zero or above one. Student provides an interpretation that is inconsistent with the calculation.	Student treats the two lights separately rather than finding a combined event. Student provides an interpretation of the union that is not complete or is unclear.	Student demonstrates recognition of the need to find the union but assumes the outcomes in the sample space are equally likely or adds $P(\text{first}) + P(\text{second})$, ignoring the intersection. Student correctly interprets the probability of the union.	Student uses an appropriate probability rule (e.g., addition rule) or solution method (e.g., table, Venn diagram) to find the probability of interest. Student correctly interprets the probability of the union.

	d S-CP.B.7	Student misinterprets the event (e.g., the union rather than an intersection) or determines the probability incorrectly and obtains a value below zero or above one.	Student demonstrates understanding of the correct event but is not able to calculate the probability.	Student demonstrates recognition of the need to find the intersection but assumes the outcomes in the sample space are equally likely or incorrectly applies a probability rule.	Student uses an appropriate probability rule (e.g., addition rule) or solution method (e.g., table, Venn diagram) to find the probability of interest.
	e S-CP.B.3	Student is not able to address the issue of independence.	Student includes only one of the three components listed.	Student includes only two of the three components listed.	Student describes an appropriate method for checking independence, includes appropriate calculations, and draws a correct conclusion.
	f S-IC.A.2	Student is not able to calculate a probability.	Student calculates an incorrect probability.	Student calculates the probability correctly but does not relate it to the scenario.	Student uses the multiplication rule to calculate the probability and evaluates the probability as small.
2	a S-CP.A.2	Student is not able to calculate a probability.	Student incorrectly adds the two given probabilities rather than multiplying.	Student knows to multiply probabilities but makes an error that results in an incorrect probability.	Student uses the multiplication rule to calculate a correct probability.
	b S-CP.A.2	Student is not able to address the issue of independence.	Student indicates that the events are not independent, but the explanation confuses independent events and mutually exclusive events.	Student correctly calculates the probability of the intersection to be 0.08 but then explains that the events are independent because 0.08 is close to 0.1.	Student indicates that the events are not independent with an explanation based on the probability of the intersection not being equal to the product of the two individual probabilities.
3	a S-CP.B.6	Student does not use information from the table.	Student reports the proportion of on-time flights or the proportion of flights arriving in Houston.	Student sets up the calculation correctly but transcribes a number incorrectly or has a small calculation error.	Student reads the correct proportion from the table.
	b S-CP.A.5	Student provides a definition that is not related to probabilities.	Student confuses independence with mutually exclusive or equiprobable.	Student describes a correct generic interpretation of probability but does not relate it to this context.	Student correctly interprets *independence* in the context of the data.

	c S-CP.A.4	Student is not able to address the issue of independence.	Student includes only one of the three components listed.	Student includes only two of the three components listed.	Student describes an appropriate method for checking independence, includes appropriate calculations, and draws a correct conclusion.
4	S-CP.A.5	Student does not think there is enough information given in the problem or does not include sufficient reasoning.	Student is not able to differentiate between the two conditional probabilities and considers them similar.	Student correctly identifies the two conditional probabilities but does not make a reasonable comparison.	Student correctly compares the two conditional proportions, $P(\text{tall} \mid \text{NBA})$ and $P(\text{NBA} \mid \text{tall})$, and correctly identifies the $P(\text{tall} \mid \text{NBA})$ as larger.
5	a S-ID.A.4	Student does not include the sketch of a normal curve.	Student draws a normal curve but with annotation.	Student draws a sketch that is missing either complete scaling (e.g., the mean but no other values) or does not include an axis label.	Student creates a reasonable sketch with scaling to represent the mean and standard deviation and includes a horizontal axis label.
	b S-ID.A.4	Student estimates the probability, but it is not based on the sketch.	Student shades at 12, but it is not consistent with the labeling and scaling of the graph. OR Student estimates a probability that is not consistent with the shading.	Student shades for a probability of less than 12 minutes or does not address the probability question.	Student shades an appropriate region given the graph. Student provides an estimate of the probability that is consistent with the shaded area.
	c S-ID.A.4	Student provides a response that does not indicate a problem with using the normal distribution.	Student recognizes the lower limit on the call length but is not able to relate that to the mean and standard deviation.	Student considers the mean and the standard deviation but does not address the lower limit of zero. OR Student draws an incorrect conclusion as to which researcher.	Student considers the shape of the distribution and indicates that extending $3SD$ below the mean would predict negative call lengths (agreeing with researcher 2).

Name ______________________________ Date ______________

1. On his way to work every day, Frank passes through two intersections with traffic signals. Sometimes the lights are green when he arrives at the light; sometimes they are red, and he must stop. The probability that he must stop at the first signal is $P(\text{first}) = 0.4$. The probability that he must stop at the second signal is $P(\text{second}) = 0.5$. The probability that he must stop at both lights is 0.3. Suppose we randomly select one morning that he travels to work, and we look at the outcomes of the two lights.

   a. Describe the event "first, not second" in words.

   This event corresponds to stopping at the first light but not the second light. (If the student reverses these, allow for credit in remaining problems if subsequent responses are consistent.)

   b. List all of the outcomes in the sample space.

   The outcome (first, second) represents stopping at both lights

   The outcome (first, not second) represents stopping at the first light but not at the second.

   The outcome (not first, second) represents not stopping at the first light but stopping at the second.

   The outcome (not first, not second) represents not stopping at either light.

   c. Calculate $P(\text{first or second})$, and interpret your result in context.

   $P(\text{first or second}) = P(\text{first}) + P(\text{second}) - P(\text{first and second}) = 0.4 + 0.5 - 0.3 = 0.60$ by the addition rule. This is the probability that Frank must stop at at least one of the lights.

d. Calculate the probability of the event described in part (a).

Using a table of, say, 100 trips:

	Stop at First Light	No Stop at First Light	Total
Stop at Second Light	30	20	50
No Stop at Second Light	10	40	50
Total	40	60	100

So, P(stop at first but not at second) $= \frac{10}{100} = 0.10$

Alternative solutions: P(stopping at first light) = P(stopping at first and second) + P(stopping at first and not second) $0.4 = 0.3 + x$; *therefore,* $x = 0.10$.

Looking at a Venn diagram:

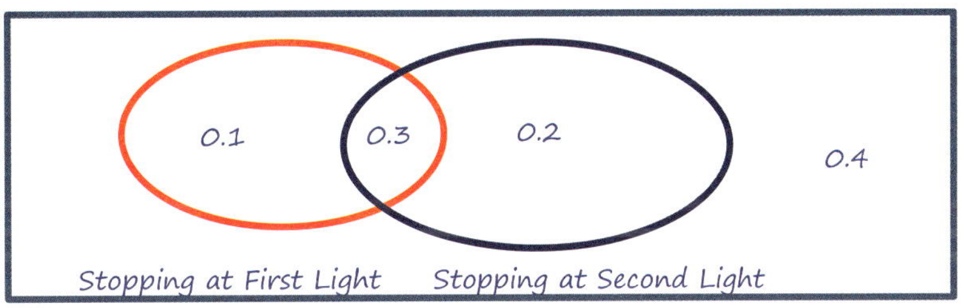

The probability of stopping at the first light but not the second light is 0.10.

e. Are the events "stopping at first light" and "stopping at second light" independent? Explain your answer.

We need to determine whether P(first and second) = P(first)P(second).

However, $0.30 \neq 0.4(0.5)$, *so these events are not independent.*

f. Assuming the probability of stopping at the first signal does not change from day to day, how surprising would it be for Frank to have to stop at the first light five days in a row? If he does have to stop at the first light five days in a row, would you question the model that assigns a probability of 0.5 to the first light each day?

Because the probability is not changing, we can multiply the probabilities together.

P(stop and stop and stop and stop and stop) = P(stop)5 $= (0.4)(0.4)(0.4)(0.4)(0.4) \approx 0.0102$

This is a small probability and indicates that this would be a surprising observation assuming the model is correct.

2. An online bookstore sells both print books and e-books (books in an electronic format). Customers can pay with either a gift card or a credit card.

   a. Suppose that the probability of the event "print book is purchased" is 0.6 and that the probability of the event "customer pays using gift card" is 0.2. If these two events are independent, what is the probability that a randomly selected book purchase is a print book paid for using a gift card?

   P(print book and pays by gift card) = P(print book)P(pays by gift card)

   $= (0.6)(0.2)$

   $= 0.12$

   b. Suppose that the probability of the event "e-book is purchased" is 0.4; the probability of the event "customer pays using gift card" is 0.2; and the probability of the event "e-book is purchased and customer pays using a gift card" is 0.1. Are the two events "e-book is purchased" and "customer pays using a gift card" independent? Explain why or why not.

   The two events are not independent. If they were independent, the probability of the event "e-book is purchased and customer pays using a gift card" would be (0.4)(0.2), which is 0.08. Because this probability is actually 0.1, the events are not independent.

3. Airlines post the estimated arrival times for all of their flights. However, sometimes the flights arrive later than expected. The following data report the number of flights that were "on time" or "late" for two different airlines in November 2012 for all flights to Houston, Chicago, and Los Angeles:

	Houston		Chicago		Los Angeles	
	On Time	**Late**	**On Time**	**Late**	**On Time**	**Late**
Airline A	7,318	1,017	466	135	544	145
Airline B	598	70	8,330	1,755	2,707	566

   a. Use the data to estimate the probability that a randomly selected flight arriving in Houston will be on time.

$$\frac{7318 + 598}{7318 + 58 + 1017 + 70} \approx 0.935$$

Consider only the flights to Chicago and Los Angeles for these two airlines combined.

	Chicago	Los Angeles
On Time	8,796	3,251
Late	1,890	711

b. Explain what it means to say that the events "arriving on time" and "Chicago" are independent.

It would mean that the probability of arriving on time is the same for Chicago as it is for Los Angeles. In other words, the likelihood of arriving on time does not depend on which city the plane is arriving.

c. Do the events "arriving on time" and "Chicago" appear to be approximately independent? Explain your answer.

$$P(\text{on time} \mid \text{Chicago}) = \frac{8796}{8796 + 1890} \approx 0.823$$

$$P(\text{on time} \mid \text{Los Angeles}) = \frac{3251}{3251 + 711} \approx 0.821$$

The on-time percentage is pretty similar for either city, so the events are roughly independent.

4. The average height of the 140 million U.S. males is 5 ft. and 10 in. Some males from the U.S. become professional basketball players. The average height of the 350–450 professional basketball players in the NBA (National Basketball Association) is about 6 ft. and 7 in. Which of the following probabilities should be larger? Or would they be similar?

- The probability that a U.S. male over 6 ft. tall is a professional basketball player
- The probability that a professional basketball player is over 6 ft. tall

Explain your reasoning.

$P(\text{tall} \mid \text{NBA player})$ is a fairly high number as almost all NBA players are tall.
$P(\text{NBA player} \mid \text{tall})$ is a pretty small number as not many tall men are in the NBA. So, the second probability is much larger.

5. A researcher gathers data on how long teenagers spend on individual cell phone calls (in number of minutes). Suppose the research determines that these calls have a mean 10 minutes and standard deviation 7 minutes.

   a. Suppose the researcher also claims that the distribution of the call lengths follows a normal distribution. Sketch a graph displaying this distribution. Be sure to add a scale and to label your horizontal axis.

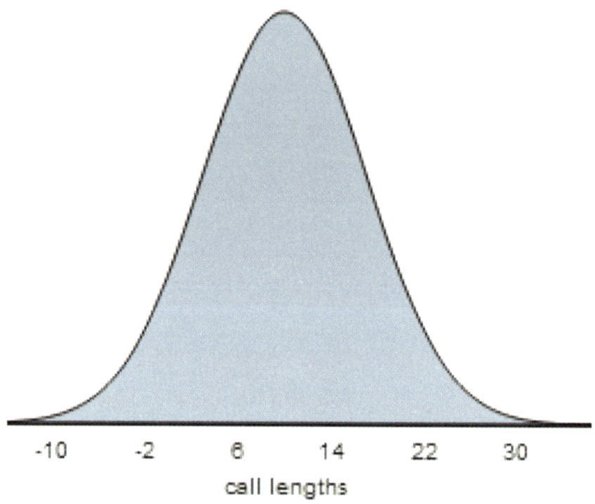

b. Using your graph, shade the area that represents the probability that a randomly selected call lasts more than 12 minutes. Is this probability closer to 0.50 or to 0.05?

Twelve minutes is very close (within one standard deviation) to the mean of 10 minutes, so the probability of being more than 12 minutes will be close to 0.5.

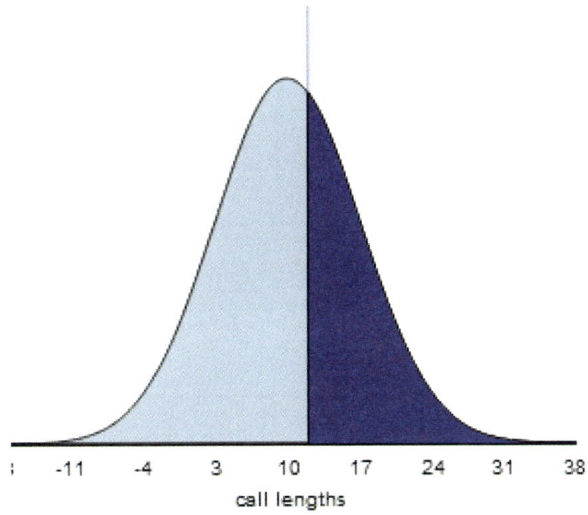

c. After looking at the above mean and standard deviation of the call length data, a second researcher indicates that she does not think that a normal distribution is an appropriate model for the call length distribution. Which researcher (the first or the second) do you think is correct? Justify your choice.

We notice from the sketch in part (a) that this implies a noticeable fraction of the calls would last less than zero minutes, which is not possible. I agree with the second researcher.

Topic C

Drawing Conclusions Using Data from a Sample

S-IC.A.1, S-IC.B.3, S-IC.B.4, S-IC.B.6

Focus Standards:	S-IC.A.1	Understand statistics as a process for making inferences about population parameters based on a random sample from that population.
	S-IC.B.3	Recognize the purposes of and differences among sample surveys, experiments, and observational studies; explain how randomization relates to each.
	S-IC.B.4	Use data from a sample survey to estimate a population mean or proportion; develop a margin of error through the use of simulation models for random sampling.
	S-IC.B.6	Evaluate reports based on data.
Instructional Days:	11	
Lesson 12:	Types of Statistical Studies (P)[1]	
Lesson 13:	Using Sample Data to Estimate a Population Characteristic (P)	
Lessons 14–15:	Sampling Variability in the Sample Proportion (E, E)	
Lessons 16–17:	Margin of Error When Estimating a Population Proportion (E, P)	
Lessons 18–19:	Sampling Variability in the Sample Mean (E, P)	
Lessons 20–21:	Margin of Error When Estimating a Population Mean (E, P)	
Lesson 22:	Evaluating Reports Based on Data from a Sample (P)	

This topic introduces different types of statistical studies (e.g., observational studies, surveys, and experiments) (**S-IC.B.3**). The role of randomization (i.e., random selection in observational studies and surveys and random assignment in experiments) is addressed. A discussion of random selection (i.e., selecting a sample at random from a population of interest) shows students how selecting participants at random provides a representative sample, thereby allowing conclusions to be generalized from the sample to the population. A discussion of random assignment in experiments, which involves assigning subjects to experimental groups at random, helps students see that random assignment is designed to create comparable groups making it possible to assess the effects of an explanatory variable on a response.

[1]Lesson Structure Key: **P**-Problem Set Lesson, **M**-Modeling Cycle Lesson, **E**-Exploration Lesson, **S**-Socratic Lesson

The distinction between population characteristics and sample statistics (first made in Grade 7) is revisited. Scenarios are introduced in which students are asked a statistical question that involves estimating a population mean or a population proportion. For example, students are asked to define an appropriate population, population characteristic, sample, and sample statistics that might be used in a study of the time it takes students to run a quarter mile or a study of the proportion of national parks that contain bald eagle nests.

In this topic, students use data from a random sample to estimate a population mean or a population proportion. Building on what they learned about sampling variability in Grade 7, students use simulation to create an understanding of margin of error. In Grade 7, students learned that the proportion of successes in a random sample from a population varies from sample to sample due to the random selection process. They understand that the value of the sample proportion is not exactly equal to the value of the population proportion. In Algebra II, they use margin of error to describe how different the value of the sample proportion might be from the value of the population proportion. Students begin by using a physical simulation process to carry out a simulation. Starting with a population that contains 40% successes (using a bag with 40 black beans and 60 white beans), they select random samples from the population and calculate the sample proportion. By doing this many times, they are able to get a sense of what kind of differences are likely. Their understanding should then extend to include the concept of margin of error. Students then proceed to use technology to carry out a simulation. Once students understand the concept of margin of error, they go on to learn how to calculate and interpret it in context (**S-IC.A.1**, **S-IC.B.4**). Students also evaluate reports from the media in which sample data are used to estimate a population mean or proportion (**S-IC.B.6**).

Lesson 12: Types of Statistical Studies

Student Outcomes

- Students distinguish between observational studies, surveys, and experiments.
- Students explain why random selection is an important consideration in observational studies and surveys and why random assignment is an important consideration in experiments.
- Students recognize when it is reasonable to generalize the results of an observational study or survey to some larger population and when it is reasonable to reach a cause-and-effect conclusion about the relationship between two variables.

Lesson Notes

A statistical study is a four-step process that begins by asking a question that can be answered with data. The next steps are to (1) collect appropriate data, (2) organize and analyze the data, and (3) arrive at a conclusion in the context of the original question. Grade 6 (Module 6 Lesson 1) introduced students to the first step in asking a statistical question while this lesson focuses on the second step on collecting data. The primary resource for the four-step process is *The GAISE Report (Guidelines in Assessment and Instruction for Statistics Education).* A free download can be found at https://www.amstat.org/education/gaise/GAISEPreK-12_Full.pdf.

This lesson discusses the three main types of statistical studies: observational studies, surveys, and experiments. It defines each type, gives examples for each, and asks students to distinguish between them.

- An *observational study* records the values of variables for members of a sample. There are several types of observational studies. Observational studies are designed to observe subjects as they are, without any manipulation by the researcher.
- A *survey* is a type of observational study that gathers data by asking people a number of questions.
- An *experiment* assigns subjects to treatments to see what effect the treatments have on some response.

Classwork

Opening Exercise (2 minutes)

Read through the opening text as a class. Discuss each of the three types of statistical studies: observational studies, surveys, and experiments, making sure that students understand the differences between the types of studies. A visual organizer might be helpful here, especially for English language learners.

> *Scaffolding:*
>
> A graphic organizer for statistical studies could resemble the Venn diagram below.
>
> 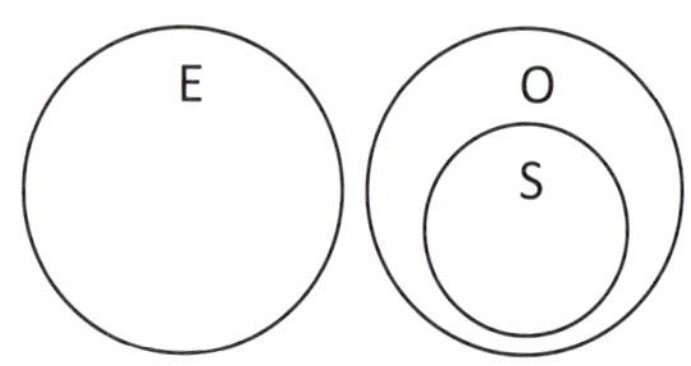
>

Opening Exercise

You want to know what proportion of the population likes rock music. You carefully consider three ways to conduct a study. What are the similarities and differences between the following three alternatives? Do any display clear advantages or disadvantages over the others?

> a. You could pick a random sample of people and ask them the question, "Do you like rock music?" and record their answers.
>
> b. You could pick a random sample of people and follow them for a period of time, noting their music purchases, both in stores and online.
>
> c. You could pick a random sample of people, separate them into groups, and have each group listen to a different genre of music. You would collect data on the people who display an emotional response to the rock music.
>
> A statistical study begins by asking a question that can be answered with data. The next steps are to collect appropriate data, organize and analyze them, and arrive at a conclusion in the context of the original question. This lesson focuses on the three main types of statistical studies: observational studies, surveys, and experiments. The objective of an observational study and a survey is to learn about characteristics of some population, so the data should be collected in a way that would result in a representative sample. This speaks to the importance of random selection of subjects for the study. The objective of an experiment is to answer such questions as "What is the effect of treatments on a response variable?" Data in an experiment need to be collected in a way that does not favor one treatment over another. This demonstrates the importance of random assignment of subjects in the study to the treatments.
>
> An observational study is one in which the values of one or more variables are observed with no attempt to affect the outcomes. One kind of observational study is a survey. A survey requires asking a group of people to respond to one or more questions. (A poll is one example of a survey.) An experiment differs from an observational study. In an experiment, subjects are assigned to treatments for the purpose of seeing what effect the treatment has on some response while an observational study makes no attempt to affect the outcomes (i.e., no treatment is given). Note that subjects could be people, animals, or any set of items that produce variability in their responses. Here is an example of an observational study: In a random sample of students, it was observed that those students who played a musical instrument had better grades than those who did not play a musical instrument. In an experiment, a group of students who do not currently play a musical instrument would be assigned at random to having to play a musical instrument or not having to play a musical instrument for a certain period of time. Then, at the end of the period of time, we would compare academic performance.
>
> Classify each of the three study methods about rock music as an observational study, a survey, or an experiment.

Example 1 (5–7 minutes): Survey

Discuss each part of the example as a class.

A survey asks people to respond to one or more questions. First graders have no doubt done this in asking their classmates, "What is your favorite color?" But constructing clear questions may not be so easy. A poorly worded question may confuse the person answering it.

For example, take the question, "Do you like your school's cafeteria food?" How do you answer if you bring your own lunch? Or what if you like the salad but do not like the turkey sandwich? Assuming that the person answering eats the cafeteria food, a better survey approach would be to say, "For each of the following items offered by our cafeteria, check the appropriate box."

Example 1: Survey

Item	I like the item.	I do not like the item.	I have never tried the item.
Salad			
Vegetable Pizza			
Turkey Sandwich			
Raspberry Tea			

a. It is easy to determine if a study is a survey. A survey asks people to respond to questions. But surveys can be flawed in several ways. Questions may be confusing. For example, consider the following question:

What kind of computer do you own? (*Circle one*) Mac IBM-PC

How do you answer that question if you do not own a computer? How do you answer that question if you own a different brand? A better question would be

Do you own a computer? (*Circle one*) Yes No

If you answered yes, what brand of computer is it? ________________

Now consider the question, "Do you like your school's cafeteria food?"

Rewrite the question in a better form. Keep in mind that not all students may use the school's cafeteria, and even if they do, there may be some foods that they like and some that they do not like.

A survey should not be generalized beyond the set of people who responded unless participants are randomly selected from some identifiable group. How to select a random sample was covered in Grade 7, but it may be necessary to review how to use a random number table with students.

b. Something else to consider with surveys is how survey participants are chosen. If the purpose of the survey is to learn about some population, ideally participants would be randomly selected from the population of interest. If people are not randomly selected, misleading conclusions from the survey data may be drawn. There are many famous examples of this. Perhaps the most famous case was in 1936 when *The Literary Digest* magazine predicted that Alf Landon would beat incumbent President Franklin Delano Roosevelt by 370 electoral votes to 161. Roosevelt won 523 to 8.

Ten million questionnaires were sent to prospective voters (selected from the magazine's subscription list, automobile registration lists, phone lists, and club membership lists), and over two million questionnaires were returned. Surely such a large sample should represent the whole population. How could *The Literary Digest* prediction be so far off the mark?

The sample was biased toward the wealthy, and in 1936, during the heart of the Great Depression, the wealthy were not representative of the whole population of voters. Another difficulty was that not everyone who received the survey chose to return it. That resulted in what is called voluntary response bias.

c. Write or say to your neighbor two things that are important about surveys.

Scaffolding:

Nobody appreciated the power of random sampling more than George Gallup who developed the Gallup Poll. Have students do research on how Gallup got started in the polling business in 1936. How did he correctly predict the 1936 election while *The Literary Digest* was so far off?

Example 2 (7–9 minutes): Observational Study

Work through each part of the example as a class.

A primary difference between an observational study and an experiment is that an experiment imposes treatments on subjects to see what effect those treatments might have on their responses. Observational studies do not impose treatments on subjects. There is no attempt to influence the responses. Sometimes situations are encountered in which an observational study needs to be done because an experiment cannot be performed. For example, it would be unethical (to say the least) to impose a dangerous treatment on people, such as assigning them to smoke a pack of cigarettes per day or exposing them purposely to asbestos.

A word of warning: There are many examples of famous unethical experiments that are incredibly disturbing. So, if assigning students to find and report on an unethical experiment, consider providing them with a prescreened list rather than letting them loose on the Internet.

Example 2: Observational Study

a. **An observational study records the values of variables for members of a sample but does not attempt to influence the responses. For example, researchers investigated the link between the use of cell phones and brain cancer. There are two variables in this study: One is the extent of cell phone usage, and the second is whether a person has brain cancer. Both variables were measured for a group of people. This is an observational study. There was no attempt to influence peoples' cell phone usage to see if different levels of usage made any difference in whether or not a person developed brain cancer.**

Why would studying any relationship between asbestos exposure and lung cancer be an observational study and not an experiment?

Random sampling of subjects is important in observational studies and surveys in order to eliminate bias so that sample results may be generalized to the population from which the sample was taken.

b. **In an observational study (just as in surveys), the people or objects to be observed would ideally be selected at random from the population of interest. This would eliminate bias and make it possible to generalize from a sample to a population. For example, to determine if the potato chips made in a factory contain the desired amount of salt, a sample of chips would be selected randomly so that the sample can be considered to be representative of the population of chips.**

Discuss how a random sample of 100 chips might be selected from a conveyor belt of chips.

MP.3 Cause-and-effect is a very important issue. Allow ample time to discuss this issue, and have students critique several examples. Students might find some examples to be humorous. For example, there is a strong positive relationship between the number of television sets in a country and the life expectancies of its residents. Could one conclude that to increase the life expectancies of the residents of some country, they need only to increase the number of television sets in that country? Surely not. But why not? A "rich" country would have many television sets but would also have access to health care and fresh water, which provide a better quality of life and, hence, a higher life expectancy. On the other hand, a poor country would have fewer television sets but would also have less access to health care and fresh water, leading to a lower life expectancy. The economic status of a country is what explains the relationship. Economic status is a lurking variable.

A lurking variable is one that causes two variables to have a high relationship even though there is no real direct relationship between the two variables.

Ice cream sales and the number of drowning accidents are positively related. Therefore, can it be concluded that ice cream sales cause drowning accidents? The lurking variable here is time of year. More ice cream is sold in the summer than in the winter, and there is more opportunity for drowning in the summer than winter. Internet research may provide some more examples for study.

MP.2 Often people are interested in knowing what causes something to happen. The main disadvantage of an observational study is that a cause-and-effect conclusion cannot be drawn from any relationship observed in such a study. Understanding this disadvantage is important for students to identify as they analyze statistical studies. For example, an observational study indicating that extensive use of a cell phone is indeed linked with brain cancer does not mean that extensive cell phone usage causes brain cancer because there may be many other variables related to cell phone usage that may also relate to brain cancer. For example, maybe people who are heavy cell phone users tend to also have very stressful jobs, and stress may be a factor contributing to poor health. An experiment must be done in order to establish causality.

c. Suppose that an observational study establishes a link between asbestos exposure and lung cancer. Based on that finding, can we conclude that asbestos exposure causes lung cancer? Why or why not?

Answers will vary. Sample response: No. It is possible there are other variables that have not been examined, such as whether or not the subject smokes cigarettes.

d. Write or say to your neighbor two things that are important about observational studies.

Example 3 (5–7 minutes): Experiment

Discuss each part of the example as a class. Make sure that students understand why the study described in part (a) is an experiment and that they can identify the explanatory variable and the response variable.

The cause-and-effect difficulty discussed in observational studies can be rectified with an experiment. Being able to vary the treatment to see what the response is in each case enables a cause-and-effect conclusion to be made as long as the treatment groups are comparable. For instance, in the radish seedlings example, by having three fixed treatment levels of light/dark and having measured the lengths of germinated seedlings, it can be concluded that complete darkness produces the longest lengths. (See *The GAISE Report,* pages 75–79.)

Example 3: Experiment

a. **An experiment imposes treatments to see the effect of the treatments on some response. Suppose that an observational study indicated that a certain type of tree did not have as much termite damage as other trees. Researchers wondered if resin from the tree was toxic to termites. They decided to do an experiment where they exposed some termites to the resin and others to plain water and recorded whether the termites survived. The explanatory variable (treatment variable) is the exposure type (resin, plain water), and the response variable is whether or not the termites survived. We know this is an experiment because the researchers imposed a treatment (exposure type) on the subjects (termites).**

Is the following an observational study or an experiment? Why? If it is an experiment, identify the treatment variable and the response variable. If it is an observational study, identify the population of interest.

A study was done to answer the question, "What is the effect of different durations of light and dark on the growth of radish seedlings?" Three similar growth chambers (plastic bags) were created in which 30 seeds randomly chosen from a package were placed in each chamber. One chamber was randomly selected and placed in 24 hours of light, another for 12 hours of light and 12 hours of darkness, and a third for 24 hours of darkness. After three days, researchers measured and recorded the lengths of radish seedlings for the germinating seeds.

Recall that random sampling of subjects was done in observational studies and surveys in order to obtain a sample that is representative of a population. In experiments, subjects may not have been chosen randomly from a population but must be randomly assigned to treatments to create comparable treatment groups.

Note that there are more experimental designs than the one discussed in this lesson in which the treatment groups were independent of each other. For example, the seeds that were in the "24 hours of light" treatment were assigned to one and only one treatment. Clearly, they could not also be subjected to another treatment. But there are studies in which subjects undergo all of the treatments. The advantage of this type of design is to eliminate the effects that extraneous variables may have on the response. For example, consider a study investigating whether smelling roses improves student performance. One treatment would be to have a student wear an unscented mask and be timed on how fast he completes a pencil and paper maze. Another would be to time how fast a student wearing a rose-scented mask completes a pencil and paper maze. To eliminate the effect of extraneous variables (e.g., ability to do mazes, gender, and age), having each student in the study do both treatments is a better design than assigning different students to the treatments, which would expose students to only one treatment each. Such a design is called a *repeated measures design*. Randomizing the order in which the two masks are implemented counterbalances a potential performance effect. Half of the students would do the experiment with the unscented mask first, and the other half would do the experiment with the scented mask first. The idea is that the performance effect would be about the same in each ordering and, hence, not affect the overall conclusion.

b. In an experiment, random assignment of subjects to treatments is done to create comparable treatment groups. For example, a university biologist wants to compare the effects of two weed killers on pansies. She chooses 24 plants. If she applies weed killer A to the 12 healthiest plants and B to the remaining 12 plants, she will not know which plants died due to the type of weed killer used and which plants subjected to weed killer B were already on their last legs. Randomly selecting 12 plants to receive weed killer A and then assigning the rest to B would help ensure that the plants in each group are fairly similar.

How might the biologist go about randomly assigning 12 plants from the 24 candidates to receive weed killer A? Could she be sure to get exactly 12 plants assigned to weed killer A and 12 plants to weed killer B by tossing a fair coin for each plant and assigning "heads up" plants to weed killer A and "tails up" to weed killer B? If not, suggest a method that you would use.

c. Write or say to your neighbor two things that are important about experiments.

Exercises 1–3 (19 minutes)

Have students complete the exercises with a partner or small group. Then, discuss answers as a class.

Exercises 1–3

1. For each of the following study descriptions, identify whether the study is a survey, an observational study, or an experiment, and give a reason for your answer. For observational studies, identify the population of interest. For experiments, identify the treatment and response variables.

a. A study investigated whether boys are quicker at learning video games than girls. Twenty randomly selected boys and twenty randomly selected girls played a video game that they had never played before. The time it took them to reach a certain level of expertise was recorded.

Observational study. The children were observed, and no treatment was administered to them. The population of interest was all boys and girls who would play the video game that they had never played before. The study was to see who is quicker at achieving a certain level of expertise.

b. As your statistics project, you collect data by posting five questions on poster board around your classroom and recording how your classmates respond to them.

Survey. Questions were asked of classmates. Students may question whether this is really a sample survey since all classmates participated. They were not randomly chosen. Good point. Full marks to those who suggest that it is a census.

c. A professional sports team traded its best player. The local television station wanted to find out what the fans thought of the trade. At the beginning of the evening news program, they asked viewers to call one number if they favored the trade and a different number if they were opposed to the trade. At the end of the news program, they announced that 53.7% of callers favored the trade.

Survey. Such a sampling technique is called voluntary response. It is a very poor way of gathering data since subjects are clearly not randomly chosen.

d. The local department of transportation is responsible for maintaining lane and edge lines on its paved roads. There are two new paint products on the market. Twenty comparable stretches of road are identified. Paint A is randomly assigned to ten of the stretches of road and paint B to the other ten. The department finds that paint B lasts longer.

Experiment. The stretches of road that are the "subjects" in this study are randomly assigned to the treatments (the two types of paint). The response variable is the longevity of the paint. Students may note that other factors could affect the experiment, like weather, traffic, or other lurking variables.

e. The National Highway Traffic Safety Administration conducts annual studies on drivers' seatbelt use at a random selection of roadway sites in each state in the United States. To determine if seatbelt usage has increased, data are analyzed over two successive years.

Observational study. No treatment is administered. The population of interest was drivers observed at various randomly selected roadway sites. The study was to observe seatbelt usage over a two-year period. *(See* Making Sense of Statistical Studies*, pages 60–67.)*

f. People should brush their teeth at least twice a day for at least two to three minutes with each brushing. For a statistics class project, you ask a random number of students at your school questions concerning their tooth brushing activities.

Survey. *(See* Making Sense of Statistical Studies*, pages 93–101.)*

g. A study determines whether taking aspirin regularly helps to prevent heart attacks. A large group of male physicians of comparable health were randomly assigned equally to taking an aspirin every second day or to taking a placebo. After several years, the proportion of the study participants who had suffered heart attacks in each group was compared.

Experiment. The male physicians were randomly assigned to one of two treatments. The treatment variable is aspirin or no aspirin administered. The response variable is whether or not the subject suffered a heart attack. Note that this is a well-known experiment called the Physicians' Health Study. It has its own extensive website: http://phs.bwh.harvard.edu/.

2. For the following, is the stated conclusion reasonable? Why or why not?

A study found a positive relationship between the happiness of elderly people and the number of pets they have. Therefore, having more pets causes elderly people to be happier.

No. This is an observational study. No treatment was administered to the elderly. Note: Consider asking students to create the scenario as an experiment. For example, 40 elderly people were found who had the same degree of happiness on some "happiness scale." Ten of them were randomly assigned to no pets, ten to one pet, ten to two pets, and ten to three pets. Six months later, the happiness measure was taken on the people. If, for example, there was an increase in happiness as the number of pets increased, then the conclusion that having more pets causes elderly people to become happier would be reasonable.

3. A researcher wanted to find out whether higher levels of a certain drug given to experimental rats would decrease the time it took them to complete a given maze to find food.

a. Why would the researcher have to carry out an experiment rather than an observational study?

To make a cause-and-effect conclusion requires an experiment to be done. Cause-and-effect cannot be concluded from an observational study or survey.

b. Describe an experiment that the researcher might carry out based on 30 comparable rats and three dosage levels: 0 mg, 1 mg, and 2 mg.

The researcher should randomly assign ten rats to dosage level 0 mg (control), ten rats to dosage level 1 mg, and ten rats to dosage level 2 mg. The researcher should then record the time it takes each rat to complete the given maze to find food.

Closing (2 minutes)

Ask students to summarize the main ideas of the lesson with a neighbor or in writing. Use this opportunity to informally assess comprehension of the lesson. The Lesson Summary below offers some important ideas that should be included.

> **Lesson Summary**
>
> - **There are three major types of statistical studies: observational studies, surveys, and experiments.**
>   - **An *observational study* records the values of variables for members of a sample.**
>   - **A *survey* is a type of observational study that gathers data by asking people a number of questions.**
>   - **An *experiment* assigns subjects to treatments for the purpose of seeing what effect the treatments have on some response.**
> - **To avoid bias in observational studies and surveys, it is important to select subjects randomly.**
> - **Cause-and-effect conclusions cannot be made in observational studies or surveys.**
> - **In an experiment, it is important to assign subjects to treatments randomly in order to make cause-and-effect conclusions.**

Exit Ticket (5 minutes)

Name ______________________________ Date ______________

Lesson 12: Types of Statistical Studies

Exit Ticket

Is the following an observational study or an experiment? Explain your answer.

Also, if it is an experiment, then identify the treatment variable and the response variable in the context of the problem. If it is an observational study, identify the population of interest.

1. A study is done to see how high soda will erupt when mint candies are dropped into two-liter bottles of soda. You want to compare using one mint candy, five mint candies, and 10 mint candies. You design a cylindrical mechanism, which drops the desired number of mint candies all at once. You have 15 bottles of soda to use. You randomly assign five bottles into which you drop one candy, five into which you drop five candies, and five into which you drop 10 candies. For each bottle, you record the height of the eruption created after the candies are dropped into it.

2. You want to see if fifth-grade boys or fifth-grade girls are faster at solving Ken-Ken puzzles. You randomly select twenty fifth-grade boys and twenty fifth-grade girls from fifth graders in your school district. You time and record how long it takes each student to solve the same Ken-Ken puzzle correctly.

EUREKA MATH™

Exit Ticket Sample Solutions

Is the following an observational study or an experiment? Explain your answer.

Also, if it is an experiment, then identify the treatment variable and the response variable in the context of the problem. If it is an observational study, identify the population of interest.

1. A study is done to see how high soda will erupt when mint candies are dropped into two-liter bottles of soda. You want to compare using one mint candy, five mint candies, and 10 mint candies. You design a cylindrical mechanism, which drops the desired number of mint candies all at once. You have 15 bottles of soda to use. You randomly assign five bottles into which you drop one candy, five into which you drop five candies, and five into which you drop 10 candies. For each bottle, you record the height of the eruption created after the candies are dropped into it.

   Experiment. There are three treatments: one, five, and ten mint candies that constitute the treatment variable. The bottles were randomly assigned to the treatments. The response variable is the height the soda erupts.

2. You want to see if fifth-grade boys or fifth-grade girls are faster at solving Ken-Ken puzzles. You randomly select twenty fifth-grade boys and twenty fifth-grade girls from fifth graders in your school district. You time and record how long it takes each student to solve the same Ken-Ken puzzle correctly.

   Observational study. Students are randomly selected from the population of fifth-grade boys and girls. The population of interest is the fifth-grade boys and girls in your school district. The study is to observe whether boys or girls are faster at solving Ken-Ken puzzles. No treatment was administered to the students.

Problem Set Sample Solutions

1. State if the following is an observational study, a survey, or an experiment, and give a reason for your answer.

   Linda wanted to know if it is easier for students to memorize a list of common three-letter words (such as *fly, pen,* and *red*) than a list of three-letter nonsense words (such as *vir, zop,* and *twq*). She randomly selected 28 students from all tenth graders in her district. She put 14 blue and 14 red chips in a jar, and without looking, each student chose a chip. Those with red chips were given the list of common words; those with blue chips were given the list of nonsense words. She gave all students one minute to memorize their lists. After the minute, she collected the lists and asked the students to write down all the words that they could remember. She recorded the number of correct words recalled.

   Experiment. The students were randomly assigned to memorize a list of common words or a list of nonsense words. There were two treatments—common words and nonsense words. The response variable is the number of correct words recalled.

2. State if the following is an observational study, a survey, or an experiment, and give a reason for your answer.

   Ken wants to compare how many hours a week that sixth graders spend doing mathematics homework to how many hours a week that eleventh graders spend doing mathematics homework. He randomly selects ten sixth graders and ten eleventh graders and records how many hours each student spent on mathematics homework in a certain week.

   Observational study. Students were randomly selected from the two populations of interest. But no treatment was administered. The population of interest is sixth graders and eleventh graders. The study was to observe the number of hours spent on mathematics homework in a certain week.

3. Suppose that in your health class you read two studies on the relationship between eating breakfast and success in school for elementary school children. Both studies concluded that eating breakfast causes elementary school children to be successful in school.

   a. Suppose that one of the studies was an observational study. Describe how you would recognize that they had conducted an observational study. Were the researchers correct in their causal conclusion?

   Possible answer: The researchers had gone to several elementary schools in a district and randomly sampled many students. They determined whether or not the students ate breakfast on a regular basis and had the district provide the students' academic performance (names of students were withheld). Their findings were not causal since the study was an observational one. The students were not randomly assigned to a treatment (eating or not eating breakfast).

   b. Suppose that one of the studies was an experiment. Describe how you would recognize that they had conducted an experiment. Were the researchers correct in their causal conclusion?

   Possible answer: The researchers had found 100 students of comparable abilities. They randomly assigned 50 of them to eat breakfast for the academic year and the other 50 to not eat breakfast. After a year, academic achievement was determined for each student. The conclusion was causal since random assignment constituted an experiment. (Note that there is an ethical issue involved in the treatment of keeping breakfast from one group of students.)

4. Data from a random sample of 50 students in a school district showed a positive relationship between reading score on a standardized reading exam and shoe size. Can it be concluded that having bigger feet causes one to have a higher reading score? Explain your answer.

   No. This is an observational study. The main lurking variable is that age is the link between reading score and shoe size.

Use the following scenarios for Problems 5–7.

A. Researchers want to determine if there is a relationship between whether or not a woman smoked during pregnancy and the birth weight of her baby. Researchers examined records for the past five years at a large hospital.

B. A large high school wants to know the proportion of students who currently use illegal drugs. Uniformed police officers asked a random sample of 200 students about their drug use.

C. A company develops a new dog food. The company wants to know if dogs would prefer its new food over the competition's dog food. One hundred dogs, who were food deprived overnight, were given equal amounts of the two dog foods: the new food versus the competitor's food. The proportion of dogs preferring the new food versus the competitor's was recorded.

5. Which scenario above describes an experiment? Explain why.

   Scenario C is the experiment. The dogs were given a treatment, the choice between the new food and the competitor's food, and their preferences were recorded.

6. Which scenario describes a survey? Will the results of the survey be accurate? Why or why not?

   Scenario B is the survey. No, the results would likely not be accurate because students would not answer honestly about drug usage to a uniformed officer.

7. The remaining scenario is an observational study. Is it possible to perform an experiment to determine if a relationship exists? Why or why not?

   It is not possible to perform an experiment to determine if a relationship exists between mothers' smoking habits and the birth weights of babies because mothers would be randomly assigned to smoke or not smoke during pregnancy. It is unethical, and thus illegal, to tell people that they must smoke.

Lesson 13: Using Sample Data to Estimate a Population Characteristic

Student Outcomes

- Students differentiate between a population and a sample.
- Students differentiate between a population characteristic and a sample statistic.
- Students recognize statistical questions that are answered by estimating a population mean or a population proportion.

Lesson Notes

This lesson reviews and extends students' previous work from Grade 7 Module 5 Lessons 18–20. The topics covered in this lesson include the distinction between a population and a sample and between population characteristics and sample statistics. Population characteristics of interest are the mean and the proportion. Because generalizing from a sample to a population requires a random sample, selecting a random sample using a random number table (or calculator if available) is also reviewed. Students new to the curriculum may need extra support in understanding review topics and using random number tables.

Classwork

Example 1 (5 minutes): Population and Sample

Let students answer the questions posed in the example and then share their responses with a neighbor. Then, have them answer the following in writing:

- What is a population?
  - *A population is the entire set of subjects in which there is an interest.*
- What is a sample?
  - *A sample is a part of the population from which information (data) is gathered, often for the purpose of generalizing from the sample to the population.*

When students think of a population and sample, they likely think only of people. Convey to students that populations and samples are not always just composed of people. Provide other examples of populations and samples. In biology, the subjects of interest could be plants or insects. In psychology, the subjects could be rats or mice. Television sets could be the subjects in a study to determine brand quality.

Scaffolding:

- If students have trouble differentiating between a population and a sample, use a relevant example from class.
- The principal of our school is interested in determining the average number of days students are absent.
- The population being studied is all of the students enrolled at the school.
- He could use the daily attendance recorded for our math class as a sample.

Whether a set of people or objects is a population or a sample depends on the context of the situation. For instance, if the players on a specific baseball team were studied to determine, for example, the team's most valuable player for that year, then that team's players would be considered a population. There would be no need to generalize beyond that set of players. But in a study concerning the whole league, those players could be considered a sample.

Example 1: Population and Sample

Answer the following questions, and then share your responses with a neighbor.

a. **A team of scientists wants to determine the average length and weight of fish in Lake Lucerne. Name a sample that can be used to help answer their question.**

Answers will vary. The scientists could spend one day catching fish and then record the length and weight of each fish caught. The fish measured that day form a sample of the total fish population in the lake.

b. **Golf balls from different manufacturers are tested to determine which brand travels the farthest. What is the population being studied?**

The population is all golf balls made by the manufacturers.

Exercise 1 (5–7 minutes)

The Gettysburg Address is mentioned in Exercise 1(g) and on the Exit Ticket and is the focus of Example 3. Remind students that the Gettysburg Address is a famous speech given by President Abraham Lincoln at the dedication of the Soldier's National Cemetery in Gettysburg, PA, during the American Civil War in 1863. Let students work independently and confirm their answers with a neighbor.

Exercise 1

For each of the following, does the group described constitute a population or a sample? Or could it be considered to be either a population or a sample? Explain your answer.

a. **The animals that live in Yellowstone National Park**

Population. The subjects are all the animals that live in Yellowstone National Park.

b. **The first-run movies released last week that were shown at the local theater complex last weekend**

The subjects are first-run movies. If all of those released last week were acquired by our local theater management and were shown last weekend, then the first-run movies released last week would be considered a population. If our local theater management chose some of the first-run movies, then they would be a sample.

c. **People who are asked how they voted in an exit poll**

If all of the people who exited the polling place were asked how they voted, then they would be considered the population of people who exited a certain polling place on a certain day. If, however, only some of them were asked how they voted, then they would be considered a sample.

d. **Some cars on the lot of the local car dealer**

Sample. For example, a customer might only be interested in two-door models.

e. **The words of the Gettysburg Address**

Population. Note that, the number of words varies depending on what version of the address is used. For example, "can not" is two words whereas "cannot" is one.

f. **The colors of pencils available in a 36-count packet of colored pencils**

Population. The 36 colors available in the packet constitute the population of colors.

> g. **The students from your school who attended your school's soccer game yesterday**
>
> ***If the entire student body attended the game, then the students who attended the game would be considered a population. If just some of the student body attended, then they would be considered a sample of students at the school.***

Example 2 (5 minutes): Representative Sample

For a sample result to be generalized to the population from which the sample was taken, the sample values need to be representative of the population.

- Explain why the scientists from Example 1 have to use a sample of fish instead of studying the whole population.
  - *It would be impossible to measure the length and weight of every fish in the lake.*

Give students a moment to think about and respond to the question posed in the example. Ask students to share their answers for a whole-class discussion.

There is no way to guarantee that a sample exactly reflects characteristics of the population. Point out that one way that is likely to result in a representative sample is to make sure that everyone in the population has the same chance of getting into the sample. Random selection accomplishes this. Students often think of random selection as haphazard selection. So, emphasize that random selection is a formal procedure that must be followed. If a population is small enough, then number each individual (person or object) in the population from 1 to as many as there are, and then use random numbers from a calculator or from a random digit table to select the individuals for the sample.

Note that if a table is used, then the number of digits to use is the number of digits in the population size (e.g., a district with $2{,}446$ students requires each number to have four digits). The number 1 would be 0001, and the number 240 would be 0240. The advantage of using a calculator with a random number generator key is that leading zeros are not used. For example, to generate 35 random numbers from a population of $2{,}446$, the typical syntax would be random $(1, 2446, 35)$.

If students have access to graphing calculators, demonstrate how random numbers can be generated using technology.

Calculator note: To generate random integers on a TI 83 or TI 84 calculator, go to the MATH menu, and highlight PRB. Choose option 5: randInt. For example, to generate 30 random integers from a population of 314, use the syntax *randInt*$(1, 314, 30)$.

> **Example 2: Representative Sample**
>
> **If a sample is taken for the purpose of generalizing to a population, the sample must be representative of the population. In other words, it must be similar to the population even though it is smaller than the population. For example, suppose you are the campaign manager for your friend who is running for senior class president. You would like to know what proportion of students would vote for her if the election was held today. The class is too big to ask everyone (314 students). What would you do?**
>
> **Comment on whether or not each of the following sampling procedures should be used. Explain why or why not.**
>
> a. **Poll everyone in your friend's math class.**
>
> ***This may not be the best option because everyone in the class may be biased toward your friend. For example, your friend may be great at math and help a lot of students prepare for tests. Everyone may look at her favorably.***

b. Assign every student in the senior class a number from 1 to 314. Then, use a random number generator to select 30 students to poll.

This is a good option because it ensures that the sample looks like the rest of the senior class.

c. Ask every student who is going through the lunch line in the cafeteria who they will vote for.

Answers may vary. This may or may not be a good option. The students going through the lunch line might be freshmen, sophomores, or juniors who do not vote for senior class president, so their opinions do not matter. However, if only seniors in the lunch line are asked how they will vote, this could be a good sample of the senior class.

Exercise 2 (3–4 minutes)

MP.5 This exercise reviews use of a random number table that students encountered in Grade 7. If students have access to graphing calculators or appropriate software, random numbers can be generated using technology. If time allows, students should explain both procedures. If students do not have access to a table of random numbers, display the following line of random numbers on the front board of the classroom for students to use.

$$54580\ 81507\ 27102\ 54537\ 55894\ 33006\ 04229\ 91828$$

Let students work in small groups to complete the exercise. Assist students who are new to the curriculum and have not used this type of table before.

Exercise 2

There is no procedure that guarantees a representative sample. But the best procedure to obtain a representative sample is one that gives every different possible sample an equal chance to be chosen. The sample resulting from such a procedure is called a *random sample.*

Suppose that you want to randomly select 60 employees from a group of 625 employees.

Explain how to use a random number table or a calculator with a random number generator to choose 60 different numbers at random and include the students with these numbers in the sample.

Random number generator: Assign each employee a number from 1 to 625. Then, generate 60 random numbers from the random number generator. If there are any duplicate numbers, generate additional numbers as needed.

Random number table: To do so, you could number all the employees from 001 to 625. Then, you could use a random number table. Any three-digit number beyond 625 is ignored, and any duplicated three-digit number is ignored. The starting point on a random number table does not matter. For the purposes of this example, start at the far left. For example, from the following line of random digits, the first four employees randomly selected would be those with identification numbers 545, 150, 102, and 375.

Example 3 (3–5 minutes): Population Characteristics and Sample Statistics

MP.1 A statistical study begins with asking a question that can be answered by data. Students should be familiar with this since it was the first Common Core Statistics Standard in Grade 6.

Let students work independently on the example and then share their responses with a neighbor. Then ask students to summarize the following summary measures in their own words:

- What is a population characteristic?
  - *A summary measure calculated using all the individuals in a population is called a population characteristic. A population proportion and a population mean are two examples of population characteristics.*

- What is a sample statistic?
  - *If the summary measure is calculated using data from a random sample, it is called a sample statistic. For example, a sample proportion or a sample mean are sample statistics.*

A summary measure is used to analyze the data and arrive at a conclusion. Population summary measures are called *population characteristics* (or parameters), and sample summary measures are called *statistics*. Examples include a total (e.g., gross movie receipts), a mean (e.g., average time to run a mile), a proportion (e.g., the proportion of voters who favor a particular candidate), and a correlation coefficient (e.g., the correlation coefficient for the amount of vitamin C taken and the number of colds contracted).

Example 3: Population Characteristics and Sample Statistics

A statistical study begins with a question of interest that can be answered by data. Depending on the study, data could be collected from all individuals in the population or from a random sample of individuals selected from the population. Read through the following, and identify which of the summary measures represents a *population characteristic* and which represents a *sample statistic*. Explain your reasoning for each.

Suppose the population of interest is the words of the Gettysburg Address. There are 269 of them (depending on the version).

a. **The proportion of nouns in all words of the Gettysburg Address**

   It is a population characteristic because it is determined from all of the words (the entire population) of the Gettysburg Address.

b. **The proportion of nouns or the proportion of words containing the letter *e* in a random sample of words taken from the Gettysburg Address**

   It is a sample statistic because it is determined from a sample of the words (the entire population) of the Gettysburg Address.

c. **The mean length of the words in a random sample of words taken from the Gettysburg Address**

   It is a sample statistic because it is determined from a sample of the words (the entire population) of the Gettysburg Address.

d. **The proportion of all words in the Gettysburg Address that contain the letter *e***

   It is a population characteristic because it is determined from all of the words (the entire population) of the Gettysburg Address.

e. **The mean length of all words in the Gettysburg Address**

   It is a population characteristic because it is determined from all of the words (the entire population) of the Gettysburg Address.

Exercise 3 (7–9 minutes)

Let students work in small groups. If time allows, let each group present one of their answers.

Exercise 3

For the following items of interest, describe an appropriate population, population characteristic, sample, and sample statistic. Explain your answer.

Answers will vary. The following are examples of sample responses:

a. **Time it takes students to run a quarter mile**

Population: all students in the eleventh grade at your school

Population characteristic: mean time to run a quarter mile for all students in the eleventh grade at your school

Sample: a random selection of students in the eleventh grade at your school

Sample statistic: mean time to run a quarter mile for the sampled students in the eleventh grade at your school

b. **National forests that contain bald eagle nests**

Population: all national forests in the United States

Population characteristic: the proportion of all national forests in the U.S. that contain bald eagle nests

Sample: a random selection of national forests in the U.S.

Sample statistic: the proportion of the sampled national forests in the U.S. that contain bald eagle nests

c. **Curfew time of boys compared to girls**

Population: all boys and girls in the eleventh grade at your school

Population characteristic: the difference in mean curfew time of all eleventh grade boys at your school and the mean curfew time of all eleventh grade girls at your school

Sample: a random selection of some boys and girls sampled from the eleventh grade at your school

Sample statistic: the difference in mean curfew time of the sampled eleventh grade boys at your school and the mean curfew time of the sampled eleventh grade girls at your school

d. **Efficiency of electric cars**

Population: all electric cars currently being marketed

Population characteristic: the mean number of kilowatt-hours to drive 100 miles for all electric cars currently being marketed

Sample: a random selection of electric cars currently being marketed

Sample statistic: the mean number of kilowatt-hours to drive 100 miles for the sampled electric cars currently being marketed

Scaffolding:

If students struggle with the exercise, provide the following on the front board as a sample response to use as a guide.

For example, an item of interest is SAT scores.

- The population might be all students in your school who take the SAT exam in mathematics this year.
- One population characteristic is the proportion of all students who take the mathematics exam this year who score 700 or higher, and another is the mean score of all students who take the mathematics exam this year.
- A sample of 40 students might be selected from those who take the SAT exam this year.
- Two examples of sample statistics are the proportion of the 40 students who score 700 or higher and the mean score for the 40 students who take the SAT exam in mathematics this year.

Exercise 4 (5–7 minutes)

Let students continue to work in small groups. Several statistical questions about the population of students at the school or all students in a particular grade (e.g., the eleventh grade) are suggested. Discuss the questions with the whole class. Consider asking students to add questions to this list. Ask each group to select a question and write a paragraph as directed in the exercise. After three to five minutes, allow as many groups as possible to discuss their paragraphs. Focus on each group's responses that indicate how the sample is obtained and how the sample mean or proportion is used to generalize to the students in the population (e.g., the whole school or the students in the eleventh grade).

Exercise 4

Consider the following questions:

- **What proportion of eleventh graders at our high school are taking at least one advanced placement course?**
- **What proportion of eleventh graders at our high school have a part-time job?**
- **What is the typical number of hours an eleventh grader at our high school studies outside of school hours on a weekday (Monday, Tuesday, Wednesday, or Thursday)?**
- **What is the typical time (in minutes) that students at our high school spend getting to school?**
- **What is the proportion of students at our high school who plan to attend a college or technical school after graduation?**
- **What is the typical amount of time (in hours per week) that students at our high school are involved in community service?**

Select one of these questions (or a different statistical question that has been approved by your teacher). Working with your group, write a paragraph that:

- **States the statistical question of interest pertaining to the students in the population for the statistical question selected.**
- **Identifies a population characteristic of interest.**
- **Identifies the appropriate statistic based on a sample of 40 students.**
- **States what property your sample must have for you to be able to use its results to generalize to all students in your high school.**
- **Includes the details on how you would select your sample.**

Possible responses for two of the above questions: We are interested in answering the statistical question, "What proportion of eleventh graders at our high school are taking at least one advanced placement course?" We would go to our high school counselors and obtain a list of the 403 eleventh graders at our high school. We would number the students from 001 to 403. We decided to randomly select 40 students for our sample. To do so, we would find a random number table, choose a place to start, and then generate 40 different random numbers between 001 and 403. If we find that 14 of the 40 students in the sample are taking at least one AP course this year, we would use that information to calculate the sample proportion of students who are taking at least one AP course. Since our sample was a random sample, we can generalize our estimate to the population of eleventh graders at our school and estimate that the proportion of all eleventh graders who are taking at least one AP course this year is 0.35.

Also consider the following: We are interested in the question about community service. The statistical question is, "What is the typical amount of time (in hours per week) that students at our school are involved in community service?" We would go to our school's guidance department and ask them to help us. The guidance counselor has an alphabetical list of the $1,245$ students enrolled in our school. Listed students are numbered from 0001 to 1245. We decided we would select 40 students at random for our study. Using our graphing calculator, we would obtain 40 different random numbers between 1 and 1245. We would ask the guidance counselor to identify the student associated with each of the 40 numbers. We would also ask for the location of the student during the last period of the day, so we could quickly ask them a question before leaving school. We would ask, "How many hours of community service did you complete last week?" We would find the mean of the 40 numbers reported by the students in our sample. The mean is our statistic for this study. If our sample was randomly selected, we can generalize to the population of all students at the school, and we would use the sample mean to estimate the mean number of hours of community service per week for students at our school. For example, if the mean of the 40 randomly selected students is 14.5 hours, we would say that our best estimate of the mean number of hours students in our school are involved in community service is 14.5 hours per week.

Closing (2 minutes)

MP.3

- Present students with a scenario, and ask them to critique whether or not the sample chosen is representative of the population and provides a good estimate for a population characteristic. Allow for multiple responses and arguments. For example, suppose that an elementary school principal is interested in how many hours of sleep students get each night. He decides to use Mr. Ross's fifth-grade class as a sample. Do you think that this sample will provide a good estimate of the population characteristic (average number of hours of sleep for all students in the school)? Explain.
    - *Answers will vary. This may not be a representative sample. Older students may require less sleep than younger students. The estimate may be too low and not represent the sleep times of students in the lower grades.*

Ask students to summarize the main ideas of the lesson in writing or with a neighbor. Use this as an opportunity to informally assess comprehension of the lesson. The Lesson Summary below offers some important ideas that should be included.

> **Lesson Summary**
>
> **We refer to summary measures calculated using data from an entire population as *population characteristics.* We refer to summary measures calculated using data from a sample as *sample statistics.* To generalize from a sample to the corresponding population, it is important that the sample be a random sample from the population. A *random sample* is one that is selected in a way that gives every different possible sample an equal chance of being chosen.**

Exit Ticket (5 minutes)

Name ______________________________ Date ______________

Lesson 13: Using Sample Data to Estimate a Population Characteristic

Exit Ticket

Indicate whether each of the following is a summary measure from a population or from a sample. Choose the one that is more realistically the case. If it is from a population, identify the population characteristic. If it is from a sample, identify the sample statistic. Explain your reasoning.

a. 88% of the more than 300 million automobile tires discarded per year are recycled or used for fuel.

b. The mean number of words that contain the letter *e* in the Gettysburg Address

c. 64% of respondents in a recent poll indicated that they favored building a proposed highway in their town.

Exit Ticket Sample Solutions

Indicate whether each of the following is a summary measure from a population or from a sample. Choose the one that is more realistically the case. If it is from a population, identify the population characteristic. If it is from a sample, identify the sample statistic. Explain your reasoning.

a. 88% of the more than 300 million automobile discarded tires per year are recycled or used for fuel.

Population. The statement concerns all discarded tires in a year.

Characteristic: percentage of all discarded tires that are recycled or used for fuel

b. The mean number of words that contain the letter *e* in the Gettysburg Address

Population. The statement refers to the entire document.

Characteristic: mean number of words that contain the letter "e"

c. 64% of respondents in a recent poll indicated that they favored building a proposed highway in their town.

Sample. Not everyone in the town contributed.

Characteristic: percentage of respondents who favor the highway construction

Problem Set Sample Solutions

Use this space to describe specific details about the Problem Set for teacher reference.

1. In the following, identify whether the subjects being measured are the sample or the population. In some cases, they could be considered a sample or a population. Explain each answer.

Subjects	What Is Being Measured	Sample or Population? Explain
Some students in your class	Number of books in backpack	*Sample* *Only some of the students in the class were in the study.*
AA batteries of a certain brand	Lifetime	*Sample* *Realistically a population does not make sense since if all the batteries are tested, there would be none left.*
Birds in Glacier National Park	Number of species	*Sample* *It would be impossible to catch and catalog every bird in a national park.*
Students in your school	Number absent or present today	*Population* *All students in your school are being recorded as being present or being absent today.*
Words in the Constitution of the U.S.	Whether a noun or not	*Population* *All the words in the Constitution are being considered.*
Americans of voting age	Opinion on an issue	*Sample* *It is unrealistic to think that all Americans of voting age could be asked their opinions.*

2. For the following items of interest, describe an appropriate population, a population characteristic, a sample, and a sample statistic.

   a. Whether or not a driver is speeding in your school zone during school hours in a day

      Answers will vary. A sample response is given.

      Population: all drivers who drive through your school zone during school hours on a certain day

      Population characteristic: proportion of all drivers who exceed the speed limit in your school zone during school hours on a certain day

      Sample: some drivers who drive in your school zone from noon to 1:00 p.m. on a certain day

      Sample statistic: proportion of the sampled drivers who exceed the speed limit in your school zone from noon to 1:00 p.m. on a certain day

   b. Seatbelt usage of men compared to women

      Answers will vary. A sample response is given.

      Population: all men and women drivers in the state of New York

      Population characteristic: the difference in the proportion of all male drivers in the state of New York who always wear a seatbelt and the proportion of all female drivers in the state of New York who always wear a seatbelt

      Sample: some men and women drivers in the state of New York

      Sample statistic: the difference in the proportion of the male drivers sampled who always wear a seatbelt and the proportion of the female drivers sampled who always wear a seatbelt

   c. Impact of a new antidepressant on people with severe headaches

      Answers will vary. A sample response is given.

      Population: all people in New York with severe headaches

      Population characteristic: the mean time it takes the new antidepressant to eliminate the headaches for all people in New York with severe headaches

      Sample: some people in New York with severe headaches

      Sample statistic: the mean time it takes the new antidepressant to eliminate the headaches for the sampled people in New York with severe headaches

3. What are the identification numbers for ten students chosen at random from a population of 78 students based on the following string of random digits? Start at the left.

   27816 78416 01822 73521 37741 016312 68000 53645 56644 97892 63408 77919 44575

   27 67 16 01 27 35 21 37 74 10

Table of Random Digits

Row																				
1	6	6	7	2	8	0	0	8	4	0	0	4	6	0	3	2	2	4	6	8
2	8	0	3	1	1	1	1	2	7	0	1	9	1	2	7	1	3	3	5	3
3	5	3	5	7	3	6	3	1	7	2	5	5	1	4	7	1	6	5	6	5
4	9	1	1	9	2	8	3	0	3	6	7	7	4	7	5	9	8	1	8	3
5	9	0	2	9	9	7	4	6	3	6	6	3	7	4	2	7	0	0	1	9
6	8	1	4	6	4	6	8	2	8	9	5	5	2	9	6	2	5	3	0	3
7	4	1	1	9	7	0	7	2	9	0	9	7	0	4	6	2	3	1	0	9
8	9	9	2	7	1	3	2	9	0	3	9	0	7	5	6	7	1	7	8	7
9	3	4	2	2	9	1	9	0	7	8	1	6	2	5	3	9	0	9	1	0
10	2	7	3	9	5	9	9	3	2	9	3	9	1	9	0	5	5	1	4	2
11	0	2	5	4	0	8	1	7	0	7	1	3	0	4	3	0	6	4	4	4
12	8	6	0	5	4	8	8	2	7	7	0	1	0	1	7	1	3	5	3	4
13	4	2	6	4	5	2	4	2	6	1	7	5	6	6	4	0	8	4	1	2
14	4	4	9	8	7	3	4	3	8	2	9	1	5	3	5	9	8	9	2	9
15	6	4	8	0	0	0	4	2	3	8	1	8	4	0	9	5	0	9	0	4
16	3	2	3	8	4	8	8	6	2	9	1	0	1	9	9	3	0	7	3	5
17	6	6	7	2	8	0	0	8	4	0	0	4	6	0	3	2	2	4	6	8
18	8	0	3	1	1	1	1	2	7	0	1	9	1	2	7	1	3	3	5	3
19	5	3	5	7	3	6	3	1	7	2	5	5	1	4	7	1	6	5	6	5
20	9	1	1	9	2	8	3	0	3	6	7	7	4	7	5	9	8	1	8	3
21	9	0	2	9	9	7	4	6	3	6	6	3	7	4	2	7	0	0	1	9
22	8	1	4	6	4	6	8	2	8	9	5	5	2	9	6	2	5	3	0	3
23	4	1	1	9	7	0	7	2	9	0	9	7	0	4	6	2	3	1	0	9
24	9	9	2	7	1	3	2	9	0	3	9	0	7	5	6	7	1	7	8	7
25	3	4	2	2	9	1	9	0	7	8	1	6	2	5	3	9	0	9	1	0
26	2	7	3	9	5	9	9	3	2	9	3	9	1	9	0	5	5	1	4	2
27	0	2	5	4	0	8	1	7	0	7	1	3	0	4	3	0	6	4	4	4
28	8	6	0	5	4	8	8	2	7	7	0	1	0	1	7	1	3	5	3	4
29	4	2	6	4	5	2	4	2	6	1	7	5	6	6	4	0	8	4	1	2
30	4	4	9	8	7	3	4	3	8	2	9	1	5	3	5	9	8	9	2	9
31	6	4	8	0	0	0	4	2	3	8	1	8	4	0	9	5	0	9	0	4
32	3	2	3	8	4	8	8	6	2	9	1	0	1	9	9	3	0	7	3	5
33	6	6	7	2	8	0	0	8	4	0	0	4	6	0	3	2	2	4	6	8
34	8	0	3	1	1	1	1	2	7	0	1	9	1	2	7	1	3	3	5	3
35	5	3	5	7	3	6	3	1	7	2	5	5	1	4	7	1	6	5	6	5
36	9	1	1	9	2	8	3	0	3	6	7	7	4	7	5	9	8	1	8	3
37	9	0	2	9	9	7	4	6	3	6	6	3	7	4	2	7	0	0	1	9
38	8	1	4	6	4	6	8	2	8	9	5	5	2	9	6	2	5	3	0	3
39	4	1	1	9	7	0	7	2	9	0	9	7	0	4	6	2	3	1	0	9
40	9	9	2	7	1	3	2	9	0	3	9	0	7	5	6	7	1	7	8	7

Lesson 14: Sampling Variability in the Sample Proportion

Student Outcomes

- Students understand the term *sampling variability* in the context of estimating a population proportion.
- Students understand that the standard deviation of the sampling distribution of the sample proportion offers insight into the accuracy of the sample proportion as an estimate of the population proportion.

Lesson Notes

This lesson and the next revisit the concept of sampling variability in the sample proportion, introduced in Grade 7 Module 5 Lessons 17–19. Students use simulation to approximate the sampling distribution of the sample proportion and explore how to use that simulation to anticipate estimation error. In this lesson, students use a physical simulation process. (In the next lesson, they use technology to carry out a simulation.)

Together, this and Lesson 15 should span a total of two class periods.

Materials needed:

- Large bag of white dried beans
- Large bag of black dried beans
- Brown paper bags, each containing 60 white and 40 black dried beans
- One brown paper bag with beans for each group of two students

Classwork

Example 1 (3 minutes): Polls

Read and discuss Example 1 as a class. Ask students the following question posed in the text and summarized below. Discuss their answers as a class.

- If you were to take a random sample of 20 Americans, how many would you predict would say that they pay a great deal attention to nutritional information?
  - $0.40(20) \approx 8$ *or about* 8 *people*

The above answer is based on the statement that 40% of the public said they pay "a great deal" of attention to nutritional information.

> *Scaffolding:*
>
> - For struggling students, poll the class asking how many pay attention to calories when making lunch choices.
> - For advanced students, have them develop a statistical question and describe how it could be answered using a poll.

Example 1: Polls

A recent poll stated that 40% of Americans pay "a great deal" or a "fair amount" of attention to the nutritional information that restaurants provide. This poll was based on a random sample of $2,027$ adults living in the United States.

The 40% corresponds to a proportion of 0.40, and 0.40 is called a *sample proportion.* It is an estimate of the proportion of all adults who would say they pay "a great deal" or a "fair amount" of attention to the nutritional information that restaurants provide. If you were to take a random sample of 20 Americans, how many would you predict would say that they pay attention to nutritional information? In this lesson, you will investigate this question by generating distributions of sample proportions and investigating patterns in these distributions.

Your teacher will give your group a container of dried beans. Some of the beans in the container are black. With your classmates, you are going see what happens when you take a sample of beans from the container and use the proportion of black beans in the sample to estimate the proportion of black beans in the container (a *population proportion*).

Exploratory Challenge 1/Exercises 1–9 (20 minutes)

MP.2

In these exercises, students use data from a sample to estimate a population proportion and generalize from a sample to the population. Hand out the paper bags full of dried beans to each group of two students. Instruct each student to take a random sample of 20 beans. (This sampling should be done with replacement.) Ideally, the class calculates between 25 to 30 sample proportions.

While students take their random samples, draw a number line on the board. (Make the line long enough to accommodate students' sticky notes.) The scale on the line should range from approximately 0.1 to 0.7 in increments of 0.05.

Let students work in their groups on Exercises 1–3. After the class graph has been constructed, have students work on Exercises 4–9 in their groups. Discuss the answers as a class.

Exploratory Challenge 1/Exercises 1–9

1. Each person in the group should randomly select a sample of 20 beans from the container by carefully mixing all the beans and then selecting one bean and recording its color. Replace the bean, mix the bag, and continue to select one bean at a time until 20 beans have been selected. Be sure to replace each bean and mix the bag before selecting the next bean. Count the number of black beans in your sample of 20.

   Answers will vary, but the number of black beans will center around 8.

2. What is the proportion of black beans in your sample of 20? (Round your answer to 2 decimal places.) This value is called the *sample proportion* of black beans.

   Answers will vary, but the sample proportions will center around 0.4.

3. Write your sample proportion on a sticky note, and place the note on the number line that your teacher has drawn on the board. Place your note above the value on the number line that corresponds to your sample proportion.

   Class data will vary. One possible sampling distribution is shown below.

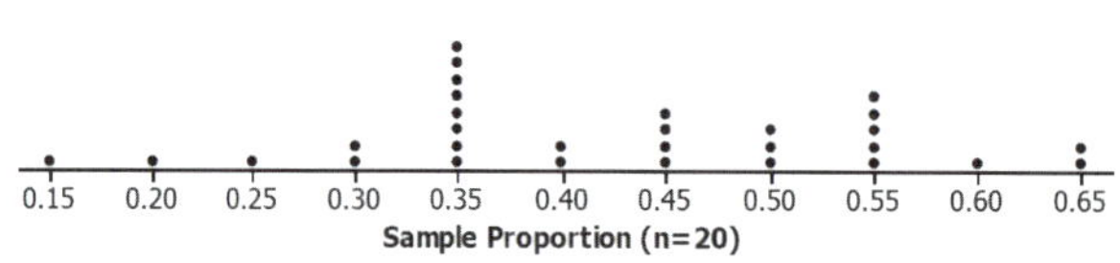

The graph of all the students' sample proportions is called the *sampling distribution* of the samples' proportions. This sampling distribution is an approximation of the actual sampling distribution of all possible samples of size 20.

4. Describe the shape of the distribution.

   Answers might vary, but the shape is generally mound shaped.

5. What was the smallest sample proportion observed?

   Answers will vary. Based on the sample graph: 0.15

6. What was the largest sample proportion observed?

   Answers will vary. Based on the sample graph: 0.65

7. What sample proportion occurred most often?

   Answers will vary, but the sample proportion should be around 0.4*. Based on the sample graph:* 0.35

Scaffolding:

- Depending on the context, the term *approximate* is used as either a noun or a verb (or an adjective).
- For English language learners, making it clear that this same term can be used in related but different ways, with different linguistic functions, can be useful.
- Choral repetition or rehearsal and a graphic organizer may help students who are struggling.

8. Using technology, find the mean and standard deviation of the sample proportions used to construct the sampling distribution created by the class.

   Answers will vary, but the mean will be approximately 0.4*, and the standard deviation will be approximately* 0.11.

9. How does the mean of the sampling distribution compare with the population proportion of 0.40?

   Answers will vary, but the two values should be about the same. In theory, the mean of the sampling distribution of sample proportions is equal to the population proportion.

Example 2 (2 minutes): Sampling Variability

MP.3 Pose the question in the example to the class. Allow for multiple responses. The following exercises provide students the opportunity to test their conjectures about treatment differences in the context of a statistical experiment:

Example 2: Sampling Variability

What do you think would happen to the sampling distribution if everyone in class took a random sample of 40 beans from the container? To help answer this question, you will repeat the process described in Example 1, but this time you will draw a random sample of 40 beans instead of 20.

Exploratory Challenge 2/Exercises 10–21 (15 minutes)

Let students continue to work in groups to complete the remaining exercises.

Exploratory Challenge 2/Exercises 10–21

10. Take a random sample with replacement of 40 beans from the container. Count the number of black beans in your sample of 40 beans.

    Answers will vary, but the number of black beans will center around 16.

11. What is the proportion of black beans in your sample of 40? (Round your answer to 2 decimal places.)

Answers will vary, but the sample proportions will center around 0.40.

12. Write your sample proportion on a sticky note, and place it on the number line that your teacher has drawn on the board. Place your note above the value on the number line that corresponds to your sample proportion.

Class data will vary. One possible sampling distribution is shown below.

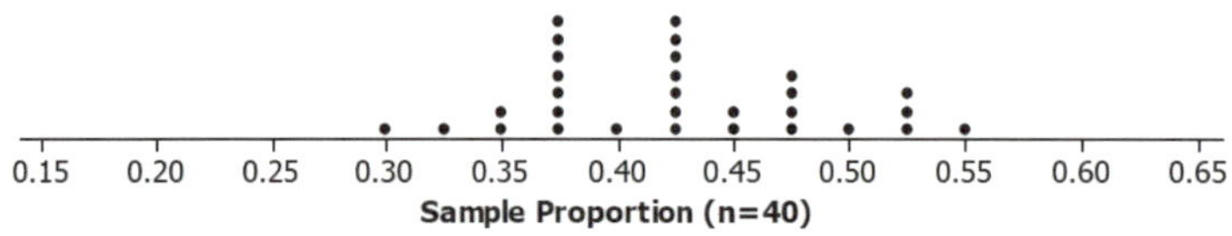

13. Describe the shape of the distribution.

Answers may vary, but the shape is generally mound shaped.

14. What was the smallest sample proportion observed?

Answers will vary. Based on the sample graph: 0.30

15. What was the largest sample proportion observed?

Answers will vary. Based on the sample graph: 0.55

16. What sample proportion occurred most often?

Answer will vary but will be approximately 0.4.

17. Using technology, find the mean and standard deviation of the sample proportions used to construct the sampling distribution created by the class.

Answers will vary, but the mean will be approximately 0.4 *and the standard deviation approximately* 0.08.

18. How does the mean of the sampling distribution compare with the population proportion of 0.40?

Answers will vary, but the two values should be about the same. In theory, the mean of the sampling distribution of sample proportions is equal to the population proportion.

19. How does the mean of the sampling distribution based on random samples of size 20 compare to the mean of the sampling distribution based on random samples of size 40?

The two means are approximately the same, about 0.4.

20. As the sample size increased from 20 to 40, describe what happened to the sampling variability (standard deviation of the distribution of sample proportions)?

The standard deviation of the distribution of the sample proportions based on a sample size of 40 *is less than the standard deviation of the distribution of the sample proportions based on a sample size of* 20.

21. **What do you think would happen to the variability (standard deviation) of the distribution of sample proportions if the sample size for each sample was 80 instead of 40? Explain.**

Because the standard deviation decreased as sample size increased from 20 to 40, I expect that the standard deviation will decrease further when the sample size is 80.

Closing (2 minutes)

Ask students to summarize the main ideas of the lesson in writing or with a neighbor. Use this as an opportunity to informally assess comprehension of the lesson. The Lesson Summary below offers some important ideas that should be included.

Lesson Summary

The sampling distribution of the sample proportion can be approximated by a graph of the sample proportions for many different random samples. The mean of the sampling distribution of the sample proportions will be approximately equal to the value of the population proportion.

As the sample size increases, the sampling variability in the sample proportion decreases; in other words, the standard deviation of the sampling distribution of the sample proportions decreases.

Exit Ticket (3 minutes)

Name ______________________________ Date ______________

Lesson 14: Sampling Variability in the Sample Proportion

Exit Ticket

A group of eleventh graders wanted to estimate the population proportion of students in their high school who drink at least one soda per day. Each student selected a different random sample of 30 students and calculated the proportion that drink at least one soda per day. The dot plot below shows the sampling distribution. This distribution has a mean of 0.51 and a standard deviation of 0.09.

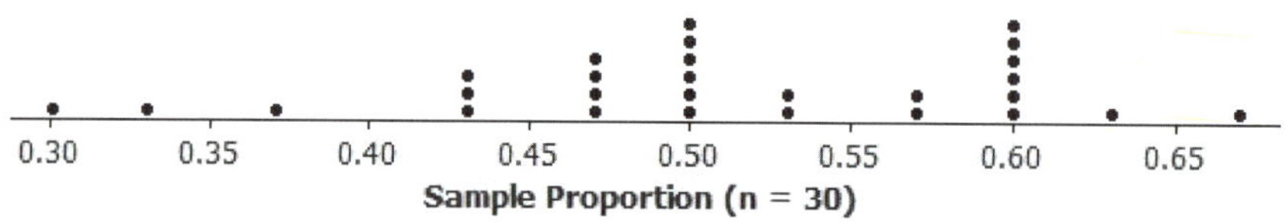

1. Describe the shape of the distribution.

2. What is your estimate for the proportion of *all* students who would report that they drink at least one soda per day?

3. If, instead of taking random samples of 30 students in the high school, the eleventh graders randomly selected samples of size 60, describe what will happen to the standard deviation of the sampling distribution of the sample proportions.

Exit Ticket Sample Solutions

A group of eleventh graders wanted to estimate the population proportion of students in their high school who drink at least one soda per day. Each student selected a different random sample of 30 students from the high school and calculated the proportion that drink at least one soda per day. The dot plot below shows the sampling distribution. This distribution has a mean of 0.51 and a standard deviation of 0.09.

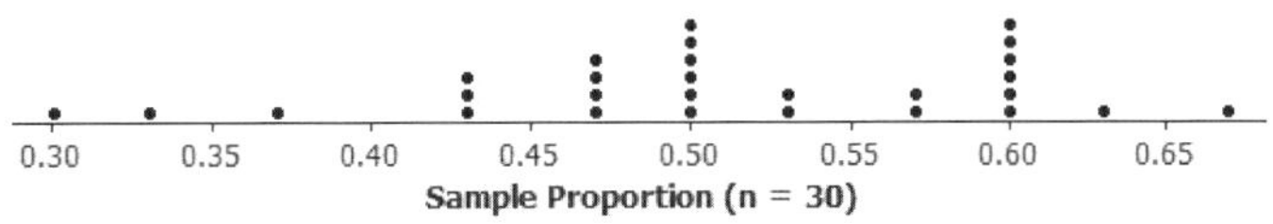

1. Describe the shape of the distribution.

   Approximately symmetric centered around 0.50

2. What is your estimate for the proportion of *all* students who would report that they drink at least one soda per day?

   0.51*, which is the mean of the sampling distribution*

3. If, instead of taking random samples of 30 students in the high school, the eleventh graders randomly selected samples of size 60, describe what will happen to the standard deviation of the sampling distribution of the sample proportions.

   The standard deviation will decrease.

Problem Set Sample Solutions

1. A class of 28 eleventh graders wanted to estimate the proportion of all juniors and seniors at their high school with part-time jobs after school. Each eleventh grader took a random sample of 30 juniors and seniors and then calculated the proportion with part-time jobs. Following are the 28 sample proportions.

   $$0.7, 0.8, 0.57, 0.63, 0.7, 0.47, 0.67, 0.67, 0.8, 0.77, 0.4, 0.73, 0.63, 0.67, 0.6, 0.77, 0.77, 0.77, 0.53, 0.57,$$
   $$0.73, 0.7, 0.67, 0.7, 0.77, 0.57, 0.77, 0.67$$

   a. Construct a dot plot of the sample proportions.

   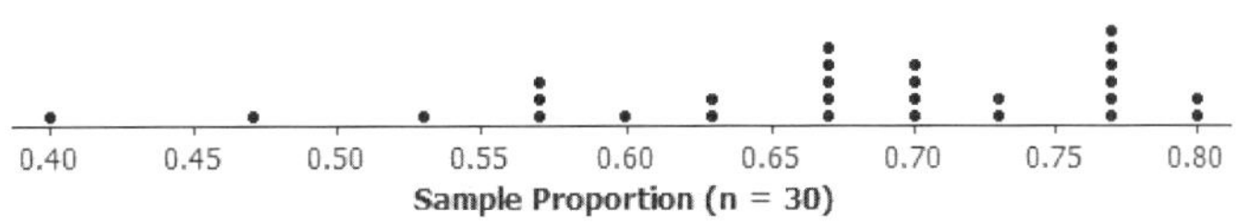

   b. Describe the shape of the distribution.

   Skewed to the left

c. Using technology, find the mean and standard deviation of the sample proportions.

Mean $= 0.67$

Standard deviation $= 0.1$

d. Do you think that the proportion of all juniors and seniors at the school with part-time jobs could be 0.7? Do you think it could be 0.5? Justify your answers based on your dot plot.

It is likely that the proportion of all juniors and seniors with part-time jobs could be 0.70 *since* 0.70 *is near the center of the dot plot. It is unlikely that the proportion of all juniors and seniors is* 0.5 *since there are very few samples with a sample proportion of* 0.5 *or less.*

e. Suppose the eleventh graders had taken random samples of size 60. How would the distribution of sample proportions based on samples of size 60 differ from the distribution for samples of size 30?

The sampling distribution would be mound shaped with approximately the same mean as the sampling distribution based on size 30*, but the standard deviation of the sampling distribution based on size* 60 *would be smaller than one based on samples of size* 30*.*

2. A group of eleventh graders wanted to estimate the proportion of all students at their high school who suffer from allergies. Each student in one group of eleventh graders took a random sample of 20 students, while each student in another group of eleventh graders each took a random sample of 40 students. Below are the two sampling distributions (shown as histograms) of the sample proportions of high school students who said that they suffer from allergies. Which histogram is based on random samples of size 40? Explain.

Histogram A is based on random samples of size 40 *because it has less variability than Histogram B.*

Histogram A

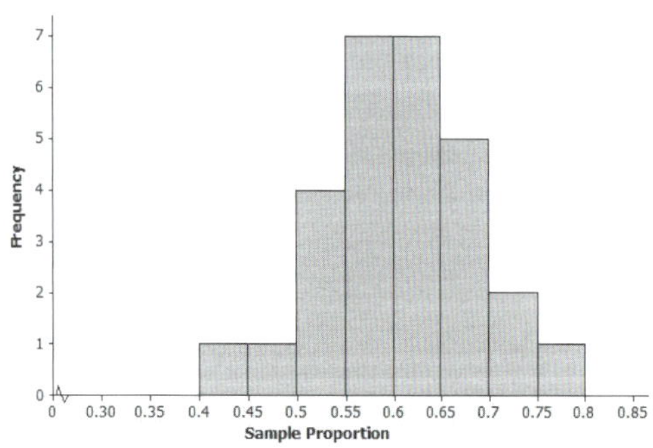

Histogram B

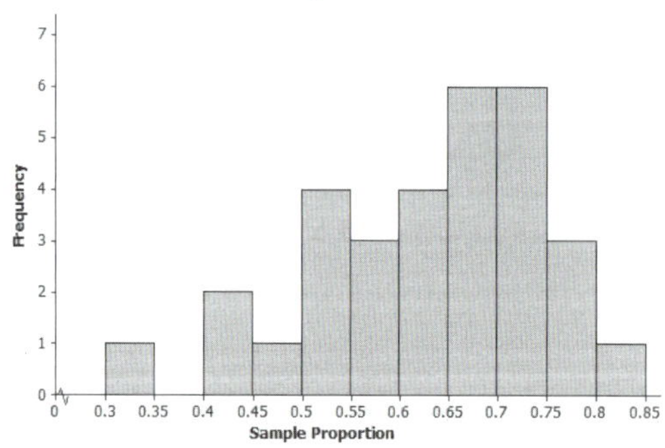

3. The nurse in your school district would like to study the proportion of all high school students in the district who usually get at least eight hours of sleep on school nights. Suppose each student in your class takes a random sample of 20 high school students in the district and each calculates their sample proportion of students who said that they usually get at least eight hours of sleep on school nights. Below is a histogram of the sampling distribution.

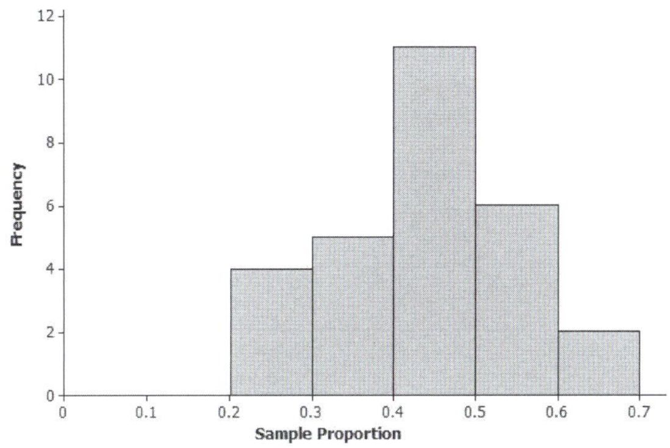

a. Do you think that the proportion of all high school students who usually get at least eight hours of sleep on school nights could have been 0.4? Do you think it could have been 0.55? Could it have been 0.75? Justify your answers based on the histogram.

The proportion of all high school students who usually get at least eight hours of sleep is likely around 0.4 *since that is near the center of the sampling distribution. The proportion could be* 0.55 *since that is still close to the center of the distribution. It is unlikely that the proportion of all high school students is* 0.75 *since none of the samples produced sample proportions as large as* 0.75.

b. Suppose students had taken random samples of size 60. How would the distribution of sample proportions based on samples of size 60 differ from those of size 20?

The means of the two distributions would be relatively close, but the standard deviation of the distribution based on samples of size 60 *would be smaller than the standard deviation of the distribution based on sample sizes of* 20.

Lesson 15: Sampling Variability in the Sample Proportion

Student Outcomes

- Students understand the term *sampling variability* in the context of estimating a population proportion.
- Students understand that the standard deviation of the sampling distribution of the sample proportion offers insight into the accuracy of the sample proportion as an estimate of the population proportion.

Lesson Notes

This lesson has the same Student Outcomes as Lesson 14, which investigated the effect of sample size on the variability of a sampling distribution of sample proportions. This lesson uses technology—either the website www.rossmanchance.com/applets/CoinTossing/CoinToss.html or a graphing calculator—to construct a sampling distribution of the proportion of heads for a different number of coin flips. Students are asked to describe the effect on the variability of the sampling distribution as the number of flips increases.

Classwork

Example 1 (5 minutes)

Introduce the following scenario to students. Before showing the steps of the simulation, discuss how students could use the beans used in Lesson 14 to design a simulation. For this scenario, 50% of the beans would be black. Students would then randomly select 40 beans (with replacement) to calculate the proportion of black beans in the sample.

MP.1 It is also important that students understand that they are assuming the principal's claim is correct. They are trying to determine if a sample proportion of 0.40 is a likely result when the population proportion is 0.50.

MP.3

Example 1

A high school principal claims that 50% of the school's students walk to school in the morning. A student attempts to verify the principal's claim by taking a random sample of 40 students and asking them if they walk to school in the morning. Sixteen of the sampled students say they usually walk to school in the morning, giving a sample proportion of $\frac{16}{40} = 0.40$, which seems to dispel the principal's claim of 50%. But could the principal be correct that the proportion of all students who walk to school is 50%?

a. **Make a conjecture about the answer.**

b. **Develop a plan for how to respond.**

Help the student make a decision on the principal's claim by investigating what kind of sample proportions you would expect to see if the principal's claim of 50% is true. You will do this by using technology to simulate the flipping of a coin 40 times.

At this point, teachers may choose to explain how to use technology to perform the simulation. Depending on available technology, students can use either a graphing calculator or a coin-tossing applet to perform simulations. Using the applet may save time during the lesson, allowing opportunities for more class discussion.

Using the Coin-Tossing Applet at the Website:

1. Use the Coin-Tossing Simulation—Rossman/Chance Applet Collection.	Go to www.rossmanchance.com/applets/CoinTossing/CoinToss.html.
2. For this example: Enter 0.5 for the probability of heads, 40 for the number of tosses, and 1 for the number of repetitions. Then, click on "Toss Coins."	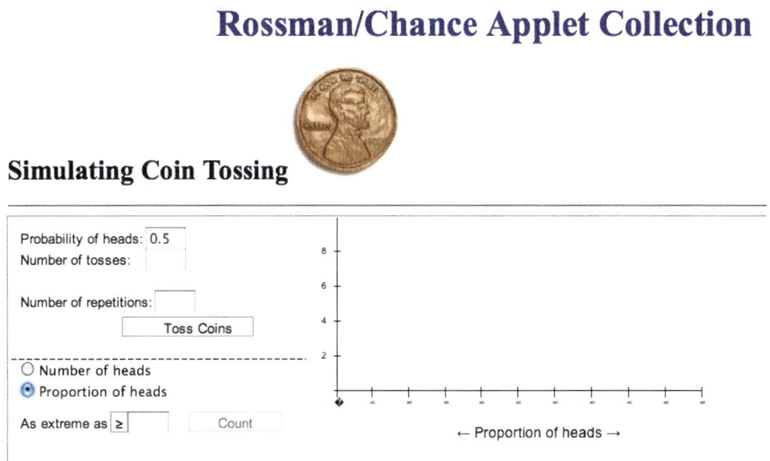

Using a TI-83 or TI-84 Graphing Calculator:

The following steps show how to use a TI-83 or TI-84 Graphing Calculator to simulate the flipping of a coin 40 times and then to calculate the sample proportion: 1. Select MATH, and highlight <PRB>.	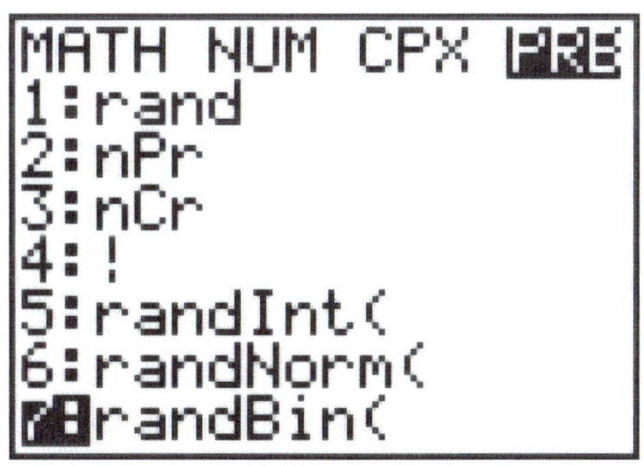

2. Choose: 7: randBin(.	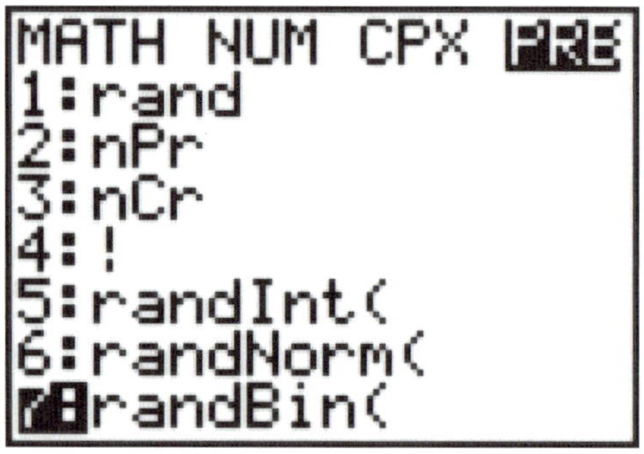
3. Input: randBin(40,0.5)/40.	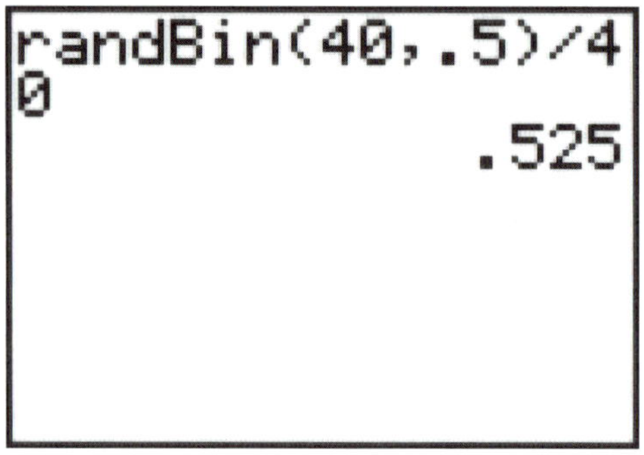
4. Press ENTER.	In this example, the 0.525 is the proportion of heads that were observed in the simulated flip of 40 coins.

Exploratory Challenge 1/Exercises 1–9 (15 minutes)

Students should work independently or in pairs on Exercises 1–3. As students report their proportions of heads, enter the reported proportions into a list if using a graphing calculator or into a list using computer-graphing software. Once all students have reported two sample proportions, construct a graph (histogram or dot plot) of the class data. Students should work on Exercises 4–9 in groups. Discuss answers as a class.

Even though each student could construct his own sampling distribution, in this first example, each student is generating just two sample proportions of heads. This way, it mirrors what was done in Lesson 14. The sampling distribution is constructed using sample proportions from the entire class.

Exploratory Challenge 1/Exercises 1–9

In Exercises 1–9, students should assume that the principal is correct that 50% of the population of students walk to school. Designate heads to represent a student who walks to school.

1. **Simulate 40 flips of a fair coin. Record your observations in the space below.**

   Answers will vary. Sample response:

   T T T H H H T H H H T T T H T T H H H

   H T H H T H T H T T H T T T H T H T T

   T T

2. **What is the sample proportion of heads in your sample of 40? Report this value to your teacher.**

   Answers will vary. Sample response: $\frac{18}{40} = 0.45$

3. **Repeat Exercises 1 and 2 to obtain a second sample of 40 coin flips.**

   Answers will vary.

Your teacher will display a graph of all the students' sample proportions of heads.

The following is an example of a sampling distribution of sample proportions of heads in 40 flips of a coin:

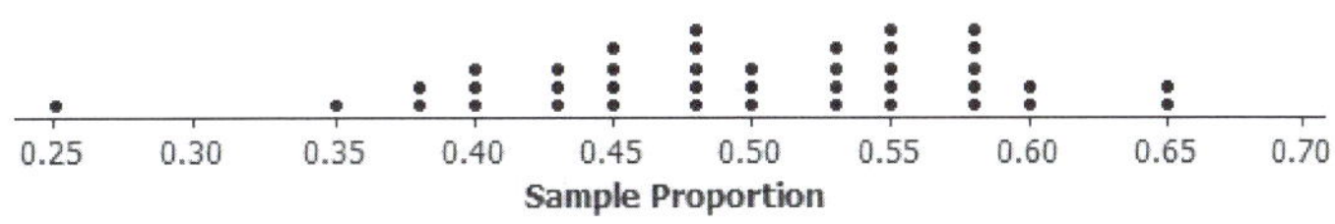

4. **Describe the shape of the distribution.**

   Answers will vary. The shape of the distribution shown above is slightly skewed.

5. **What was the smallest sample proportion observed?**

   Answers will vary. In the sample graph, 0.25

6. **What was the largest sample proportion observed?**

   Answers will vary. In the sample graph, 0.65

7. **Estimate the center of the distribution of sample proportions.**

   Answers will vary. In the sample graph, about 0.50

Your teacher will report the mean and standard deviation of the sampling distribution created by the class.

Answers will vary. The mean will be approximately 0.5, and the standard deviation will be approximately 0.079. From the sample graph, the mean is 0.493, and the standard deviation is 0.085.

8. **How does the mean of the sampling distribution compare with the population proportion of 0.50?**

   Answers will vary. From the sample response, the population proportion of 0.50 is very close to the mean of the sampling distribution.

9. **Recall that a student took a random sample of 40 students and found that the sample proportion of students who walk to school was 0.40. Would this have been a surprising result if the actual population proportion was 0.50 as the principal claims?**

   Answers will vary. Based on the sample responses, the value of 0.40 is about one standard deviation from the mean. There were quite a few samples in the simulation that resulted in sample proportions that were 0.40 or smaller. Hence, a value of 0.40 would not be a surprising result if the population was 0.50.

Example 2 (3 minutes): Sampling Variability

Give students a moment to think, write, and/or speak about the question posed in the example.

Example 2: Sampling Variability

What do you think would happen to the sampling distribution you constructed in the previous exercises had everyone in class taken a random sample of size 80 instead of 40? Justify your answer. This will be investigated in the following exercises.

Answers will vary. The results would be more accurate because there are more samples.

Now, explain how students can generate their own sampling distributions using technology. The steps are slightly different from before as students are now performing the simulation 40 times. Again, teachers should consider using the applet in order to save time.

Using the Coin-Tossing Applet at the Website:

MP.3

<table>
<tr><td>1. Use the Coin-Tossing Simulation—Rossman/Chance Applet Collection.</td><td>Go to www.rossmanchance.com/applets/CoinTossing/CoinToss.html.</td></tr>
<tr><td>2. For this example, enter 0.5 for the probability of heads, 80 for the number of tosses, and 40 for the number of repetitions. Also, select proportion of heads and summary statistics.</td><td>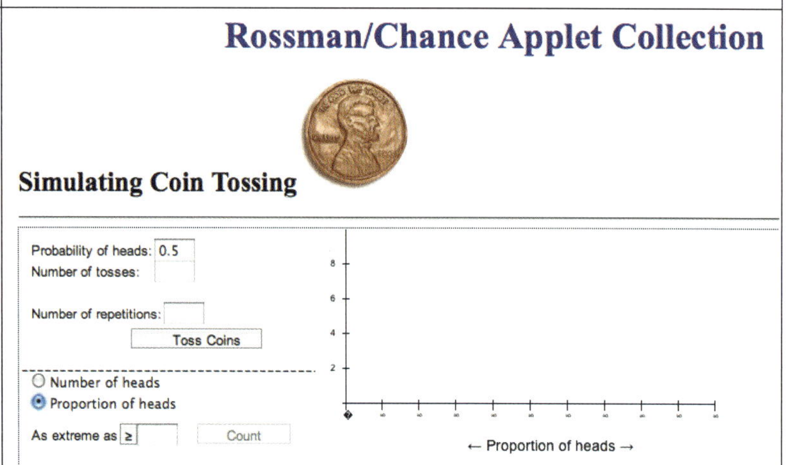
</td></tr>
</table>

Using a TI-83 or TI-84 Graphing Calculator:

MP.3

The following steps show how to use a TI-83 or TI-84 Graphing Calculator to simulate the flipping of a coin 80 times to calculate the sample proportion, repeating the process for a total of 40 times: 1. Select MATH, and highlight <PRB>.	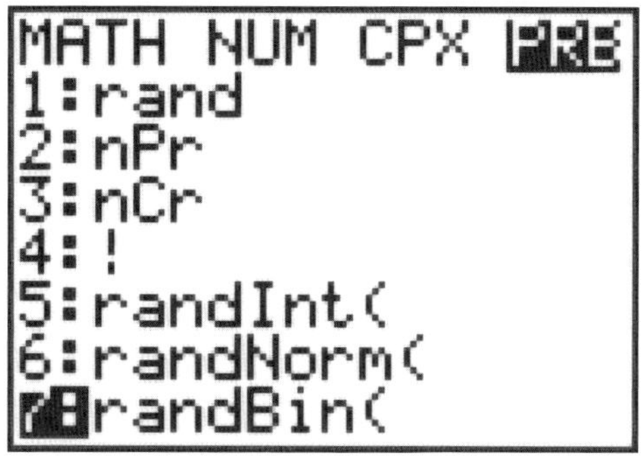
2. Choose: 7: randBin(.	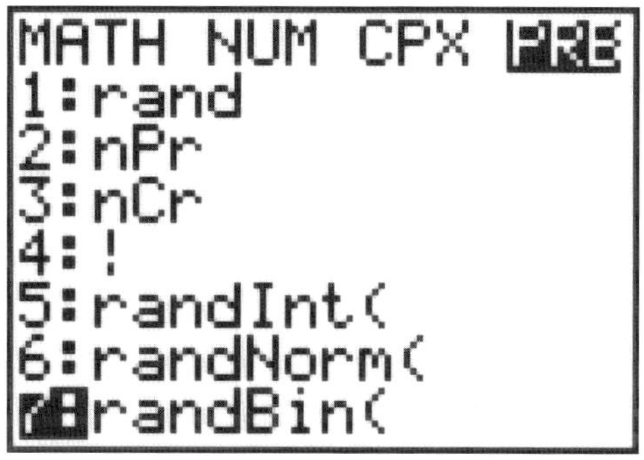
3. Input: randBin(80,0.5,40)/80→L_1. Use the randBin operation, and input 80 for the sample size and 0.5 for the population proportion, divide by 80, store the results in L_1, and then repeat 40 times.	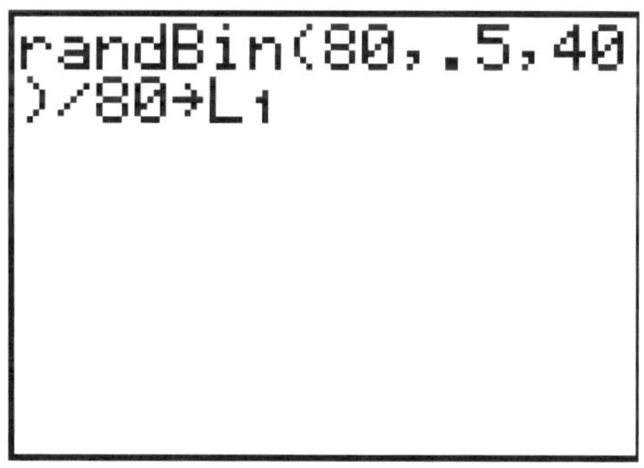

4. Press ENTER, and view the results in L_1.

   To view the relevant statistics, go to the STAT menu, and highlight <CALC>.

   Choose option 1: 1–Var Stats.

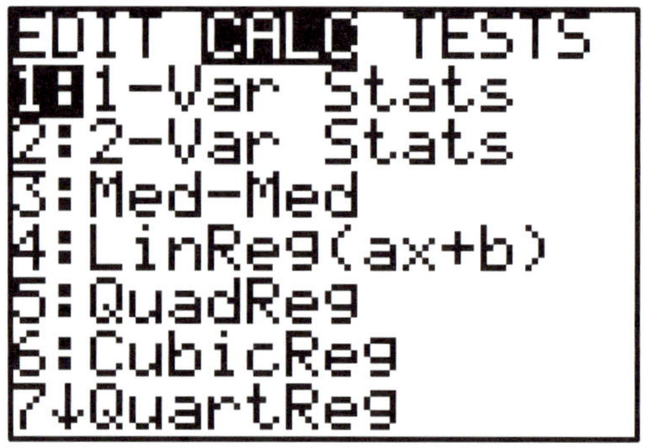

MP.3

5. Students may also choose to view a histogram of the sample proportions in L_1.

   Go to STAT PLOT. <2nd> and <Y=>

   Make sure Plot 1 is turned on, and highlight the histogram.

   The following window is useful in viewing the histogram.

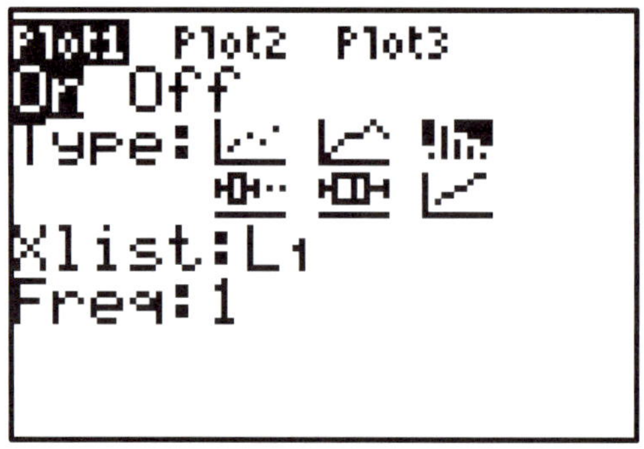

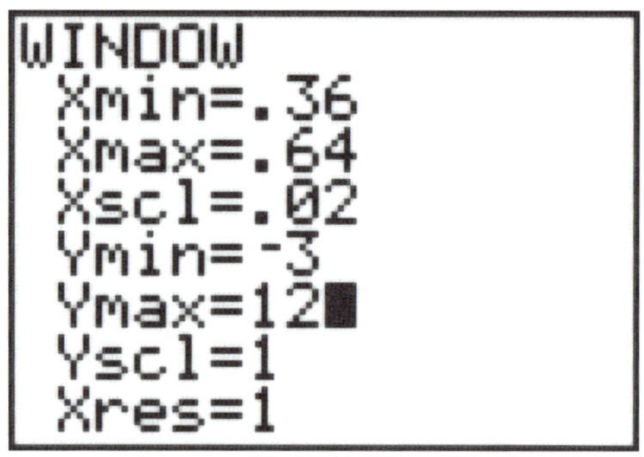

Exploratory Challenge 2/Exercises 10–22 (15 minutes)

In this set of exercises, students use technology to carry out simulations in order to study sampling variability. Students should work independently or in pairs on Exercises 10–22.

MP.5

Exploratory Challenge 2/Exercises 10–22

10. **Use technology and simulate 80 coin flips. Calculate the proportion of heads. Record your results in the space below.**

    Answers will vary. Sample response: $\frac{39}{80} = 0.4875$

EUREKA MATH™

11. Repeat flipping a coin 80 times until you have recorded a total of 40 sample proportions.

    Answers will vary. See Exercise 12 for a dot plot of the sampling distribution of the proportion of heads in 80 *flips of a coin.*

12. Construct a dot plot of the 40 sample proportions.

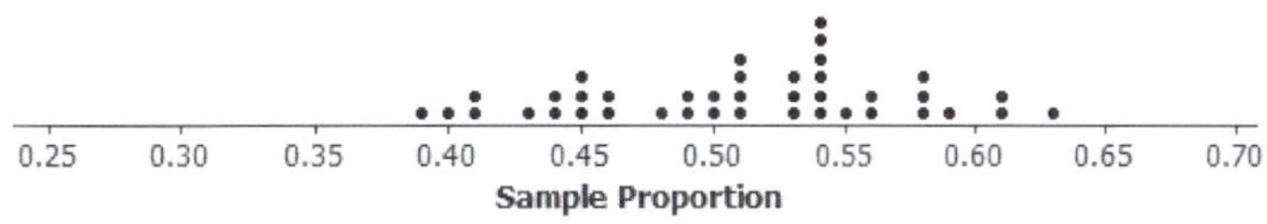

13. Describe the shape of the distribution.

    Answers will vary. From the sample response, the distribution is symmetric and mound shaped.

14. What was the smallest proportion of heads observed?

    Answers will vary. From the sample response, 0.39

15. What was the largest proportion of heads observed?

    Answers will vary. From the sample response, 0.63

MP.5

16. Using technology, find the mean and standard deviation of the distribution of sample proportions.

    Answers will vary. The mean will be approximately 0.5*, and the standard deviation will be approximately* 0.055*. In the example above, the mean is* 0.508*, and the standard deviation is* 0.061*.*

17. Compare your results with the others in your group. Did you have similar means and standard deviations?

    Answers will vary. All the groups should have similar means and standard deviations.

18. How does the mean of the sampling distribution based on 40 simulated flips of a coin (Exercise 1) compare to the mean of the sampling distribution based on 80 simulated coin flips?

    Both of the means will be approximately equal to 0.50*.*

19. Describe what happened to the sampling variability (standard deviation) of the distribution of sample proportions as the number of simulated coin flips increased from 40 to 80.

    The standard deviation decreased as the number of coin flips went from 40 *to* 80*.*

20. What do you think would happen to the variability (standard deviation) of the distribution of sample proportions if the sample size for each sample was 200 instead of 80? Explain.

    The standard deviation will decrease as the sample size increases.

MP.5

21. Recall that a student took a random sample of 40 students and found that the sample proportion of students who walk to school was 0.40. If the student had taken a random sample of 80 students instead of 40, would this have been a surprising result if the actual population proportion was 0.50 as the principal claims?

Answers will vary. The value of 0.40 is about two standard deviations from the mean. Only two of the 40 simulated samples resulted in a sample proportion of 0.40 or smaller. A sample proportion of 0.40 would be a fairly surprising result.

22. What do you think would happen to the sampling distribution you constructed in the previous exercises if everyone in class took a random sample of size 80 instead of 40? Justify your answer.

Answers will vary. The more samples, the more accurate the simulation will be because the standard deviation decreases as sample size increases.

Closing (2 minutes)

Ask students to summarize the main ideas of the lesson in writing or with a neighbor. Use this as an opportunity to informally assess comprehension of the lesson. The Lesson Summary below offers some important ideas that should be included.

Lesson Summary

The sampling distribution of the sample proportion can be approximated by a graph of the sample proportions for many different random samples. The mean of the sample proportions will be approximately equal to the value of the population proportion.

As the sample size increases, the sampling variability in the sample proportion decreases; in other words, the standard deviation of the sample proportions decreases.

Exit Ticket (5 minutes)

Name ______________________________ Date ______________

Lesson 15: Sampling Variability in the Sample Proportion

Exit Ticket

Below are three dot plots of the proportion of tails in 20, 60, or 120 simulated flips of a coin. The mean and standard deviation of the sample proportions are also shown for each of the three dot plots. Match each dot plot with the appropriate number of flips. Clearly explain how you matched the plots with the number of simulated flips.

Dot Plot 1

Mean: 0.502
Standard deviation: 0.046

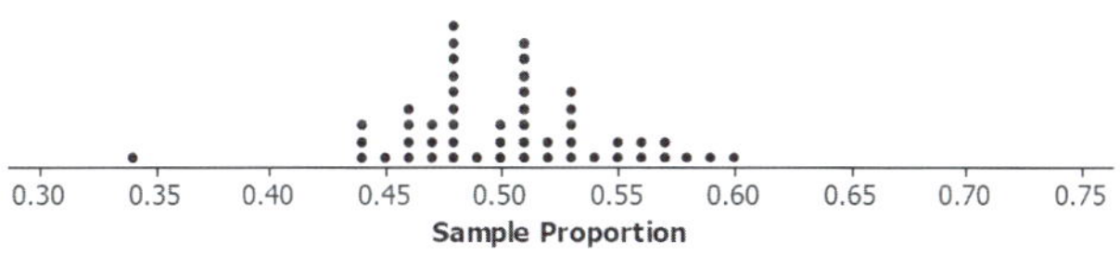

Sample Size: ________

Explain:

Dot Plot 2

Mean: 0.518
Standard deviation: 0.064

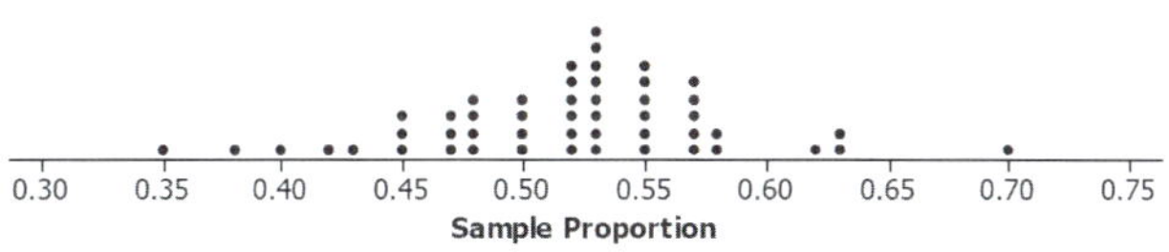

Sample Size: ________

Explain:

Dot Plot 3	
Dot Plot 3 Mean: 0.498 Standard deviation: 0.110 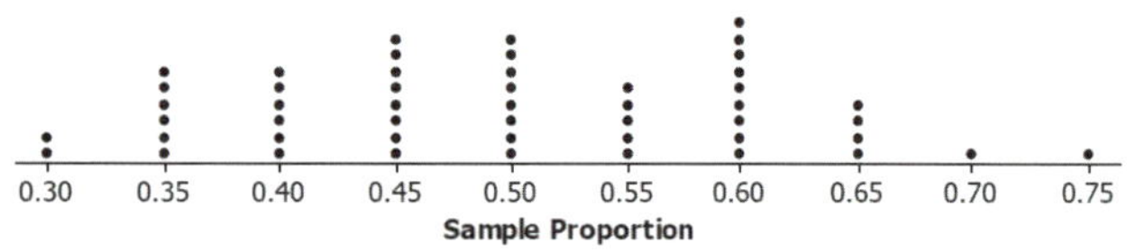	Sample Size: _________ Explain:

Exit Ticket Sample Solutions

Below are three dot plots of the proportion of tails in 20, 60, or 120 simulated flips of a coin. The mean and standard deviation of the sample proportions are also shown for each of the three dot plots. Match each dot plot with the appropriate number of flips. Clearly explain how you matched the plots with the number of simulated flips.

Dot Plot 1
Mean: 0.502
Standard deviation: 0.046

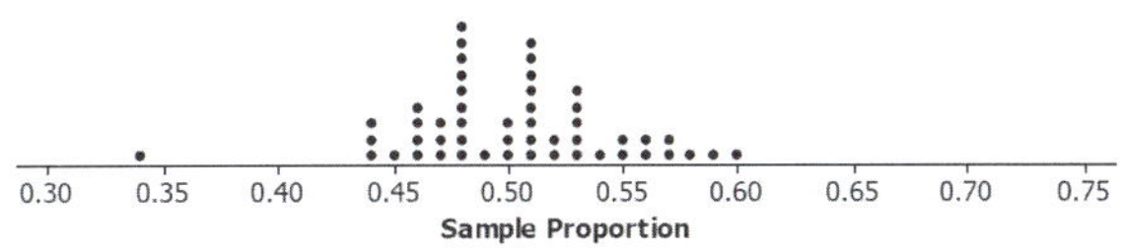

Sample Size: 120 *flips of the coin*

Explain: *As the number of flips increases, the standard deviation decreases. The sampling distribution based on* 120 *flips has the smallest standard deviation.*

Dot Plot 2
Mean: 0.518
Standard deviation: 0.064

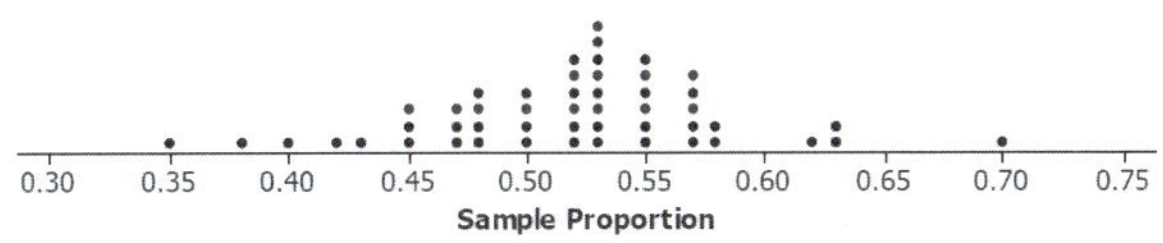

Sample Size: 60 *flips of the coin*

Explain: *As sampling size increases, the standard deviation decreases. Because this sample size falls between the other two, its standard deviation will be between the standard deviations of the other sample sizes.*

Dot Plot 3
Mean: 0.498
Standard deviation: 0.110

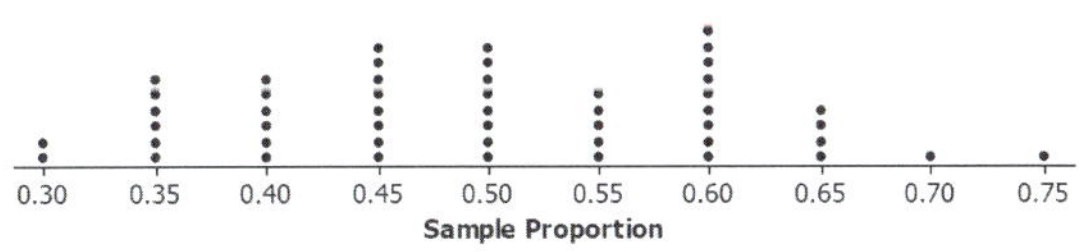

Sample Size: 20 *flips of the coin*

Explain: *This standard deviation is the largest, which means that the sample size must be the smallest.*

Problem Set Sample Solutions

1. A student conducted a simulation of 30 coin flips. Below is a dot plot of the sampling distribution of the proportion of heads. This sampling distribution has a mean of 0.51 and a standard deviation of 0.09.

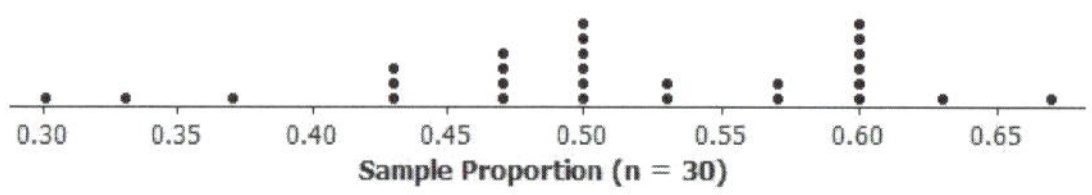

   a. Describe the shape of the distribution.

   The distribution is approximately symmetric. Some students may respond that the distribution is slightly skewed to the left.

   b. Describe what would have happened to the mean and the standard deviation of the sampling distribution of the sample proportions if the student had flipped a coin 50 times, calculated the proportion of heads, and then repeated this process for a total of 30 times.

   The mean would be approximately equal to 0.51, and the standard deviation would be less than 0.09.

2. What effect does increasing the sample size have on the mean of the sampling distribution?

   Increasing the sample size does not affect the mean of the sampling distribution. The mean of the sampling distribution is approximately equal to the population mean for any sample size.

3. What effect does increasing the sample size have on the standard deviation of the sampling distribution?

   Increasing the sample size decreases the standard deviation of the sampling distribution.

4. A student wanted to decide whether or not a particular coin was fair (i.e., the probability of flipping a head is 0.5). She flipped the coin 20 times, calculated the proportion of heads, and repeated this process a total of 40 times. Below is the sampling distribution of sample proportions of heads. The mean and standard deviation of the sampling distribution are 0.379 and 0.091, respectively. Do you think this was a fair coin? Why or why not?

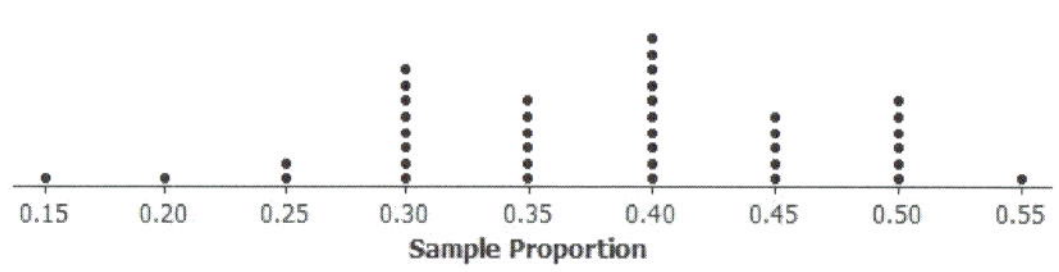

   If the coin was fair, the sampling distribution should be centered at about 0.50. Here, the sampling distribution is centered pretty far to the left of 0.50. Hence, it is unlikely that the probability of heads for this coin would be 0.50.

5. The same student flipped the coin 100 times, calculated the proportion of heads, and repeated this process a total of 40 times. Below is the sampling distribution of sample proportions of heads. The mean and standard deviation of the sampling distribution are 0.405 and 0.046, respectively. Do you think this was a fair coin? Why or why not?

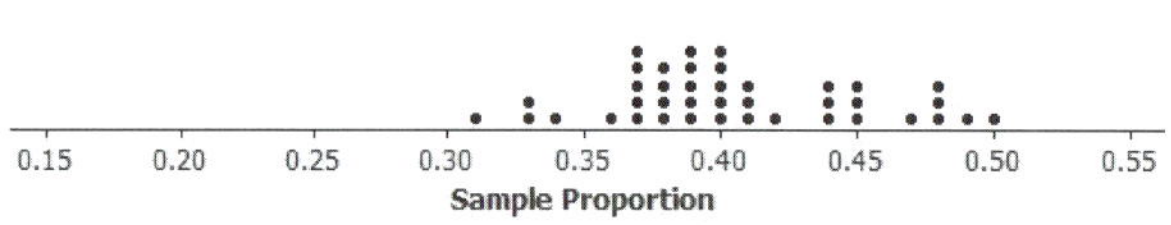

   If the coin was fair, the sampling distribution should be centered at about 0.50. Here, the sampling distribution is centered pretty far to the left of 0.50. Hence, it is unlikely that the probability of heads for this coin would be 0.50.

Lesson 16: Margin of Error When Estimating a Population Proportion

Student Outcomes

- Students use data from a random sample to estimate a population proportion.
- Students calculate and interpret margin of error in context.
- Students know the relationship between sample size and margin of error in the context of estimating a population proportion.

Lesson Notes

From prior lessons, students should recognize that the values of statistics calculated from samples selected from known populations vary from sample to sample. In this lesson, students learn what variability can tell about an unknown population. Students use data from a random sample drawn from a mystery bag to estimate a population proportion and then find and interpret a margin of error for the estimate. Comparing an observed proportion of successes from a random sample drawn from a population with an unknown proportion of successes to these sampling distributions provides information about what populations might produce a random sample like the one observed. This lesson can be done as a class investigation with students actually drawing samples and simulating sampling distributions using random number generators or by working through a set of questions that relate to the kinds of activities they would actually do in a hands-on investigation.

Classwork

Exploratory Challenge 1/Exercises 1–4 (25 minutes): Mystery Bag

For a Whole-Class Hands-On Investigation:

Prepare at least nine small bags of colored chips, one bag for each pair of students. If chips are not available, use pieces of paper with the letter R written on them. Each bag should have 20 chips with the following numbers of red chips. If the class is large, prepare duplicate bags for the percentages.

2 red chips—10% of the chips are red.
4 red chips—20% of the chips are red.
6 red chips—30% of the chips are red.
8 red chips—40% of the chips are red.
10 red chips—50% of the chips are red.
12 red chips—60% of the chips are red.
14 red chips—70% of the chips are red.
16 red chips—80% of the chips are red.
18 red chips—90% of the chips are red.

To learn about an unknown population, it is easiest to start by understanding how samples from a known population would behave. Consider asking students the following question to activate their prior knowledge and remind them of the variability inherent in different samples from the same population:

MP.3

- Suppose you know that 20% of the chips in a bag are red. Write down an estimate of the number of red chips you are likely to see in a random sample of 30 chips from the bag. Have students write or speak responses to each other or as a class.

Drawing a red chip constitutes a *success.* The proportion of red chips in each bag should be clearly written on the bottom of the bag. (Several pairs of students may have the same proportion if the class is large.) The other chips can be any color other than red. These chips represent *failures.*

Give each pair of students a bag with the proportion of reds marked on the bottom. Monitor the class, and when most students have two sets of 30 observations, bring the class together, and suggest they use technology to generate their random samples to speed up the process. Each pair of students should generate a set that corresponds to the percentage of red chips in their bags; for example, the 10% can be represented by a set consisting of $s = \{1,0,0,0,0,0,0,0,0,0\}$, 20% by $\{1,1,0,0,0,0,0,0,0,0\}$, and so on. They should draw a sample of size 30 with replacement from their sets. For some random number generators, the command would be randsamp $(s, 30)$ to generate 30 elements from the set, and then a sum command can be used to find the number of successes in the sample:

$s{:}=\{1,0,0,0,0,0,0,0,0,0\}$ ▸ $\{1.,0.,0.,0.,0.,0.,0.,0.,0.,0.\}$

$r{:}=$ randSamp$(s,30)$ ▸ $\{0.,0.,0.,1.,0.,1.,0.,0.,0.,0.,0.,0.,1.,0.,0.,1.,0.,0.,0.,0.,0.,0.,0.,1.,0.,0.,0.,0.,0.,0.\}$

sum(r) ▸ $5.$

Have students record the number of successes in a frequency tally and then select additional samples. Repeating the process about 40 or 50 times gives them a simulated sampling distribution of the number of red chips in random samples of size 30 drawn from a population that is known to have a certain percentage of red chips. Emphasize that they are drawing samples from a population with *known* proportions of successes—in this case, red chips.

Once most students have generated about 50 random samples of size 30 from their bags and recorded the number of reds in each sample in a frequency tally, bring the class together. Instruct students to do the following:

- Look at your simulated sampling distributions for the number of red chips in your samples, and write down an interval that seems to describe the number of reds you would typically get for the proportion of red chips in the bag.
  - *If I create a frequency tally or dot plot, it looks like about half of the samples resulted in having between* 10 *and* 14 *red chips.*
- Then, return to the mystery bag, and respond to Exercises 2 and 3.

For a Lesson With and Without the Hands-On Activities:

Prepare one mystery bag that has 20 chips with an unknown (to the class) proportion of red chips. Perhaps put eight red chips in the bag for a mystery proportion of 40%. Try not to use six red chips (30%) as students may confuse the percent with the sample of 30 chips that are drawn with replacement from the bag. Students do not actually have to know there are 20 chips in any of the bags because the samples are drawn with replacement.

MP.2 Begin the class with the introduction below. Lessons without the hands-on activities should begin with Exercise 1. Lessons with the hands-on activities should begin with Exercise 2. In Exercise 3, students understand that a margin of error is the interval that marks off the proportions of red chips from the expected proportion that are unlikely to occur based on the simulated sampling distribution. If the proportion from the sample did not show up in the sampling distribution nor was one of the more extreme proportions identified in the interval called the *margin of error,* then it is unlikely the proportion of red chips in the mystery bag is equal to the proportion stated in the investigation. It is important to have a class discussion on part (b) of Exercise 3 to make sure students understand that the margin of error defines an interval and not the likelihood of making a mistake.

MP.3 Bring in the mystery bag, and ask students what proportion of the chips in the bag they think are red chips. Ask students:

- How can you find the proportion of red chips in the bag?

Students should write or speak with a partner to develop a plan. (Note: Taking the chips out, examining them, and counting the red ones is not an option.)

Have one student draw a chip from the bag and a different student record whether the chip was red or not red. Return the chip to the bag, shake the bag, and have students draw and record the color of a second chip. Continue the process until they have a sample of 30 chips. Ask students to write down their predictions for the proportion of red chips in the bag based on the sample results.

> *Scaffolding:*
>
> Encourage advanced learners to develop their own plans for determining the proportion and carry it out, without scaffold questions given.
>
> Offer struggling students a simpler example (e.g., a bag with only four chips—one red—in it) that illustrates the ideas at work here. Show a visual of this, and ask questions, such as "How many red chips would you have in a sample of 10? 20? 50?"

Exploratory Challenge 1/Exercises 1–4: Mystery Bag

In this lesson, you will use data from a random sample drawn from a mystery bag to estimate a population proportion and learn how to find and interpret a margin of error for your estimate.

1. **Write down your estimate for the proportion of red chips in the mystery bag based on the random sample of 30 chips drawn in class.**

   If 15 red chips were in the sample of 30, some students might suggest that $\frac{1}{2}$ of the chips in the mystery bag were red. Others might suggest an interval around 0.5.

2. **Tanya and Raoul had a paper bag that contained red and black chips. The bag was marked 40% red chips. They drew random samples of 30 chips, with replacement, from the bag. (They were careful to shake the bag after they replaced a chip.) They had 9 red chips in their sample. They drew another random sample of 30 chips from the bag, and this time they had 12 red chips. They repeated this sampling process 50 times and made a plot of the number of red chips in each sample. A plot of their sampling distribution is shown below.**

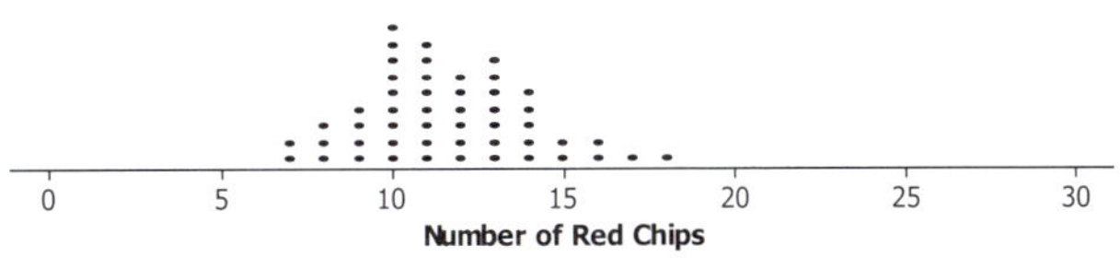

   a. **What was the most common number of red chips in the 50 samples? Does this seem reasonable? Why or why not?**

   The most common number in the samples was 10 red chips, which seems reasonable because we would expect to have about 40% successes, and 40% of 30 is 12 successes, and 10 is close to that.

b. **What number of red chips, if any, never occurred in any of the samples?**

They never got a sample with a number of red chips less than or equal to 6 *or more than or equal to* 20.

c. **Give an interval that contains the likely number of red chips in samples of size** 30 **based on the simulated sampling distribution.**

From 7 *to* 18 *red chips*

Note: Do not focus on exactly what "likely" means. The object is to see if it is at all reasonable for an outcome to occur by chance.

d. **Do you think the number of red chips in the mystery bag could have come from a sample drawn from a bag that had** 40% **red chips? Why or why not?**

This depends on the number of red chips that were observed in the random sample drawn from the mystery bag. If the observed number of red chips was 18*, the answer would be "yes" because* 18 *red chips occurred just by chance in the simulated sampling distribution. Some students might suggest* 18 *or more red chips only occurred once in* 50 *random samples, so it was not too likely but could happen by chance. The important concept is that students look at the simulated distribution to see where the observed outcome falls with respect to that distribution.*

Nine different bags of chips were distributed to small teams of students in the class. Each bag had a different proportion of red chips. Each team simulated drawing 50 **different random samples of size** 30 **from their bags and recorded the number of red chips for each sample. The graphs of their simulated sampling distributions are shown below.**

Population 10% **Red Chips**

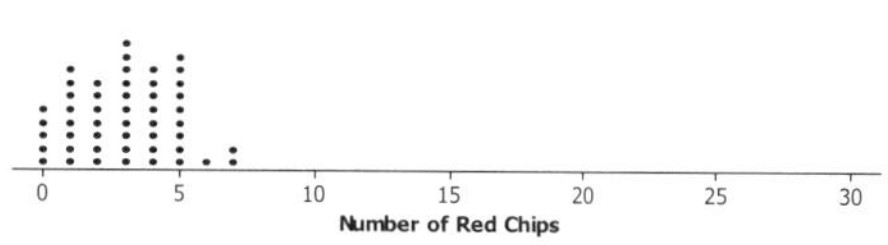

Population 20% **Red Chips**

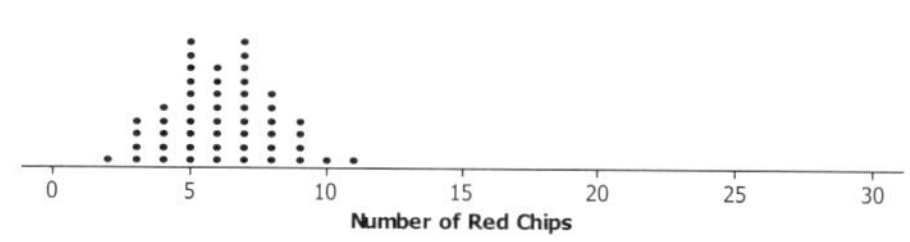

Population 30% **Red Chips**

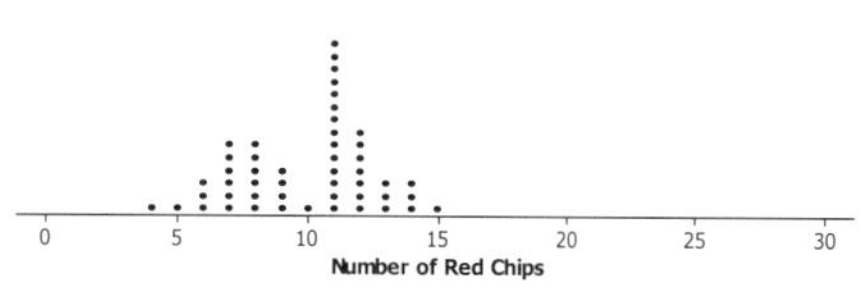

Population 40% **Red Chips**

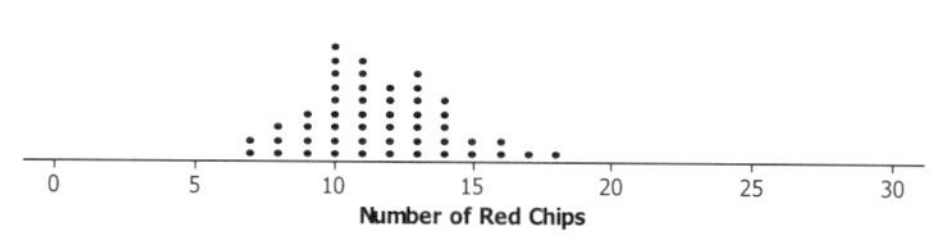

Population 50% **Red Chips**

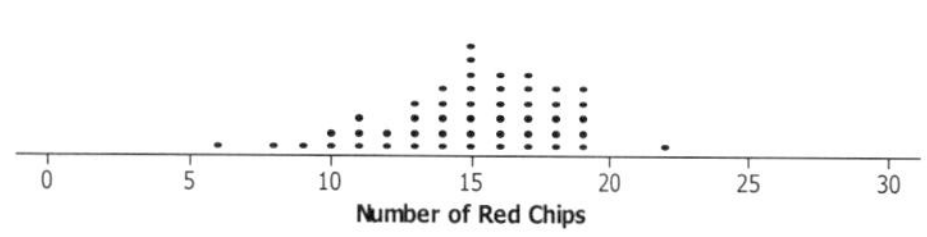

Population 60% **Red Chips**

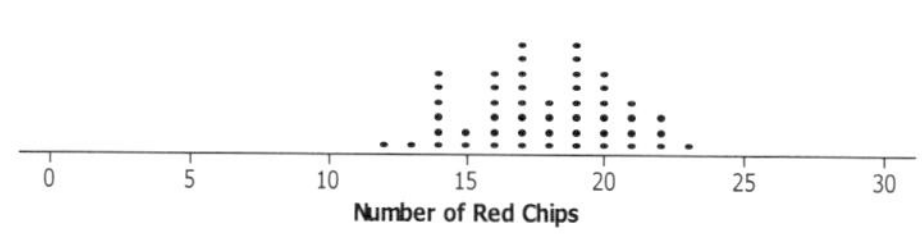

Population 70% Red Chips

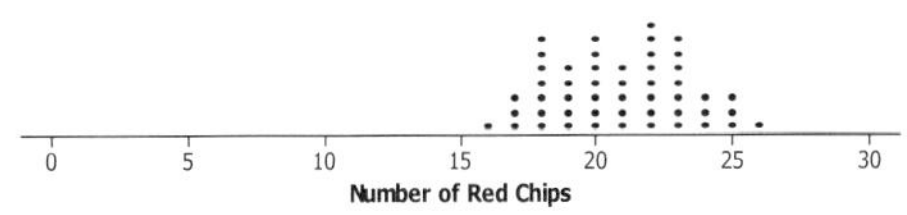

Population 80% Red Chips

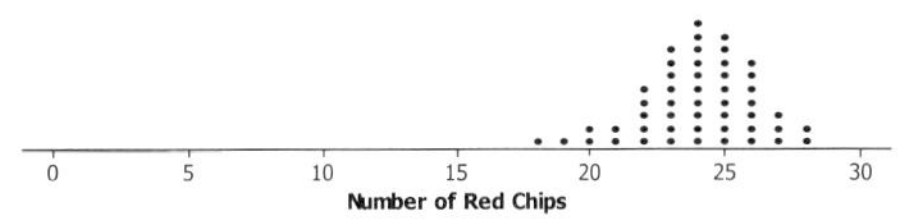

Population 90% Red Chips

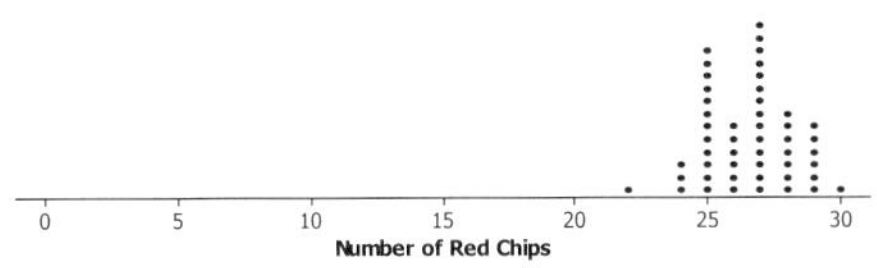

3. **Think about the number of red chips in the random sample of size 30 that was drawn from the mystery bag.**

   a. **Based on the simulated sampling distributions, do you think that the mystery bag might have had 10% red chips? Explain your reasoning.**

   If the random sample of size 30 from the mystery bag had 18 red chips, the answer would be "no" because 18 never showed up once in all of the samples.

   b. **Based on the simulated sampling distributions, which of the percentages 10%, 20%, 30%, 40%, 50%, 60%, 70%, 80%, and 90% might reasonably be the percentage of red chips in the mystery bag?**

   If the number of red chips in the sample from the mystery bag was 18, it looks like, just by chance, the sample could have been drawn from a population having from 40% to 80% red chips.

   c. **Let p represent the proportion of red chips in the mystery bag. (For example, $p = 0.40$ if there are 40% red chips in the bag.) Based on your answer to part (b), write an inequality that describes plausible values for p. Interpret the inequality in terms of the mystery bag population.**

   $0.40 \leq p \leq 0.80$

   This means that based on the simulated sampling distributions, the true proportion of red chips in the mystery bag could have been anywhere from 0.40 to 0.80. It would not have been surprising for a random sample of size 30 drawn from any of these populations to have included 18 red chips.

4. **If the inequality like the one you described in part (c) of Exercise 3 went from 0.30 to 0.60, it is sometimes written as 0.45 ± 0.15. The value 0.15 is called a *margin of error.* The margin of error represents an interval from the expected proportion that would not contain any proportions or very few proportions based on the simulated sampling distribution. Proportions in this interval are not expected to occur when taking a sample from the mystery bag.**

   a. **Write the inequality you found in Exercise 3 part (c), using this notation. What is the margin of error?**

   Using 18 as the number of red chips in the random sample from the mystery bag, the interval would be 0.60 ± 0.20. The margin of error is defined as an interval of 0.20.

Scaffolding:

- English language learners may need a quick explanation of the term *margin.*
- It can be the part of a page that is above, below, or to the side of a printed part.
- In this lesson, it can be a measure or degree of difference.
- A Frayer diagram may be used to explain margin of error.

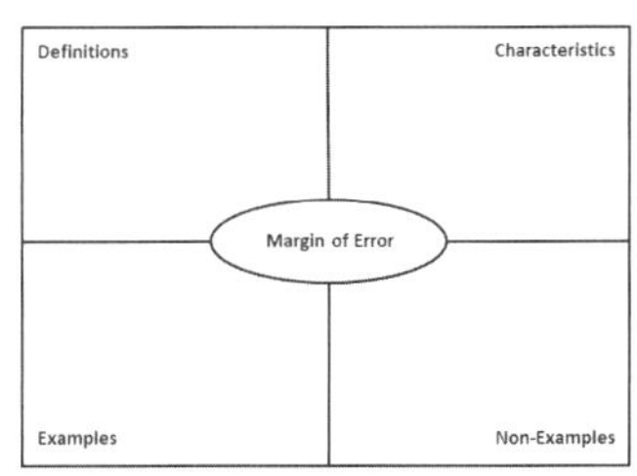

b. **Suppose Sol said, "So this means that the actual proportion of red chips in the mystery bag was 60%." Tonya argued that the actual proportion of red chips in the mystery bag was 20%. What would you say?**

They are both wrong. The notation does not mean that the center of the interval is the actual population but that it is the center of an interval made by adding and subtracting the margin of error. A random sample drawn from any proportion in the interval could have produced an outcome of 18 red chips. Tonya has mixed up the number describing the length of the interval with the population proportion.

Exploratory Challenge 2/Exercises 5–7 (10 minutes): Samples of Size 50

All students should do Exercise 5. As in the prior part of the lesson, students (or a subset of students) may actually simulate the sampling distributions. Or they can use the distributions provided in Exercise 6. The simulation would be similar to that for a sample of size 30, simply replacing 30 in the command with 50. All students should respond to Exercise 6.

Exploratory Challenge 2/Exercises 5–7: Samples of Size 50

MP.3

5. **Do you think the margin of error would be different in Exercise 4 if you had sampled 50 chips instead of 30? Try to convince a partner that your conjecture is correct.**

   They might be different but maybe only a little bit. I am not sure why the sample size would make a difference because the counts would be different, but it would still be centered around the same proportion (40% of a sample of size 30 is 12 red chips; 40% of a sample of size 50 is 20 red chips—different counts but the same proportion).

6. **Below are simulated sampling distributions of the number of red chips for samples of size 50 from populations with various percentages of red chips.**

Population with 10% Red Chips

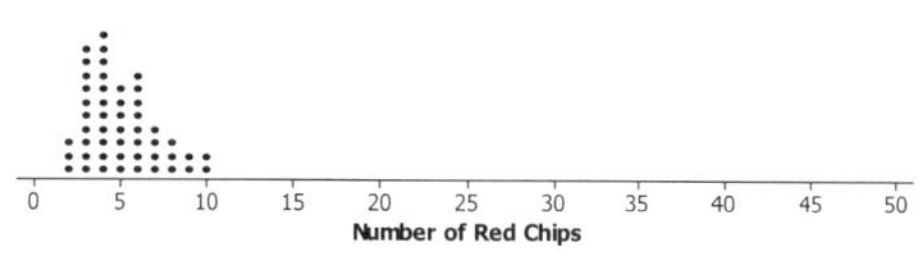

Population with 20% Red Chips

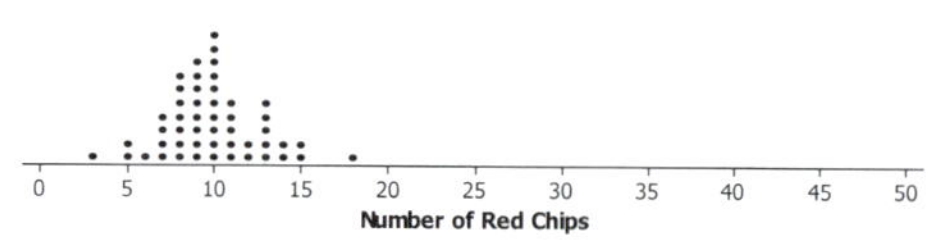

Population with 30% Red Chips

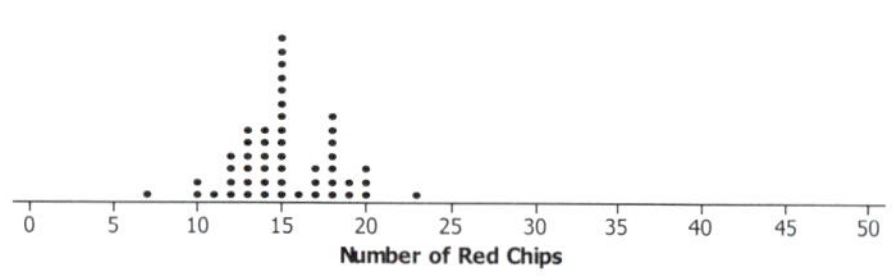

Population with 40% Red Chips

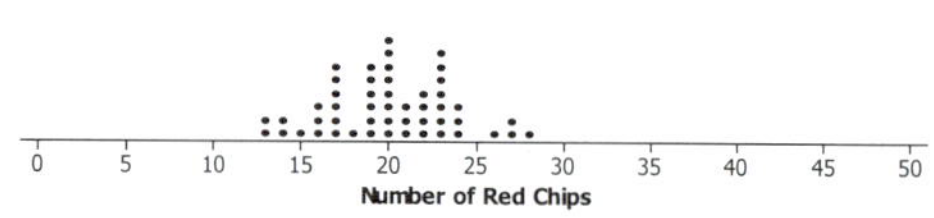

Population with 50% Red Chips

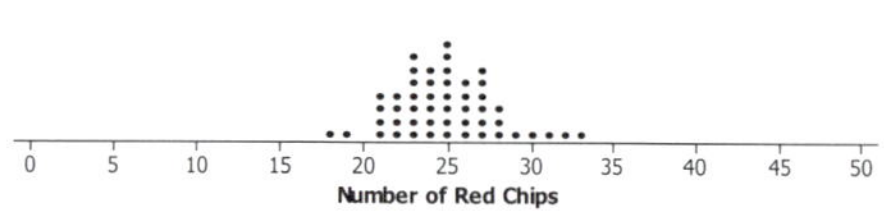

Population with 60% Red Chips

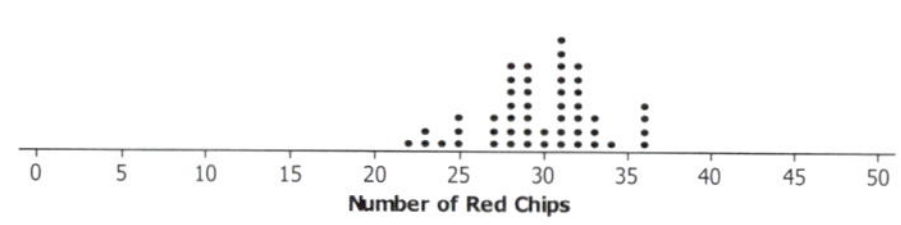

EUREKA MATH™

Population with 70% Red Chips

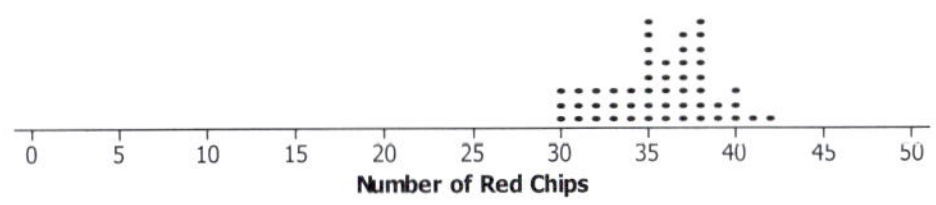

Population with 80% Red Chips

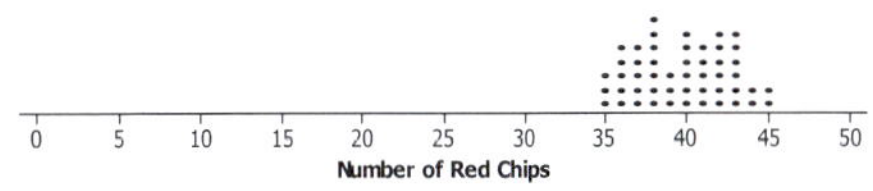

Population with 90% Red Chips

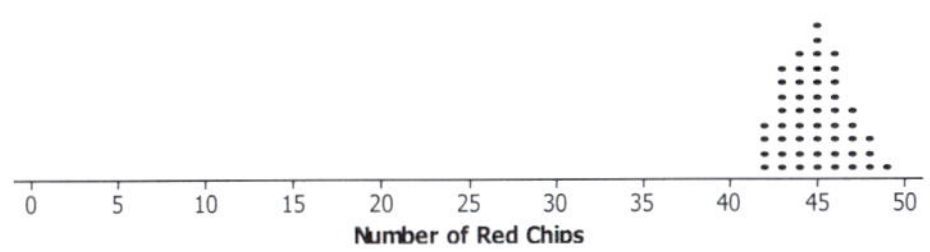

a. **Suppose you drew 30 red chips in a random sample of 50 from the mystery bag. What are plausible values for the proportion of red chips in the mystery bag? Explain your reasoning.**

Plausible population proportions are from 0.50 to 0.70.

b. **Write an expression that contains the margin of error based on your answer to part (a).**

The margin of error statement would be 0.60 ± 0.10. The margin of error would be 0.10.

7. **Remember your conjecture from Exercise 5, and compare the margin of error you found for a sample of size 30 (from Exercise 3) to the margin of error you found for a sample of size 50.**

a. **Was your reasoning in Exercise 5 correct? Why or why not?**

I forgot that the sample size might have to do with the spread of the distribution, not just the center, so my reasoning was not correct.

b. **Explain why the change in the margin of error makes sense.**

The margin of error was 0.10 for the sample size of 50, which is less than the margin of error for the sample size of 30. It makes sense that the margin of error would decrease as the sample size increases because as the sample size increases, the variability from sample to sample decreases, and the sample proportions tend to be closer to the actual population proportion.

Closing (5 minutes)

- Why do you suppose we use language like *margin of error* to define the interval describing a population that might have produced a sample proportion of red chips like the one from the mystery bag?
  - *The word* error *is misleading; it makes you think that you are off by a certain percent; it really tells you the range of possible population proportions for the population from which a random sample is drawn.*
- How could you apply the concepts from this lesson to investigate the proportion of female customers in a coffee shop on a weekday morning if you had observed 18 female customers in a random sample of 30 people during that time?
  - *The results would be the same because female customers would be like the red chips, and the sample size remains the same. Going into the coffee shop to take a random sample is the same as drawing a random sample from a mystery bag.*

- You knew the sample size and the observed outcome when you started the investigation you just did. Why were these important to know?
  - *You needed the sample size to know what size samples to generate, and you had to have an observed outcome to think about what might be plausible populations considering the simulated distributions of sample proportions for populations of known proportions.*

Ask students to summarize the main ideas of the lesson in writing or with a neighbor. Use this as an opportunity to informally assess comprehension of the lesson. The Lesson Summary below offers some important ideas that should be included.

Lesson Summary

In this lesson, you investigated how to make an inference about an unknown population proportion based on a random sample from that population.

- **You learned how random samples from populations with known proportions of successes behave by simulating sampling distributions for samples drawn from those populations.**
- **Comparing an observed proportion of successes from a random sample drawn from a population with an unknown proportion of successes to these sampling distributions gives you some information about what populations might produce a random sample like the one you observed.**
- **These plausible population proportions can be described as $p \pm M$. The value of M is called a *margin of error*.**

Exit Ticket (5 minutes)

Name ______________________________ Date ______________

Lesson 16: Margin of Error When Estimating a Population Proportion

Exit Ticket

1. Suppose you drew a sample of 12 red chips in a sample of 30 from a mystery bag. Describe how you would find plausible population proportions using the simulated sampling distributions we generated from populations with known proportions of red chips.

2. What would happen to the interval containing plausible population proportions if you changed the sample size to 60?

Exit Ticket Sample Solutions

1. Suppose you drew a sample of 12 red chips in a sample of 30 from a mystery bag. Describe how you would find plausible population proportions using the simulated sampling distributions we generated from populations with known proportions of red chips.

   Answers will vary. I would look at the simulated distributions to see which ones contained an outcome of 12 red chips. For those that did, the corresponding population proportion would be included in the set of plausible population proportions. (Students might use the actual distributions if they have them available and use the corresponding interval as an example in their answers—0.20 to 0.50 in those above.)

2. What would happen to the interval containing plausible population proportions if you changed the sample size to 60?

   The width of the interval would decrease, and the margin of error would be smaller.

Problem Set Sample Solutions

1. Tanya simulated drawing a sample of size 30 from a population of chips and got the following simulated sampling distribution for the number of red chips:

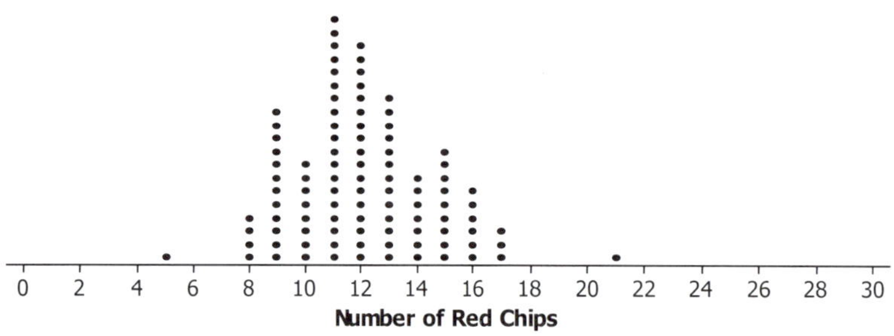

   Which of the following results seem like they might have come from this population? Explain your reasoning.

   I. 8 red chips in a random sample of size 30

   II. 12 red chips in a random sample of size 30

   III. 24 red chips in a random sample of size 30

   Samples that had 8 and 12 red chips might have come from this population because they occurred by chance in the random samples from the simulation. A sample with 24 red chips never occurred by chance, so it seems more unlikely to happen for this population.

EUREKA MATH

2. 64% of the students in a random sample of 100 high school students intended to go to college. The graphs below show the result of simulating random samples of size 100 from several different populations where the success percentage was known and recording the percentage of successes in the sample.

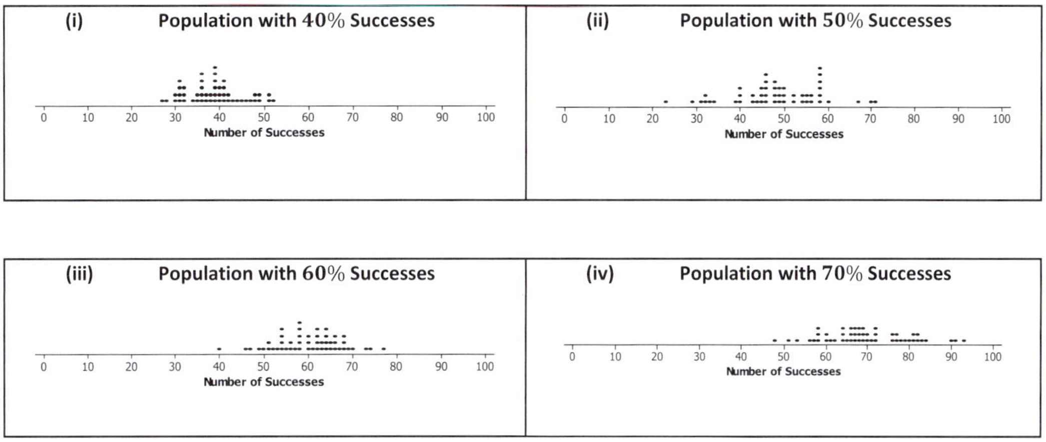

a. Based on these graphs, which of the following are plausible values for the percentage of successes in the population from which the sample was selected: 40%, 50%, 60%, or 70%? Explain your thinking.

The graphs show that 64% successes was a likely outcome for samples from populations with 60% and 70% successes. While exactly 64% did not occur in the 50% success population, it was in the range of observed sample percentages and, thus, could have happened. None of the samples from the 40% success population had a percentage of successes as large as 64%, so it would not seem likely that the sample came from this population.

b. Would you need more information to determine plausible values for the actual proportion of the population of high school students who intend to go to some postsecondary school? Why or why not?

Yes, you would need more information because you have not really looked at any simulated distributions of sample proportions larger than 70%. And 80% or 90% might turn out to be plausible as well.

3. Suppose the mystery bag had resulted in the following number of red chips. Using the simulated sampling distributions found earlier in this lesson, find a margin of error in each case.

a. The number of red chips in a random sample of size 30 was 10.

0.20 to 0.50, or 0.35 ± 0.15, for a margin of error of 0.15

b. The number of red chips in a random sample of size 30 was 21.

0.50 to 0.80, or 0.65 ± 0.15, for a margin of error of 0.15

c. The number of red chips in a random sample of size 50 was 22.

0.40 to 0.60, or 0.50 ± 0.10, for a margin of error of 0.10

4. The following intervals were plausible population proportions for a given sample. Find the margin of error in each case.

a. From 0.35 to 0.65

0.50 ± 0.15

b. From 0.72 to 0.78

0.75 ± 0.03

c. From 0.84 to 0.95

0.895 ± 0.055

d. From 0.47 to 0.57

0.52 ± 0.05

5. Decide if each of the following statements is true or false. Explain your reasoning in each case.

a. The smaller the sample size, the smaller the margin of error.

False. The smaller the sample size, the larger the margin of error.

b. If the margin of error is 0.05 and the observed proportion of red chips is 0.45, then the true population proportion is likely to be between 0.40 and 0.50.

True. The range of plausible values for the population proportion is 0.40 to 0.50.

6. Extension: The margin of error for a sample of size 30 is 0.20; for a sample of 50, it is 0.10. If you increase the sample size to 70, do you think the margin of error for the percent of successes will be 0.05? Why or why not?

No. When we simulated the sampling distributions for a sample size 100, the margin of error got smaller but was not 0.05.

Lesson 17: Margin of Error When Estimating a Population Proportion

Student Outcomes

- Students use data from a random sample to estimate a population proportion.
- Students calculate and interpret margin of error in context.
- Students know the relationship between sample size and margin of error in the context of estimating a population proportion.

Lesson Notes

A general approach for finding a margin of error involves using the standard deviation of a sample proportion. With appropriate sample sizes, 95% of all sample proportions are within *about* two standard deviations of the true population proportion. Therefore, due to natural sampling variability, 95% of all samples have a sample proportion of $\text{true proportion} \pm 2 \cdot \text{sample standard deviation}$, where 2SD is the margin of error. The first half of this lesson leads students through an example of finding and interpreting the standard deviation of a sampling distribution for a sample proportion. The focus of the second half of the lesson centers on the concept that if a sample size is large, then the sampling distribution of the sample proportion is approximately normal. To use the normal model for a sampling distribution, the *Success-Failure condition* (in which $np \geq 10$ and $nq \geq 10$) and the 10% *condition* (i.e., the sample size is no larger than 10% of the population) must both be met.

Classwork

In this lesson, you will find and interpret the standard deviation of a simulated distribution for a sample proportion and use this information to calculate a margin of error for estimating the population proportion.

Scaffolding:

Some students may have trouble moving from the count of the number of successes to the proportion. Suppose there are 15 seniors out of 20 high school students.

- To find the proportion, divide the number of successes by the sample size: $\frac{15}{20} = 0.75$.
- To find the percentage, multiply the proportion by 100%: $0.75 \times 100\% = 75\%$.

Exercises 1–6 (18 minutes): Standard Deviation for Proportions

In this set of exercises, students find and interpret the standard deviation of a sample proportion. Note that there is a shift in the notation to account for the sampling context, where the sample proportion of successes—denoted by $\hat{p}$ and read as *p-hat*—is a statistic obtained from the sample, as opposed to p, which is the proportion of successes in the entire population.

Exercises 1–6: Standard Deviation for Proportions

In the previous lesson, you used simulated sampling distributions to learn about sampling variability in the sample proportion and the margin of error when using a random sample to estimate a population proportion. However, finding a margin of error using simulation can be cumbersome and can take a long time for each situation. Fortunately, given the consistent behavior of the sampling distribution of the sample proportion for random samples, statisticians have developed a formula that will allow you to find the margin of error quickly and without simulation.

1. 30% of students participating in sports at Union High School are female (a proportion of 0.30).

   a. If you took many random samples of 50 students who play sports and made a dot plot of the proportion of female students in each sample, where do you think this distribution will be centered? Explain your thinking.

   Answers will vary. Since 30% of 50 is 15, I would expect the sampling distribution to be centered around $\frac{15}{50}$ female students.

   b. In general, for any sample size, where do you think the center of a simulated distribution of the sample proportion of female students in sports at Union High School will be?

   The sampling distribution should be centered at around 0.3. Some samples will result in a sample proportion of female students that is greater than 0.3, and some will result in a sample proportion of female students that is less than 0.3, but the sample proportions should center around 0.3.

2. Below are two simulated sampling distributions for the sample proportion of female students in random samples from all the students at Union High School.

Simulated Sampling Distribution 1

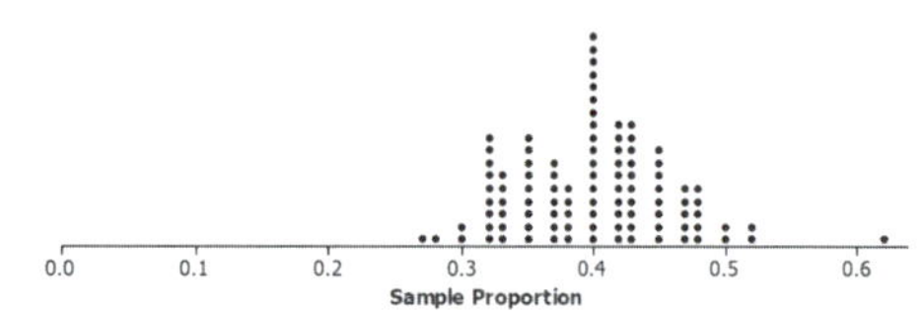

Simulated Sampling Distribution 2

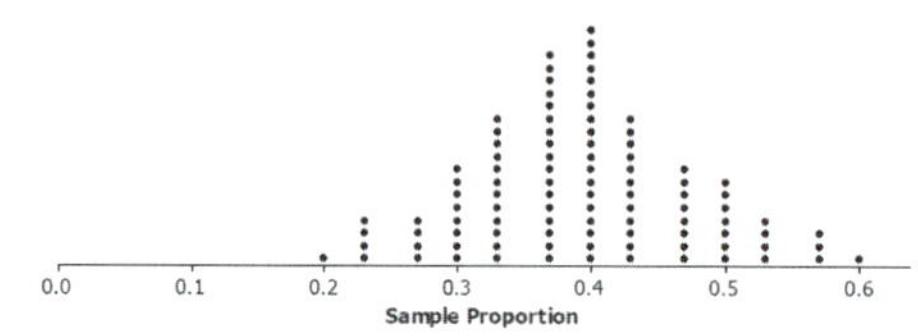

   a. Based on the two sampling distributions above, what do you think is the population proportion of female students?

   Answers will vary, but students should give an answer around 0.4.

   b. One of the sampling distributions above is based on random samples of size 30, and the other is based on random samples of size 60. Which sampling distribution corresponds to the sample size of 30? Explain your choice.

   Simulated Sampling Distribution 2 corresponds to the sample size of 30. I chose this one because it is more spread out—there is more sample-to-sample variability in Simulated Sampling Distribution 2 than in Simulated Sampling Distribution 1.

3. Remember from your earlier work in statistics that distributions were described using shape, center, and spread. How was spread measured?

   The spread of a distribution was measured with either the standard deviation or (sometimes) the interquartile range.

4. For random samples of size n, the standard deviation of the sampling distribution of the sample proportion can be calculated using the following formula:

$$\text{standard deviation} = \sqrt{\frac{p(1-p)}{n}},$$

where p is the value of the population proportion and n is the sample size.

a. If the proportion of female students at Union High School is 0.4, what is the standard deviation of the distribution of the sample proportions of female students for random samples of size 50? Round your answer to three decimal places.

$$\sqrt{\frac{(0.4)(0.6)}{50}} \approx 0.069$$

b. The proportion of males at Union High School is 0.6. What is the standard deviation of the distribution of the sample proportions of male students for random samples of size 50? Round your answer to three decimal places.

$$\sqrt{\frac{(0.6)(0.4)}{50}} \approx 0.069$$

c. Think about the graphs of the two distributions in parts (a) and (b). Explain the relationship between your answers using the center and spread of the distributions.

The two distributions are alike, but one is centered at 0.6 and the other at 0.4. The spread, as measured by the standard deviation of the two distributions, will be the same.

5. Think about the simulations that your class performed in the previous lesson and the simulations in Exercise 2 above.

a. Was the sampling variability in the sample proportion greater for samples of size 30 or for samples of size 50? In other words, does the sample proportion tend to vary more from one random sample to another when the sample size is 30 or 50?

There was more variability from sample to sample when the sample size was 30.

b. Explain how the observation that the variability in the sample proportions decreases as the sample size increases is supported by the formula for the standard deviation of the sample proportion.

You divide by n in the formula, and as n (a positive whole number) increases, the result of the division will be smaller.

6. Consider the two simulated sampling distributions of the proportion of female students in Exercise 2 where the population proportion was 0.4. Recall that you found $n = 60$ for Distribution 1 and $n = 30$ for Distribution 2.

a. Find the standard deviation for each distribution. Round your answer to three decimal places.

In Simulated Sampling Distribution 1, $n = 60$, and the standard deviation is 0.063.

In Simulated Sampling Distribution 2, $n = 30$, and the standard deviation is 0.089.

b. **Make a sketch, and mark off the intervals one standard deviation from the mean for each of the two distributions. Interpret the intervals in terms of the proportion of female students in a sample.**

Simulated Sampling Distribution 1: $n = 60$	Simulated Sampling Distribution 2: $n = 30$
The interval is from 0.337 *to* 0.463. *Typically, the proportion of female students in a sample of size* 60 *will be between* 0.337 *to* 0.463, *or about* 34%–46%. 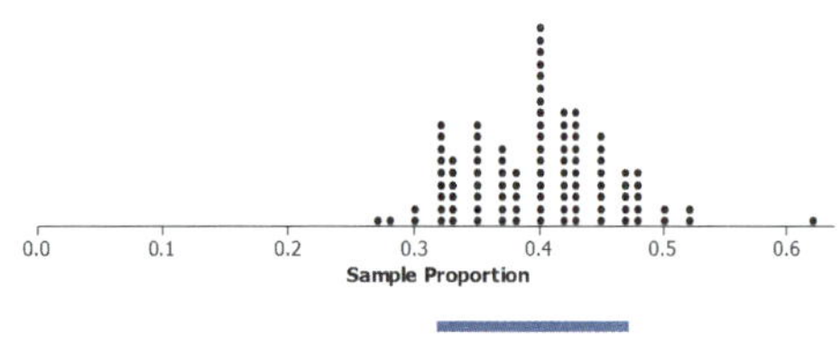	*The interval is from* 0.311 *to* 0.489. *Typically, the proportion of female students in a sample of size* 30 *will be between* 0.311 *to* 0.489, *or about* 31%–49%. 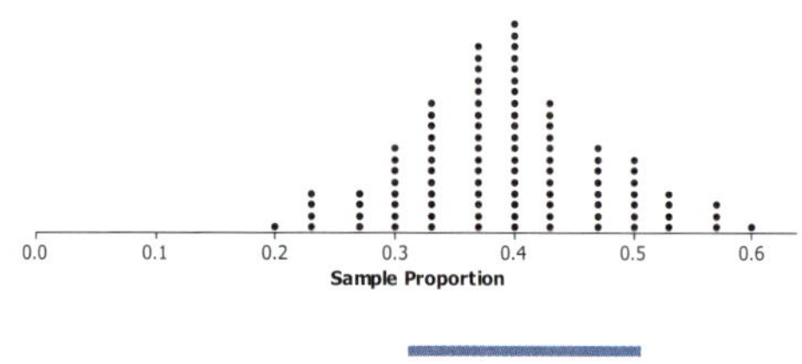

In general, three results about the sampling distribution of the sample proportion are known.

- **The sampling distribution of the sample proportion is centered at the actual value of the population proportion, p.**
- **The sampling distribution of the sample proportion is less variable for larger samples than for smaller samples. The variability in the sampling distribution is described by the standard deviation of the distribution, and the standard deviation of the sampling distribution for random samples of size n is $\sqrt{\frac{p(1-p)}{n}}$, where p is the value of the population proportion. This standard deviation is usually estimated using the sample proportion, which is denoted by $\hat{p}$ (read as p-hat), to distinguish it from the population proportion. The formula for the estimated standard deviation of the distribution of sample proportions is $\sqrt{\frac{\hat{p}(1-\hat{p})}{n}}$.**
- **As long as the sample size is large enough that the sample includes at least 10 successes and failures, the sampling distribution is approximately normal in shape. That is, a normal distribution would be a reasonable model for the sampling distribution.**

Exercises 7–12 (17 minutes): Using the Standard Deviation with Margin of Error

MP.2

The focus of this exercise set centers on the fact that if the sample size is large, the sampling distribution of the sample proportion is approximately normal. Combining this information with what students have learned about normal distributions leads to the fact that about 95% of the sample proportions are within two standard deviations of the value of the population (the mean of the sampling distribution). Note that if the population proportion is close to 0 or 1, either no one or everyone has the characteristic of interest, and the normal approximation to the sampling distribution is not appropriate unless the sample size is very, very large. To use the result based on the normal distribution, values of n and p should satisfy $np \geq 10$ and $n(1-p) \geq 10$. This is the same as saying that the sample is large enough, and at least 10 successes and failures would be expected in the sample. In addition to this *Success-Failure condition*, a 10% *condition* must also be met. The sample size must be less than 10% of the population to ensure that samples are independent.

Building from this normal approximation to the sampling distribution of $\hat{p}$, a formula can be created for the margin of error, dependent on the sample size and the proportion of successes observed in the sample.

MP.2 In Exercise 9, interested and motivated students might analyze the change in the rate at which the margin of error decreases and note that it is not constant; the margin of error is decreasing at a smaller and smaller rate as the sample size increases, which suggests a possible limiting factor.

MP.7 In Exercises 11 and 12, students look for and make use of structure as they consider the formula for margin of error and reason about how margin of error is affected by sample size and the value of the sample proportion.

Exercises 7–12: Using the Standard Deviation with Margin of Error

7. **In the work above, you investigated a simulated sampling distribution of the proportion of female students in a sample of size 30 drawn from a population with a known proportion of 0.4 female students. The simulated distribution of the proportion of red chips in a sample of size 30 drawn from a population with a known proportion of 0.4 is displayed below.**

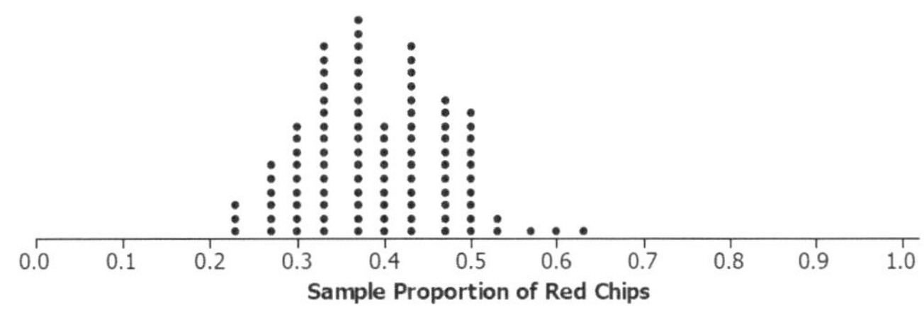

   a. **Use the formula for the standard deviation of the sample proportion to calculate the standard deviation of the sampling distribution. Round your answer to three decimal places.**

   The standard deviation should be about 0.089*.*

   b. **The distribution from Exercise 2 for a sample of size 30 is below. How do the two distributions compare?**

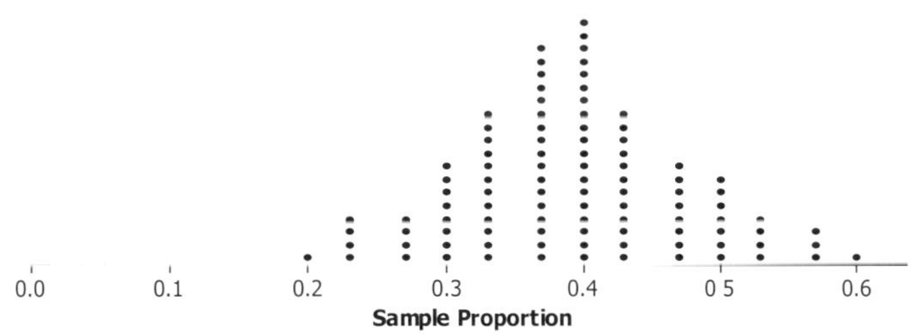

   The shapes of the two sampling distributions of the proportions are slightly different, but they both center at 0.4 *and have the same estimated standard deviation,* 0.089*.*

c. How many of the values of the sample proportions are within one standard deviation of 0.4? How many are within two standard deviations of 0.4?

The typical distance of the values from 0.4 is one standard deviation, between 0.311 and 0.489. All but two of the values are within two standard deviations from 0.4, or between 0.222 and 0.578. See the dot plot below.

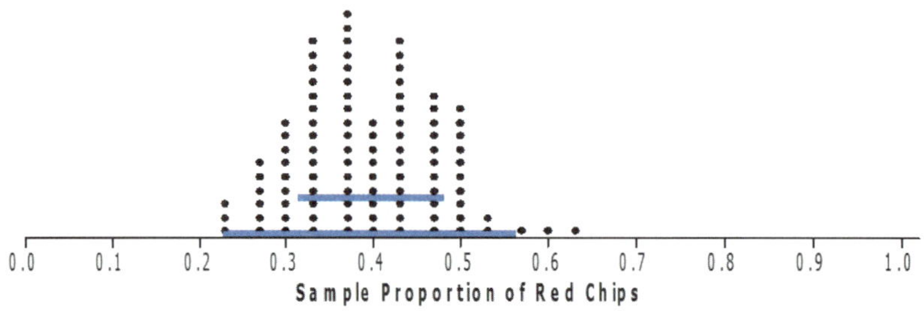

In general, for a known population proportion, about 95% of the outcomes of a simulated sampling distribution of a sample proportion will fall within two standard deviations of the population proportion. One caution is that if the proportion is close to 1 or 0, this general rule may not hold unless the sample size is very large. You can build from this to estimate a proportion of successes for an unknown population proportion and calculate a margin of error without having to carry out a simulation.

If the sample is large enough to have at least 10 of each of the two possible outcomes in the sample but small enough to be no more than 10% of the population, the following formula (based on an observed sample proportion $\hat{p}$) can be used to calculate the margin of error. The standard deviation involves the parameter p that is being estimated. Because p is often not known, statisticians replace p with its estimate $\hat{p}$ in the standard deviation formula. This estimated standard deviation is called the *standard error* of the sample proportion.

8.

a. Suppose you draw a random sample of 36 chips from a mystery bag and find 20 red chips. Find $\hat{p}$, the sample proportion of red chips, and the standard error.

$\hat{p} = \frac{20}{36} \approx 0.56$, *and the standard error is* $\sqrt{\frac{\hat{p}(1-\hat{p})}{n}} = \sqrt{\frac{(0.56)(0.44)}{36}} \approx 0.083.$

b. Interpret the standard error.

The sample proportion was 0.56, so I estimate that the proportion of red chips in the bag is 0.56. The actual population proportion probably isn't exactly equal to 0.56, but I expect that my estimate is within 0.083 of the actual value.

When estimating a population proportion, *margin of error* can be defined as the *maximum expected difference* between the value of the population proportion and a sample estimate of that proportion (the farthest away from the actual population value that you think your estimate is likely to be).

If $\hat{p}$ is the sample proportion for a random sample of size n from some population and if the sample size is large enough,

$$\text{estimated margin of error} = 2\sqrt{\frac{\hat{p}(1-\hat{p})}{n}}.$$

9. Henri and Terence drew samples of size 50 from a mystery bag. Henri drew 42 red chips, and Terence drew 40 red chips. Find the margins of error for each student.

   Henri's estimated margin of error is 0.104*; Terence's estimated margin of error is* 0.113.

10. Divide the problems below among your group, and find the sample proportion of successes and the estimated margin of error in each situation:

    a. Sample of size 20, 5 red chips

    Sample proportion of red chips is 0.25*, which is* 25%*; estimated margin of error is* 0.194*, which is* 19.4%.

    b. Sample of size 40, 10 red chips

    Sample proportion of red chips is 0.25*, which is* 25%*; estimated margin of error is* 0.137*, which is* 13.7%.

    c. Sample of size 80, 20 red chips

    Sample proportion of red chips is 0.25*, which is* 25%*; estimated margin of error is* 0.097*, which is* 9.7%.

    d. Sample of size 100, 25 red chips

    Sample proportion of red chips is 0.25*, which is* 25%*; estimated margin of error is* 0.087 *,which is* 8.7%.

11. Look at your answers to Exercise 2.

    a. What conjecture can you make about the relation between sample size and margin of error? Explain why your conjecture makes sense.

    As the sample size increases, the margin of error decreases. If you have a larger sample size, you can get a better estimate of the proportion of successes that are in the population, so the margin of error should be smaller.

    b. Think about the formula for a margin of error. How does this support or refute your conjecture?

    In the formula for the margin of error, the sample size is in the denominator of a fraction. If the sample size is large, that means the result of the division gets smaller. So, if the proportion is the same for three different-sized random samples, the smallest result would be when you divided by the largest sample size.

12. Suppose that a random sample of size 100 will be used to estimate a population proportion.

    a. Would the estimated margin of error be greater if $\hat{p} = 0.4$ or $\hat{p} = 0.5$? Support your answer with appropriate calculations.

    For $\hat{p} = 0.4$*: estimated margin of error* $= 2\sqrt{\frac{(0.4)(0.6)}{100}} \approx 0.098$

    For $\hat{p} = 0.5$*: estimated margin of error* $= 2\sqrt{\frac{(0.5)(0.5)}{100}} \approx 0.100$

    The estimated margin of error is greater when $\hat{p} = 0.5$.

b. **Would the estimated margin of error be greater if $\widehat{p} = 0.5$ or $\widehat{p} = 0.8$? Support your answer with appropriate calculations.**

For $\widehat{p} = 0.5$: *estimated margin of error* $= 2\sqrt{\frac{(0.5)(0.5)}{100}} \approx 0.100$

For $\widehat{p} = 0.8$: *estimated margin of error* $= 2\sqrt{\frac{(0.8)(0.2)}{100}} \approx 0.080$

The estimated margin of error is greater when $\widehat{p} = 0.5$.

c. **For what value of $\widehat{p}$ do you think the estimated margin of error will be greatest? (Hint: Draw a graph of $\widehat{p}(1 - \widehat{p})$ as $\widehat{p}$ ranges from 0 to 1.)**

The estimated margin of error is greater when $\widehat{p} = 0.5$. *The value of* $\widehat{p}(1 - \widehat{p})$ *is greatest when* $\widehat{p} = 0.5$. *(See graph below.) This value is in the numerator of the fraction in the formula for margin of error, so the margin of error is greatest when* $\widehat{p} = 0.5$.

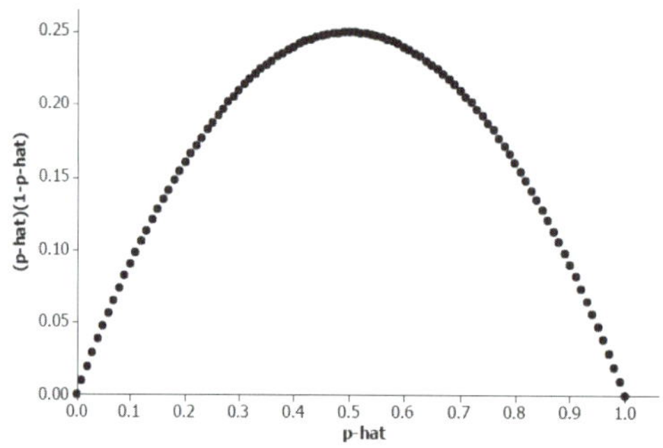

Closing (5 minutes)

- How does the work you did earlier on normal distributions relate to the margin of error?
  - *In a normal distribution, about 95% of the outcomes are within two standard deviations of the mean. We used the same thinking to get the margin of error formula.*
- How will your thinking about the margin of error change in each of the following situations?
  - A poll of 100 randomly selected people found 42% favor changing the voting age.
  - A sample of 100 red chips from a mystery bag found 42 red chips.
  - 42 cars in a random sample of 100 cars sold by a dealer were white.
    - *In all of the examples, the sample proportion and the sample size are the same and would give the same margin of error.*
- Why is it important to have a random sample when you are finding a margin of error?
  - *A random sample is important because you need to know that the behavior of the sample you choose should be fairly consistent across different samples. Having randomly selected samples is the only way to be sure of this.*

Ask students to summarize the main ideas of the lesson in writing or with a neighbor. Use this as an opportunity to informally assess comprehension of the lesson. The Lesson Summary below offers some important ideas that should be included.

Lesson Summary

- **Because random samples behave in a consistent way, a large enough sample size allows you to find a formula for the standard deviation of the sampling distribution of a sample proportion. This can be used to calculate the margin of error: $M = 2\sqrt{\frac{\hat{p}(1-\hat{p})}{n}}$, where $\hat{p}$ is the proportion of successes in a random sample of size n.**
- **The sample size is large enough to use this result for estimated margin of error if there are at least 10 of each of the two outcomes.**
- **The sample size should not exceed 10% of the population.**
- **As the sample size increases, the margin of error decreases.**

Exit Ticket (5 minutes)

Name ______________________________ Date ______________

Lesson 17: Margin of Error When Estimating a Population Proportion

Exit Ticket

1. Find the estimated margin of error when estimating the proportion of red chips in a mystery bag if 18 red chips were drawn from the bag in a random sample of 50 chips.

2. Explain what your answer to Problem 1 tells you about the number of red chips in the mystery bag.

3. How could you decrease your margin of error? Explain why this works.

Exit Ticket Sample Solutions

1. Find the estimated margin of error when estimating the proportion of red chips in a mystery bag if 18 red chips were drawn from the bag in a random sample of 50 chips.

   The margin of error would be 0.136.

2. Explain what your answer to Problem 1 tells you about the number of red chips in the mystery bag.

   The sample proportion of 0.36 *is likely to be within* 0.136 *of the actual value of the population proportion. This means that the proportion of red chips in the bag might be somewhere between* 0.22 *and* 0.496*, or about* 22%*–*50% *red chips.*

3. How could you decrease your margin of error? Explain why this works.

   The margin of error could be decreased by increasing sample size. The larger the sample size, the smaller the standard deviation, thus the smaller the margin of error.

Problem Set Sample Solutions

1. Different students drew random samples of size 50 from the mystery bag. The number of red chips each drew is given below. In each case, find the margin of error for the proportions of the red chips in the mystery bag.

   a. 10 red chips

      The margin of error will be approximately 0.113.

   b. 28 red chips

      The margin of error will be approximately 0.140.

   c. 40 red chips

      The margin of error will be approximately 0.113.

2. The school newspaper at a large high school reported that 120 out of 200 randomly selected students favor assigned parking spaces. Compute the margin of error. Interpret the resulting interval in context.

   The margin of error will be $2\sqrt{\frac{(0.6)(0.4)}{200}}$*, which is approximately* 0.069*. The resulting interval is* 0.6 ± 0.069*, or from* 0.531 *to* 0.669*. The proportion of students who favor assigned parking spaces is from* 0.531 *to* 0.669.

3. A newspaper in a large city asked 500 women the following: "Do you use organic food products (such as milk, meats, vegetables, etc.)?" 280 women answered "yes." Compute the margin of error. Interpret the resulting interval in context.

   The margin of error will be $2\sqrt{\frac{(0.56)(0.44)}{500}}$*, which is approximately* 0.044*. The resulting interval is* 0.56 ± 0.044 *or from* 0.516 *to* 0.604*. The proportion of women who use organic food products is between* 0.516 *and* 0.604.

4. The results of testing a new drug on $1,000$ people with a certain disease found that 510 of them improved when they used the drug. Assume these $1,000$ people can be regarded as a random sample from the population of all people with this disease. Based on these results, would it be reasonable to think that more than half of the people with this disease would improve if they used the new drug? Why or why not?

The margin of error would be about 0.032, which is about 3.2%, which means that the sample proportion of 0.510 is likely to be within 0.032 of the value of the actual population proportion. That means that the population proportion might be as small as 0.478, which is 47.8%. So, it is not reasonable to think that more than half of the people with the disease would improve if they used the new drug.

5. A newspaper in New York took a random sample of 500 registered voters from New York City and found that 300 favored a certain candidate for governor of the state. A second newspaper polled $1,000$ registered voters in upstate New York and found that 550 people favored this candidate. Explain how you would interpret the results.

In New York City, the proportion of people who favor the candidate is 0.60 ± 0.044, which is the range from 0.556 to 0.644. In upstate New York, the proportion of people who favor this candidate is 0.55 ± 0.031, which is the range from 0.519 to 0.581. Because the margins of error for the two candidates produce intervals that overlap, you cannot really say that the proportion of people who prefer this candidate is different for people in New York City and people in upstate New York.

6. In a random sample of $1,500$ students in a large suburban school, $1,125$ reported having a pet, resulting in the interval 0.75 ± 0.022. In a large urban school, 840 out of $1,200$ students reported having a pet, resulting in the interval 0.7 ± 0.026. Because these two intervals do not overlap, there appears to be a difference in the proportion of suburban students owning a pet and the proportion of urban students owning a pet. Suppose the sample size of the suburban school was only 500, but 75% still reported having a pet. Also, suppose the sample size of the urban school was 600, and 70% still reported having a pet. Is there still a difference in the proportion of students owning a pet in suburban schools and urban schools? Why does this occur?

The resulting intervals are as follows:

For suburban students: $0.75 \pm 2\sqrt{\frac{0.75(0.25)}{500}} \approx 0.75 \pm 0.039$, *or from* 0.711 *to* 0.789.

For urban students: $0.7 \pm 2\sqrt{\frac{0.7(0.3)}{600}} \approx 0.7 \pm 0.037$, *or from* 0.663 *to* 0.737.

No, there does not appear to be a difference in the proportion of students owning a pet in suburban and urban schools. This occurred because the margins of error are larger due to the smaller sample size.

7. Find an article in the media that uses a margin of error. Describe the situation (an experiment, an observational study, a survey), and interpret the margin of error for the context.

Students might bring in poll results from a newspaper.

Lesson 18: Sampling Variability in the Sample Mean

Student Outcomes

- Students understand the term *sampling variability* in the context of estimating a population mean.
- Students understand that the standard deviation of the sampling distribution of the sample mean offers insight into the accuracy of the sample mean as an estimate of the population mean.

Lesson Notes

MP.4

This is the first of two lessons to build on the concept of sampling variability in the sample mean first developed in Grade 7 Module 5 Lessons 17–19. Students use simulation to approximate the sampling distribution of the sample mean and explore how the simulated sampling distribution provides information about the anticipated estimation error when using a sample mean to estimate a population mean. Students learn how simulating samples gives information about how sample means vary.

Each student or small group of students should have a bag with slips of paper numbered one to 100. (Or as an alternative, have them generate the random numbers using technology.) Prepare a number line on a wall or board that goes from 1 to 5 with each unit divided into tenths. Students should have sticky notes to post the mean segment lengths in their random samples on the number line, so leave enough space for the sticky notes.

In Exercises 4 and 5, students need to share the values of the means for their individual samples. Consider having them report their means while the others record them (or enter them if they are using technology). To facilitate the process, consider giving students a copy of the values from the table below used in the simulated sampling distribution of means suggested as possible answers to Exercise 3 and Exercise 5. (Or do this as a whole-class activity with one student entering the values to avoid errors in entering the data.)

Length of Segments

2	3	3	2	4	8	2	5	2	2
5	5	4	3	1	2	3	2	2	3
1	1	2	4	1	1	4	5	4	3
2	1	1	1	2	2	3	1	8	4
4	2	1	1	3	5	1	1	4	2
1	3	7	3	3	3	1	2	3	3
1	2	4	3	1	1	7	3	1	7
2	2	3	2	3	2	2	1	2	1
3	1	8	4	2	2	1	1	5	3
1	4	1	2	5	3	3	3	5	4

Be sure students label their graphs completely in their answers to the questions. Understanding what they are graphing is an important part of understanding the concepts involved in this exercise.

Classwork

Exploratory Challenge/Exercises 1–7 (40 minutes): Random Segments

Provide each student or small group with a copy of the worksheet that is located at the end of this lesson.

Exploratory Challenge/Exercises 1–7: Random Segments

The worksheet contains 100 segments of different lengths. The length of a segment is the number of rectangles spanned on the grid. For example, segment 2 has length 5.

1. **Briefly review the sheet, and estimate the mean length of the segments. Will your estimate be close to the actual mean? Why or why not?**

   Answers will vary. Some may estimate 5, others as low as 2.

   Sample response: I estimate the mean length is 4. I believe my estimate will be close to the actual mean because it appears as if a large number of the segments are around 4, and if I average the longer and shorter lengths, the average is also around 4.

2. **Look at the sheet. With which of the statements below would you agree? Explain your reasoning.**

   The mean length of the segments is

   a. **Close to 1.**
   b. **Close to 8.**
   c. **Around 5.**
   d. **Between 2 and 5.**

   The only choice that makes sense to me is (d), between 2 and 5. The smallest segment length was 1, so it does not make sense that the mean would be the smallest. The largest segment length was 8, so again, it would not make sense to have the mean be 8 or even 5 because there are a lot of segments of lengths 1 and 2.

Some estimates for the mean are not reasonable because they are lengths of the longest segments, which would not account for lengths that are shorter. (This would also be true for the shortest segment lengths.) While an interval estimate might make sense, it is still hard to know for sure whether that interval is a good estimate. A better way to get a good estimate is to use random samples.

3. **Follow your teacher's directions to select ten random numbers between 1 and 100. For each random number, start at the upper left cell with a segment value of 2, and count down and to the right the number of cells based on the random number selected. The number in the cell represents the length of a randomly selected segment.**

   a. **On a number line, graph the lengths of the corresponding segments on the worksheet.**

      For the random numbers $\{8, 23, 35, 74, 40, 75, 9, 50, 54, 64\}$, the sample lengths would be $\{2, 2, 1, 1, 2, 1, 3, 5, 2, 3\}$. The graph would look like:

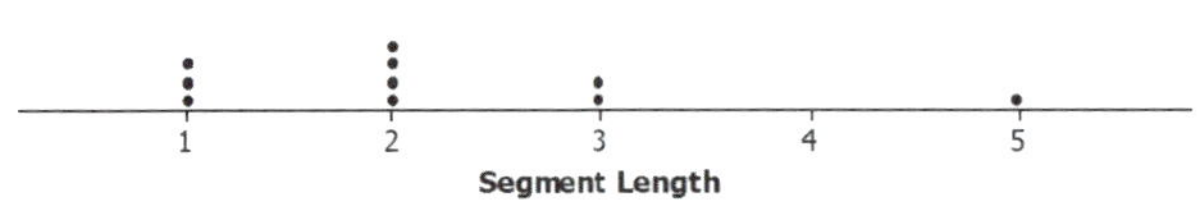

b. Find the mean and standard deviation of the lengths of the segments in your sample. Mark the mean length on your graph from part (a).

The mean sample length is 2.2 units, and the standard deviation is 1.23 units.

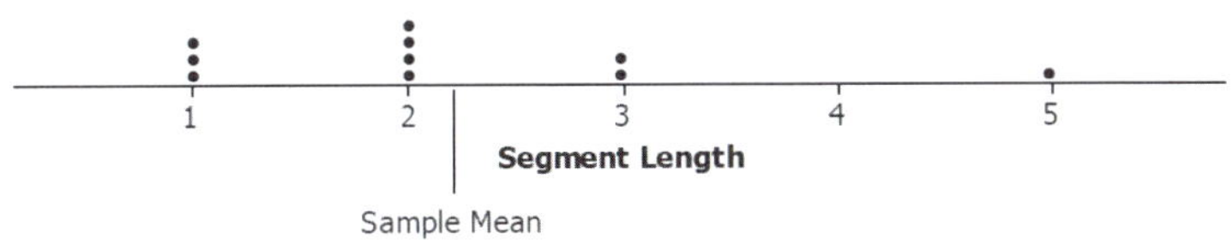

4. Your sample provides some information about the mean length of the segments in one random sample of size 10, but that sample is only one among all the different possible random samples. Let's look at other random samples and see how the means from those samples compare to the mean segment length from your random sample.

Record the mean segment length for your random sample on a sticky note, and post the note in the appropriate place on the number line your teacher set up.

Sample response (based on a class with 31 students):

Simulated sampling distribution of mean segment lengths for samples of size 10

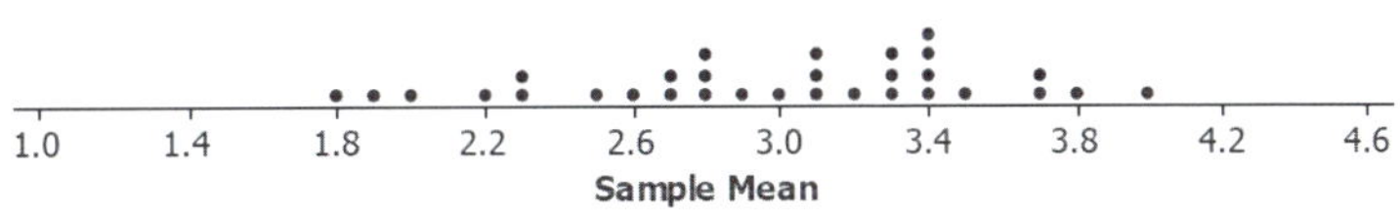

a. Jonah looked at the plot and said, "Wow. Our means really varied." What do you think he meant?

Many of the samples had mean lengths that were different. Some were the same, but most were not.

b. Describe the simulated sampling distribution of mean segment lengths for samples of size 10.

The maximum mean segment length in our samples of size 10 was 4.0, and the minimum was 1.7. The sample means seemed to center around 3, with most of the segments from about 2.5 to 3.5 units long.

c. How did your first estimate (from Exercise 1) compare to your sample mean from the random sample? How did it compare to the means in the simulated distribution of the sample means from the class?

My estimate was way off. I thought the mean length would be larger, like maybe 4.5 units. My sample mean was only 2.2 units long, which was smaller than all but three of the other sample means.

5. Collect the values of the sample means from the class.

a. Find the mean and standard deviation of the simulated distribution of the sample means.

Sample response (based on the 31 sample means used to produce the answer to Exercise 3): The mean of the simulated distribution of sample means is 2.97, and the standard deviation is 0.57.

b. Interpret the standard deviation of the simulated sampling distribution in terms of the length of the segments.

Sample response: A typical distance of a sample mean from the center of the sampling distribution is about 0.57.

c. What do you observe about the values of the means in the simulated sampling distribution that are within two standard deviations from the mean of the sampling distribution?

Sample response: All of the sample means were within two standard deviations from the mean of the sampling distribution, from 1.53 to 4.11.

6. Generate another set of ten random numbers, find the corresponding lengths on the sheet, and calculate the mean length for your sample. Put a sticky note with your sample mean on the second number line. Then, answer the following questions.

Sample response (based on a class with 31 students): second simulated sampling distribution of mean segment lengths for 33 samples of size 10

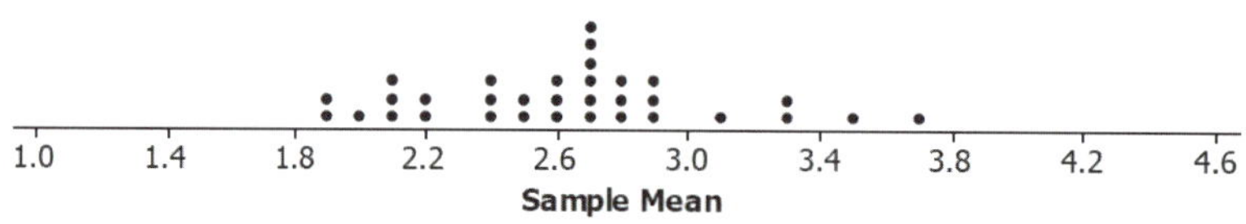

a. Find the mean and standard deviation of the simulated distribution of the sample means.

Sample response: The mean is 2.63 units, and the standard deviation is 0.44 unit.

b. Interpret the standard deviation of the simulated sampling distribution in terms of the length of the segments.

Sample response: A typical distance of a sample mean from the center of the sampling distribution is about 0.44.

c. What do you observe about the values of the means in the simulated sampling distribution that are within two standard deviations from the mean of the sampling distribution?

Sample response: Only one or two sample means were not within two standard deviations from the mean of the sampling distribution, 1.75 to 3.51.

7. Suppose that we know the actual mean of all the segment lengths is 2.78 units.

MP.2

a. Describe how the population mean relates to the two simulated distributions of sample means.

The simulated sampling distributions of sample means were both centered around values very close to the population mean.

b. Tonya was concerned that neither of the simulated distributions of sample means had a value around 5, but some of the segments on the worksheet were 5 units long, and some were as big as 8 units long. What would you say to Tonya?

The simulated sampling distribution was of the means of the samples, not of individual segment lengths. It could be possible to have a mean length of 5 from a sample of ten segment lengths, but it is not very likely.

Closing (2–3 minutes)

- Why is the concept of *random* samples important in exploring how a simulated sampling distribution provides information about the anticipated estimation error when using a sample mean to estimate a population mean?
  - *If the samples are not random, the distribution of the sample means might not have centers close to the population mean, and the standard deviations of different sampling distributions might not tell the same story about the distributions of the sample means.*
- What is the difference between a *sample mean* and a *distribution of sample means*? (You may use the segment lengths in your answer.)
  - *A sample mean is the mean of the values of the segment lengths in a given sample. A distribution of sample means is the distribution of all sorts of sample means calculated from many different samples.*

Ask students to summarize the main ideas of the lesson in writing or with a neighbor. Take this opportunity to informally assess comprehension of the lesson. The Lesson Summary below offers some important ideas that should be included.

> Lesson Summary
>
> **In this lesson, you drew a sample from a population and found the mean of that sample.**
>
> - **Drawing many samples of the same size from the same population and finding the mean of each of those samples allows you to build a simulated sampling distribution of the sample means for the samples you generated.**
> - **The mean of the simulated sampling distribution of sample means is close to the population mean.**
> - **In the two examples of simulated distributions of sample means we generated, most of the sample means seemed to fall within two standard deviations of the mean of the simulated distribution of sample means.**

Exit Ticket (3 minutes)

Name ______________________________ Date ____________

Lesson 18: Sampling Variability in the Sample Mean

Exit Ticket

Describe what a *simulated distribution of sample means* is and what the *standard deviation of the distribution* indicates. You may want to refer to the segment lengths in your answer.

Exit Ticket Sample Solutions

Describe what a *simulated distribution of sample means* is and what the *standard deviation of the distribution* indicates. You may want to refer to the segment lengths in your answer.

You draw samples of a given size from a population (here, it was the 100 segment lengths), find the mean segment length of each sample, and plot the sample mean lengths. The resulting distribution of the sample means from those random samples is the simulated distribution of sample means. The standard deviation of that distribution gives you an idea of how the sample means vary from sample to sample.

Problem Set Sample Solutions

1. The three distributions below relate to the population of all of the random segment lengths and to samples drawn from that population. The eight phrases below could be used to describe a whole graph or a value on the graph. Identify where on the appropriate graph the phrases could be placed. (For example, *segment of length* 2 could be placed by any of the values in the column for 2 on the plot labeled *Length*.)

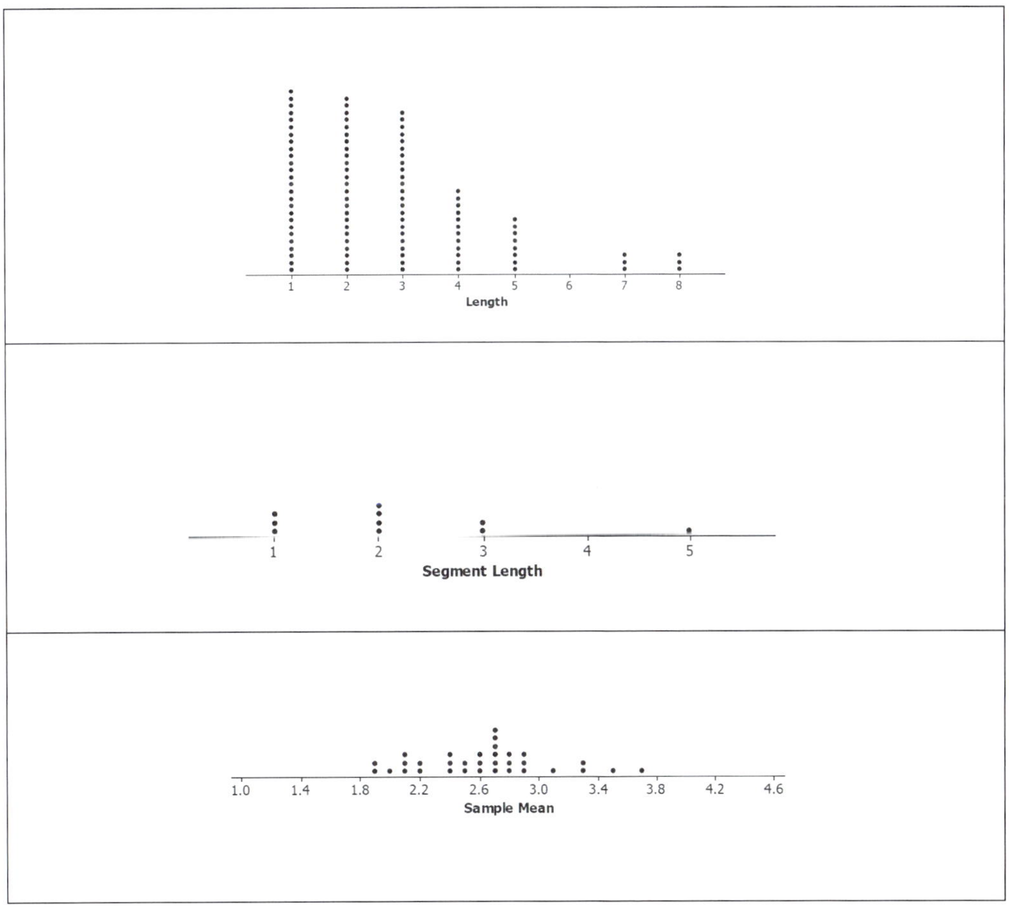

a. Random sample of size 10 of segment lengths
b. Segment of length 2
c. Sample mean segment length of 2
d. Mean of sampling distribution, 2.6
e. Simulated distribution of sample means
f. Sample segment lengths
g. Population of segment lengths
h. Mean of all segment lengths, 2.78

Possible answers are shown below. (Note that segment of length 2 could be used in more than one place and on more than one graph.)

Population of Segment Lengths

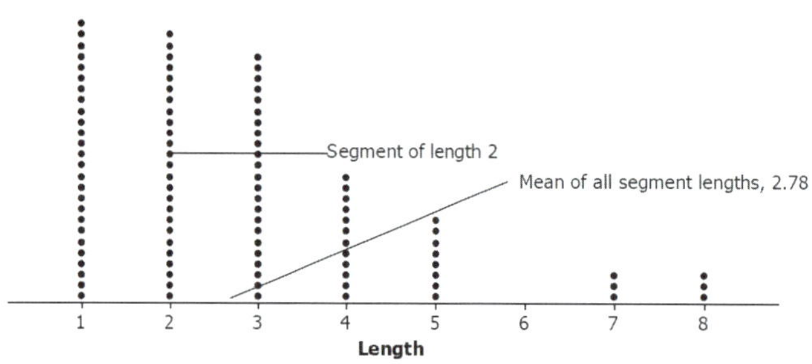

Random Sample of Size 10 of Segment Lengths

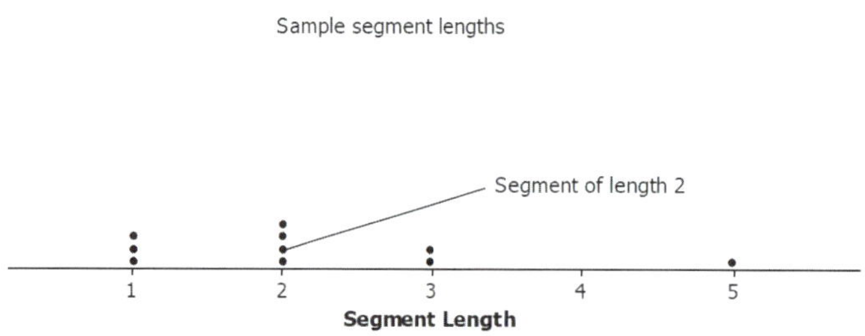

Simulated Distribution of Sample Means

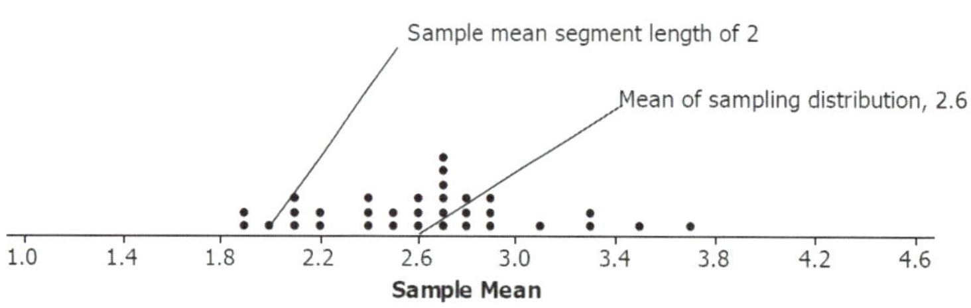

2. **The following segment lengths were selected in four different random samples of size 10.**

If students work in small groups, indicate that each group member work on a different sample.

Lengths Sample A	Lengths Sample B	Lengths Sample C	Lengths Sample D
1	1	1	2
2	3	5	2
1	1	1	7
5	2	3	2
3	1	4	5
1	5	2	2
2	3	2	3
2	4	4	5
3	3	3	5
1	3	4	4

a. **Find the mean segment length of each sample.**

Sample A mean is 2.1 (1.29)*; Sample B mean is* 2.6 (1.35)*; Sample C mean is* 2.9 (1.37)*; Sample D mean is* 3.7 (1.77)*.*

The standard deviation of each sample is noted in parentheses as reference for part (b).

b. **Find the mean and standard deviation of the four sample means.**

The mean of the sample means is 2.825*, and the standard deviation of the sample means is* 0.67*.*

c. **Interpret your answer to part (b) in terms of the variability in the sampling process.**

A typical distance of a sample mean from the mean of the four samples (2.825) *is* 0.67*.*

3. **Two simulated sampling distributions of the mean segment lengths from random samples of size 10 are displayed below.**

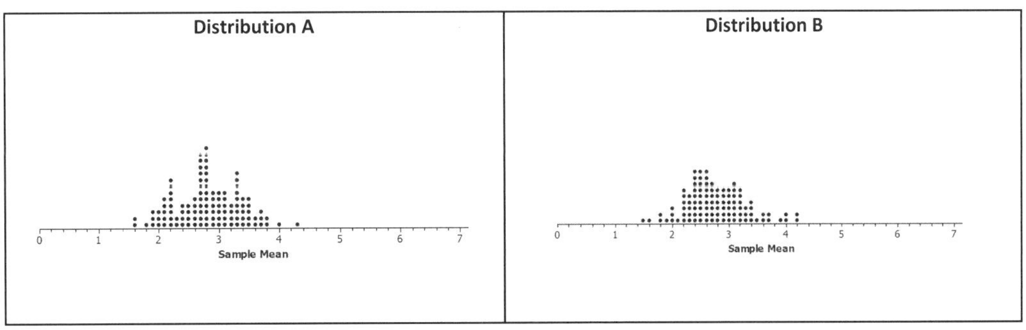

a. **Compare the two distributions with respect to shape, center, and spread.**

Both distributions are mound shaped with the center a bit below 3*, about* 2.8*. The maximum mean segment length in both is about* 4.2 *units, and the minimum is about* 1.5 *or* 1.6*. Most of the sample means in both distributions are between about* 2 *and* 4*.*

b. **Distribution A has a mean of 2.82, and Distribution B has a mean of 2.77. How do these means compare to the population mean of 2.78?**

The mean segment lengths of the two simulated distributions of sample means are very close to the actual mean segment length.

c. **Both Distribution A and Distribution B have a standard deviation 0.54. Make a statement about the distribution of sample means that makes use of this standard deviation.**

Answers will vary. Possible answers include the following:

- *Almost all of the sample means in Distribution A are within two standard deviations of the mean of the sample means, 1.74 to 3.90. The same is true for the sample means in Distribution B; the sample means are almost all from 1.69 to 3.85.*
- *A typical distance of a sample mean from the center of the sampling distribution is 0.54.*

4. **The population distribution of all the segment lengths is shown in the dot plot below. How does the population distribution compare to the two simulated sampling distributions of the sample means in Problem 3?**

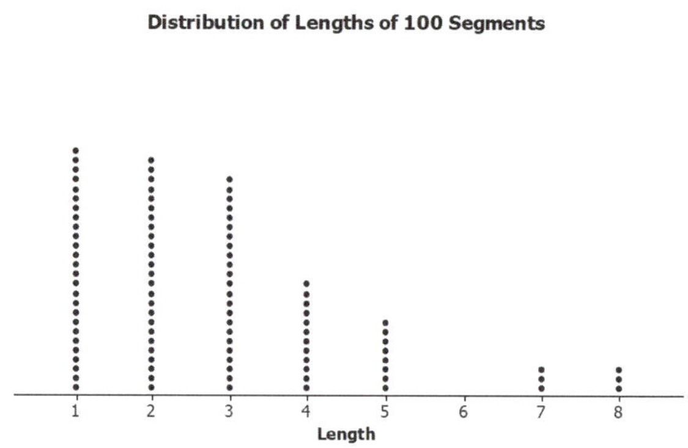

The distribution of all of the lengths is skewed right. Most of the lengths were 1, 2, or 3 units. The simulated sampling distributions of the sample means were both mound shaped, with the centers about the same, and not like the shape of the population.

Exercises 1–7: Random Segments

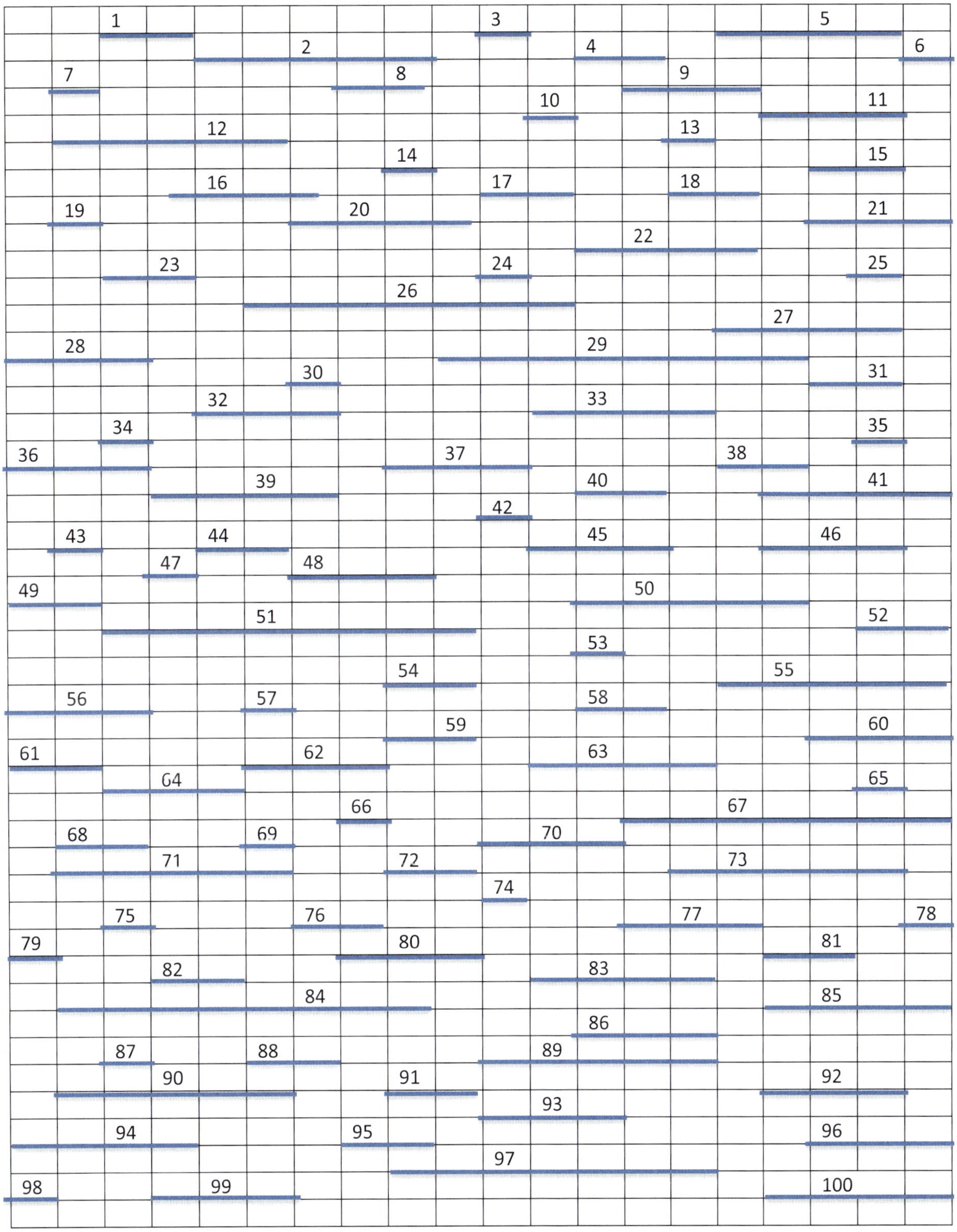

Lesson 19: Sampling Variability in the Sample Mean

Student Outcomes

- Students understand the term *sampling variability* in the context of estimating a population mean.
- Students understand that the standard deviation of the sampling distribution of the sample mean conveys information about the anticipated accuracy of the sample mean as an estimate of the population mean.

Lesson Notes

This is the second of two lessons building on the concept of sampling variability in the sample mean first developed in Grade 7 Module 5 Lessons 17–19. Students use simulation to approximate the sampling distribution of the sample mean for random samples from a population. They also explore how the simulated sampling distribution provides information about the anticipated estimation error when using a sample mean to estimate a population mean and how sample size affects the distribution of the sample mean.

Classwork

This lesson uses simulation to approximate the sampling distribution of the sample mean for random samples from a population, explores how the simulated sampling distribution provides insight into the anticipated estimation error when using a sample mean to estimate a population mean, and covers how sample size affects the distribution of the sample mean.

Exercises 1–6 (35 minutes): SAT Scores

In this lesson, consider giving students the population data, and have them use technology to take random samples (without replacement) from the population. (Copy and paste the table into a spreadsheet, and send it to students.) A typical set of commands to generate a random sample might be *randsamp(SAT_scores, 20)*, where *SAT_scores* is the name of the list containing the scores and 20 is the sample size. The random sample should refresh by clicking on the command line or by using a command such as Control R.

Note that the sample answers for the simulated distributions typically display the means from about 50 random samples. If it is possible to generate many more samples quickly with technology, students should do so. The basic characteristics of a distribution of sample means (center, spread, mound shaped) do not change much as more samples are added to the first 50 or so—which is suggested by contrasting Exercise 3 parts (a) and (b)—and students should experience this themselves by generating their own distributions with many samples.

Part (a) of both Exercises 2 and 3 is important to discuss, as it highlights the difference in the distribution of the values in the sample (the scores) and the distribution of the sample means (the mean of the scores) in a sample.

Have students share the simulated sampling distributions they generate for Exercise 3. Consider using screen capture/sharing software, or have students walk around the room with Post-it notes looking at each student's handheld or computer screen and recording what they see about the distributions. Recognizing that all of the simulated distributions have essentially the same characteristics is a key factor in understanding why it is possible to make general statements about how random samples behave in relation to the population.

MP.2

Exercises 1–6: SAT Scores

1. SAT scores vary a lot. The table on the next page displays the 506 scores for students in one New York school district for a given year.

Table 1: SAT Scores for District Students

441	395	369	350	521	691	648	521	498	413	486
440	415	481	392	800	448	603	503	486	476	500
391	359	447	550	432	158	379	394	495	442	507
395	504	399	424	456	729	356	392	514	388	518
445	436	386	493	467	493	440	387	512	431	467
499	412	457	389	323	319	550	450	517	405	506
486	519	369	373	348	532	496	488	504	444	
396	473	319	367	679	472	613	561	522	408	
451	427	369	560	602	520	567	495	473	424	
362	391	371	407	436	366	582	528	533	463	
328	613	357	438	436	713	603	525	553	446	
414	466	382	362	777	259	557	508	495	466	
409	486	627	589	749	410	639	516	520	632	
526	334	608	374	634	443	556	506	506	526	
391	497	378	358	566	442	496	568	544	546	
529	392	387	373	198	555	499	476	525	529	
529	426	470	378	345	431	613	490	548	455	
574	379	380	561	712	197	556	547	543	431	
363	382	370	379	504	254	596	489	474	386	
486	434	365	530	685	372	580	506	529	434	
418	722	674	504	645	501	605	511	566	362	
527	437	388	525	509	662	445	489	487	426	
441	395	377	561	448	503	602	523	510	404	
467	463	427	519	491	448	638	530	518	493	
387	433	446	525	352	662	570	507	515	515	
503	371	394	569	779	158	558	504	516	407	
350	392	368	484	689	691	535	522	505	409	
583	416	406	416	513	729	623	503	536	422	
370	370	350	446	624	493	465	524	547	612	
499	422	344	420	465	319	460	523	528	486	
399	532	347	446	504	532	375	524	527	394	
374	545	377	462	390	472	540	501	523	424	
372	427	391	528	576	520	564	482	540	393	
559	371	339	533	756	366	547	502	480	420	
330	390	404	543	451	713	568	503	516	415	
567	529	377	460	505	259	588	439	501	394	
371	341	469	391	540	410	502	474	452	473	
503	356	417	623	436	443	510	477	507	531	
327	351	356	587	298	442	589	458	486	469	
528	377	370	528	449	555	537	494	500	453	
447	404	355	356	352	431	410	447	507	442	
572	369	364	523	574	197	330	517	518	509	
379	396	383	404	518	460	500	457	467	435	
456	396	400	505	682	623	531	471	506	427	
406	535	404	512	474	587	509	541	509	489	
420	388	375	514	629	528	571	513	597	480	
395	370	398	516	656	523	527	441	509	516	
355	417	376	498	539	505	457	489	567	501	
423	419	451	460	553	514	552	498	509	452	
438	348	369	541	400	629	561	538	597	507	

a. Looking at the table above, how would you describe the population of SAT scores?

It is hard to tell. I see some numbers as low as 198 and others in the 700's. You cannot tell much from just looking at the individual numbers.

b. Jason used technology to draw a random sample of size 20 from all of the scores and found a sample mean of 487. What does this value represent in terms of the graph below?

Random Sample from District SAT Scores

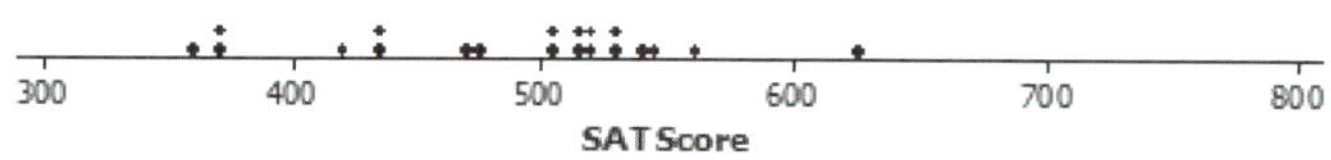

This represents the average SAT score for the sample and indicates where the scores in the dot plot are centered. If you computed the mean of the values for the SAT scores in the sample (i.e., one student had an SAT score about 350, two had scores a bit over 350, one student scored about 625, and so on), you would find the mean SAT scores for those students to be 487.

2. If you were to take many different random samples of 20 from this population, describe what you think the sampling distribution of these sample means would look like.

The sampling distribution of these samples may be centered on a value close to 487. The spread may be similar to the example above, possibly from 350 to 600.

3. Everyone in Jason's class drew several random samples of size 20 and found the mean SAT score. The plot below displays the distribution of the mean SAT scores for their samples.

Random Sample from District SAT Scores

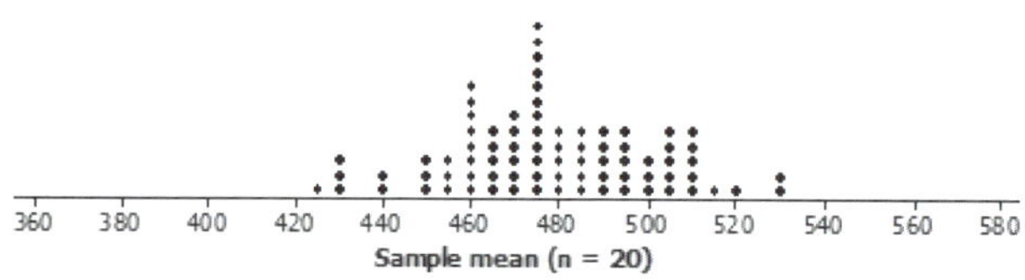

a. How does the simulated sampling distribution compare to your conjecture in Exercise 2? Explain any differences.

I estimated the mean SAT score to be a bit higher than it seems to be from the simulated distribution, but my estimate was not that far off. I thought the spread would range from 350 to 600 like in Jason's sample, but that was not a good estimate. I was thinking of the individual SAT scores in the sample and mixing them up with the means of samples of 20 scores, which is why my estimate was off.

MP.5

b. Use technology to generate many more samples of size 20, and plot the means of those samples. Describe the shape of the simulated distribution of sample mean SAT scores.

Answers will vary. The following displays an example of a simulated distribution of sample means from random samples from district SAT scores:

The distribution is normal. The mean of the sample mean SAT scores appears to be around 480, and the sample mean scores range from a little over 420 to about 530.

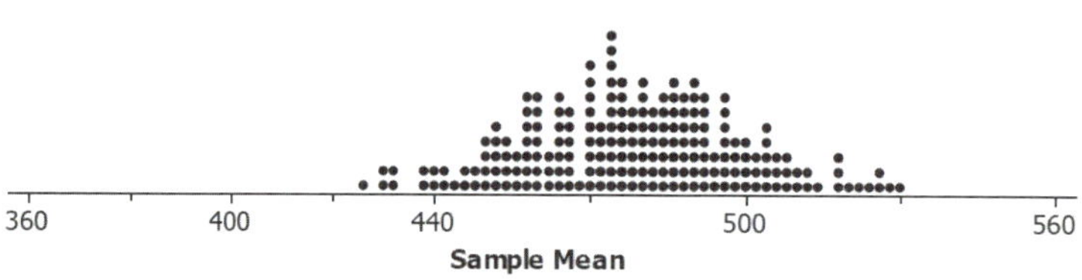

c. How did the simulated distribution using more samples compare to the one you generated in Exercise 3?

The maxima and minima were nearly the same in both of the distributions, 420 to 530 and a little over 420 to 520. The mean SAT scores of the simulated distributions of sample means in both seemed to be about 480.

d. What are the mean and standard deviation of the simulated distribution of the sample mean SAT scores you found in part (b)? (Use technology and your simulated distribution of the sample means to find the values.)

The mean SAT score of the simulated distribution of sample means is 478. The standard deviation is 21.5.

e. Write a sentence describing the distribution of sample means that uses the mean and standard deviation you calculated in part (d).

Almost all of the SAT scores are within two standard deviations from the mean, from 435 to 521.

4. Reflect on some of the simulated sampling distributions you have considered in previous lessons.

a. Make a conjecture about how you think the size of the sample might affect the distribution of the sample SAT means.

The larger the sample size, the smaller the spread of the distribution of sample means.

MP.5

b. To test the conjecture, investigate the sample sizes 5, 10, 40, 50, and the simulated distribution of sample means from Exercise 3. Divide the sample sizes among your group members, and use technology to simulate sampling distributions of mean SAT scores for samples of the different sizes. Find the mean and standard deviation of each simulated sampling distribution.

The following represent simulated distributions of sample means of district SAT scores for different size random samples:

Mean 469; standard deviation is 45.4.

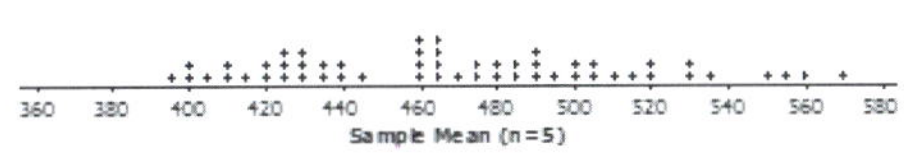

Mean 468; standard deviation is 32.9.

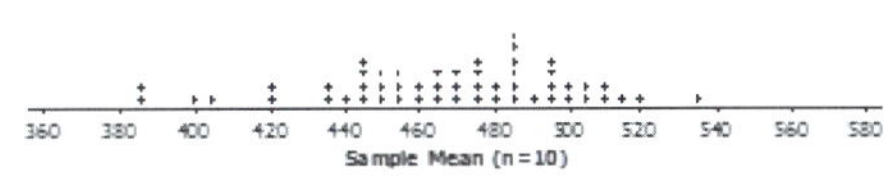

Mean 478; standard deviation is 22.8.

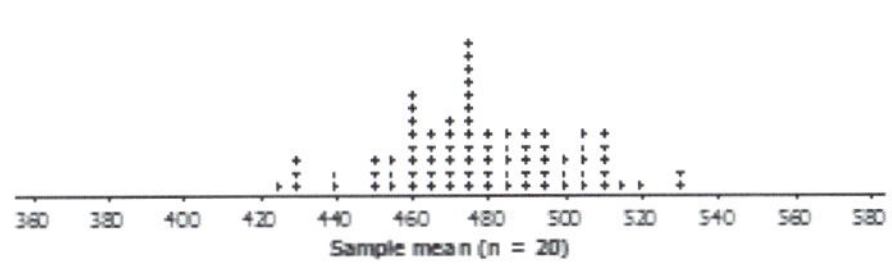

Mean 475; standard deviation is 16.1.

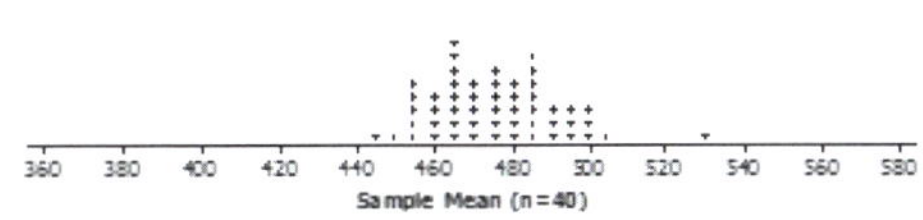

MP.5

Mean 476; standard deviation is 13.8.

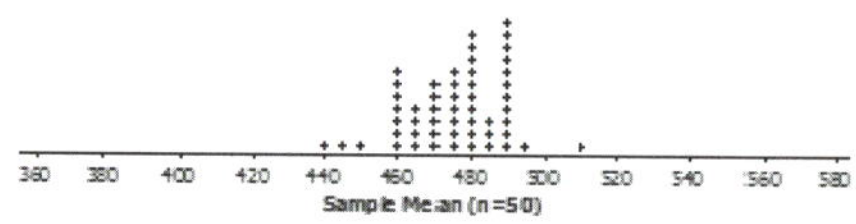

c. How does the sample size seem to affect the simulated distributions of the sample SAT mean scores? Include the simulated distribution from part (b) of Exercise 3 in your response. Why do you think this is true?

As the sample size increases, the spread decreases. The standard deviation went from 40 for a sample of size 5 to about 14 for a sample of size 50. The means of the sampling distributions of mean SAT scores varied from 468 for the distribution for samples of size 10 to 478 for samples of size 20. I would expect that a bigger sample would be more likely to look a lot like the population, and so bigger samples would not tend to be as different from one another as smaller samples. Because of this, the sample means would not differ as much from sample to sample for bigger samples.

5.

a. For each of the sample sizes, consider how the standard deviation seems to be related to the range of the sample means in the simulated distributions of the sample SAT means you found in Exercise 4.

In each case, nearly all of the sample means are within two standard deviations of the mean or are a normal distribution.

b. How do your answers to part (a) compare to the answers from other groups?

Everyone had simulated sampling distributions that looked fairly alike and were centered in about the same place. The sample means for each sampling distribution were typically within two standard deviations of the mean of the simulated sampling distributions. The distributions were normal.

6.

a. Make a graph of the distribution of the population consisting of the SAT scores for all of the students.

Possible response below of a distribution of the SAT scores for district students:

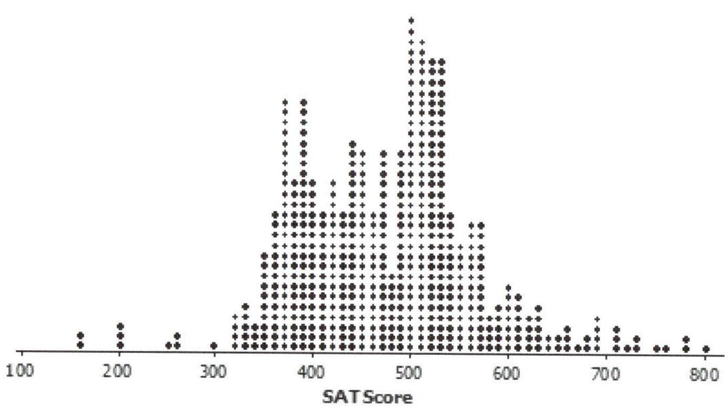

b. **Find the mean of the distribution of SAT scores. How does it compare to the mean of the sampling distributions you have been simulating?**

The mean SAT score for the students in the district is 475.1 *or* 475*. This is close to the means of the sampling distributions, even for fairly small samples.*

Closing (5minutes)

- Why do we call the sampling distributions we generated *simulated* sampling distribution of sample means rather than the *sampling distribution of sample means*?
  - *The sampling distribution of sample means is the distribution of all the possible sample means for a sample of a given size (i.e., every possible combination of the population values for that size). The simulated sampling distribution is a subset of the actual sampling distribution that, because it is random, approximates the actual sampling distribution.*
- If you had a simulated distribution of the mean SAT scores for random samples of size 100, how do you think the distribution would compare to the distribution you found for samples of size 50?
  - *The mean would be somewhere around* 475*, and the standard deviation would be smaller, so the values (the sample means) would be closer together.*

Ask students to summarize the main ideas of the lesson in writing or with a neighbor. Use this opportunity to informally assess comprehension of the lesson. The Lesson Summary below offers some important ideas that should be included.

Lesson Summary

For a given sample, you can find the sample mean.

- **There is variability in the sample mean. The value of the sample mean varies from one random sample to another.**
- **A graph of the distribution of sample means from many different random samples is a simulated sampling distribution.**
- **Sample means from random samples tend to cluster around the value of the population mean. That is, the simulated sampling distribution of the sample mean will be centered close to the value of the population mean.**
- **The variability in the sample mean decreases as the sample size increases.**
- **Most sample means are within two standard deviations of the mean of the simulated sampling distribution.**

Exit Ticket (5 minutes)

Name ______________________________ Date ______________

Lesson 19: Sampling Variability in the Sample Mean

Exit Ticket

1. Describe the difference between a population distribution, a sample distribution, and a simulated sampling distribution, and make clear how they are different.

2. Use the standard deviation and mean of the sampling distribution to describe an interval that includes most of the sample means.

Exit Ticket Sample Solutions

1. Describe the difference between a population distribution, a sample distribution, and a simulated sampling distribution, and make clear how they are different.

   The distribution of the elements in a population (the SAT scores for students in a district) is a population distribution; the distribution of the elements in a random sample from that population (a subset of a given size chosen at random from the SAT scores) is a sample distribution; a simulated distribution of sample means for many random samples of a given size chosen from the population (the means of different random samples of the same size of students' SAT scores) is a simulated sampling distribution.

   Some students might also suggest that the meaning of sampling distribution of all samples is the samples of a given size selected from a population. This would be the distribution of the means of every possible sample that might be chosen.

2. Use the standard deviation and mean of the sampling distribution to describe an interval that includes most of the sample means.

   Sample response: Typically, most of the means of the different random samples of the same size chosen from a population will be within two standard deviations of the mean or the mean ± 2 standard deviations.

Problem Set Sample Solutions

1. Which of the following will have the smallest standard deviation? Explain your reasoning.

   A sampling distribution of sample means for samples of size:

   a. 15 b. 25 c. 100

   The largest sample size, 100, will have the smallest standard deviation because as the sample size increases, the variability in the sample mean decreases.

2. In light of the distributions of sample means you have investigated in the lesson, comment on the statements below for random samples of size 20 chosen from the district SAT scores.

   a. Josh claimed he took a random sample of size 20 and had a sample mean score of 320.

      A mean score of 320 seems very unlikely. None of the samples we have investigated had a sample mean score that low.

   b. Sarfina stated she took a random sample of size 20 and had a sample mean of 520.

      This seems plausible for the simulated distributions of sample mean scores; 520 was high, but still some of the random samples had mean scores greater than 520.

   c. Ana announced that it would be pretty rare for the mean SAT score in a random sample to be more than three standard deviations from the mean SAT score of 475.

      Given that the sample means in nearly all of the simulated distributions of the sample means were usually within two standard deviations from the mean, Ana is correct. It could happen, but it would not be usual.

3. Refer to your answers for Exercise 4, and then comment on each of the following:

   a. A random sample of size 50 produced a mean SAT score of 400.

      A mean score of 400 was less than any of the sample means in the simulated sampling distribution of sample means for samples of size 50, so this seems unlikely.

   b. A random sample of size 10 produced a mean SAT score of 400.

      A mean score of 400 was within two standard deviations of the mean for random samples of size 10, so it could have come from one of the samples.

   c. For what sample sizes was a sample mean SAT score of 420 plausible? Explain your thinking.

      A sample mean of 420 occurred in the simulated sampling distributions for samples of size 5 and 10 but not at all in the simulated distributions for samples of size 20, 40, and 50. So, it seems like 420 was a plausible outcome for samples of size 5 and 10.

4. Explain the difference between the sample mean and the mean of the sampling distribution.

   Each sample of SAT scores had a mean SAT score, which is the sample mean. Then, all of those sample means formed a distribution of sample means, and we found the mean of that set, the mean of the sampling distribution of the sample mean—the mean of the means of the different samples.

Lesson 20: Margin of Error When Estimating a Population Mean

Student Outcomes

- Students use data from a random sample to estimate a population mean.
- Students calculate and interpret margin of error in context.
- Students know the relationship between sample size and margin of error in the context of estimating a population mean.

Lesson Notes

Lessons 16 and 17 introduced the concept of margin of error in the context of estimating a population proportion. The concept of *margin of error* may have been difficult to grasp for those students who see the word *error* and think "mistake." Lessons 16 and 17 showed that margin of error is interpreted as the farthest away from the value of the population proportion that an estimate is likely to be. The margin of error was also used to calculate an interval of plausible values for the population proportion.

In this lesson, margin of error is first developed visually and then estimated by twice the standard deviation of the sampling distribution of the sample proportion. This, and the next lesson, develops the idea of the margin of error when sample data are used to estimate a population mean.

Classwork

Example 1 (5 minutes): Describing a Population of Numerical Data

Provide each student the page of 100 numbered rectangles located at the end of this teacher lesson. For this example, let students work in pairs to answer the questions. Use whole-class discussion to develop those answers.

Example 1: Describing a Population of Numerical Data

The course project in a computer science class was to create 100 computer games of various levels of difficulty that had ratings on a scale from 1 (easy) to 20 (difficult). We will examine a representation of the data resulting from this project. Working in pairs, your teacher will give you a page that contains 100 rectangles of various sizes.

a. **What do you think the rectangles represent in the context of the 100 computer games?**

Each rectangle represents a computer game.

b. **What do you think the sizes of the rectangles represent in the context of the 100 computer games?**

The size of the rectangle, or the number of squares that comprise it, represents the difficulty rating of the computer game. The minimum rating is 1; the maximum is 20.

c. Why do you think the rectangles are numbered from 00 to 99 instead of from 1 to 100?

Anticipating that a random sample will be taken later in the lesson, it is easiest if all the labels have the same number of digits. So, 100 is conveniently designated as 00. The integers from 1 to 9 are represented by 01, 02, and so on.

Exploratory Challenge 1/Exercises 1–3 (5 minutes): Estimate the Population Mean Rating

Let students work with their partners to answer the questions.

Exploratory Challenge 1/Exercises 1–3: Estimate the Population Mean Rating

1. Working with your partner, discuss how you would calculate the mean rating of all 100 computer games (the population mean).

   To find the population mean, all 100 ratings would have to be added and then divided by 100. This is not hard, but it can be a tedious calculation to make. If students do not think adding 100 numbers is too bad, suggest that the number of computer games might have been $1,000$.

2. Discuss how you might select a random sample to estimate the population mean rating of all 100 computer games.

   A good answer would include stating a reasonable sample size, e.g., 10 or more. It should also state that a random number table or a calculator with a random number generator should be used to generate the 10 random two-digit numbers. The generated numbers identify the rectangles (computer games) that would be chosen for the sample.

3. Calculate an estimate of the population mean rating of all 100 computer games based on a random sample of size 10. Your estimate is called a *sample mean,* and it is denoted by $\bar{x}$. Use the following random numbers to select your sample:

   $$34\ \ 86\ \ 80\ \ 58\ \ 04\ \ 43\ \ 96\ \ 29\ \ 44\ \ 51$$

   The respective ratings for the given random numbers are $12, 5, 2, 4, 1, 4, 18, 10, 1$, and 16. Based on this sample, the estimate for the population mean rating is $\frac{73}{10}$, or 7.3.

Exploratory Challenge 2/Exercises 4–6 (10 minutes): Build a Distribution of Sample Means

Let students work with their partners to generate four sets of random numbers. Prepare a number line for the class to post their sample means. Be sure to provide enough room on the number line so the sample means do not overlap.

Exploratory Challenge 2/Exercises 4–6: Build a Distribution of Sample Means

MP.5

4. Work in pairs. Using a table of random digits or a calculator with a random number generator, generate four sets of ten random numbers. Use these sets of random numbers to identify four random samples of size 10. Calculate the sample mean rating for each of your four random samples.

   Answers will vary. To build a distribution, you should have 50 or more randomly generated sample mean estimates. So, if you have between 25 and 30 students in your class, then 12 to 15 pairs should generate about four sample mean estimates, which will provide the right number of estimates for the distribution.

> **5. Write your sample means on separate sticky notes, and post them on a number line that your teacher has prepared for your class.**

Be sure that there is enough length on the number line so that the sticky notes do not overlap.

> **6. The actual population mean rating of all 100 computer games is 7.5. Does your class distribution of sample means center at 7.5? Discuss why it does or does not.**
>
> *A possible reason why their sample means do not center at 7.5 is that they need more samples. They may suggest that they were very unlucky and got a distribution that centered well above or well below 7.5. That is possible, but it is highly unlikely.*
>
> *Note: There is a theoretical result that says, for random samples, the expected value of the sample mean is the mean value of the population from which the sample was taken. But that theory is not part of the curriculum. However, students may reason that if a population is divided into samples of equal size, then the mean of sample means is the same as the mean of the whole. They might give an example, such as the following: "Consider four samples each of size three (e.g., $4, 1, 3$; $2, 2, 7$; $5, 9, 6$; $3, 6, 5$). The respective means are $\frac{8}{3}, \frac{11}{3}, \frac{20}{3}$, and $\frac{14}{3}$ whose mean is their sum divided by 4, precisely the same as the sum of the 12 values divided by 12." Students might then argue that this would also apply to random samples and then go on to try to demonstrate conceptually that taking means produces values that tend to congregate or balance around the population mean.*

Example 2 (5 minutes): Margin of Error

This example has students visualize the concept of margin of error. Using the dot plot, they (roughly) determine the number of rating points within which almost all the sample means fall (i.e., within that number of points from the population mean 7.5).

This example should also clarify the meaning of the word *error* in the phrase *margin of error* insofar as the word does not imply mistake but refers to estimation error (i.e., the error that is made when a sample is used to estimate a population value).

Read through the example as a class, and convey the following:

- Almost all of the sample means are between 4 and 11. That is, almost all are roughly within 3.5 rating points of the population mean 7.5. The value 3.5 is a visual estimate of the *margin of error*. Highlight this description with students.
- Although not a formal definition of margin of error, it is a visual representation that motivates students' understanding of the term.
- It is not really an error in the sense of "mistake." Rather, it is how far the estimate for the population mean is likely to be from the actual value of the population mean.

Pose the question presented in the text to the class.

- Based on the class distribution of sample means, is the visual estimate of margin of error close to 3.5?
  - *Discuss this question using the distribution of the sample means from the class. It is anticipated that the margin of error is also close to 3.5. If it is very different, examine the results, and possibly examine what values seemed to make this estimate different.*

Example 2: Margin of Error

Suppose that 50 random samples, each of size ten, produced the sample means displayed in the following dot plot.

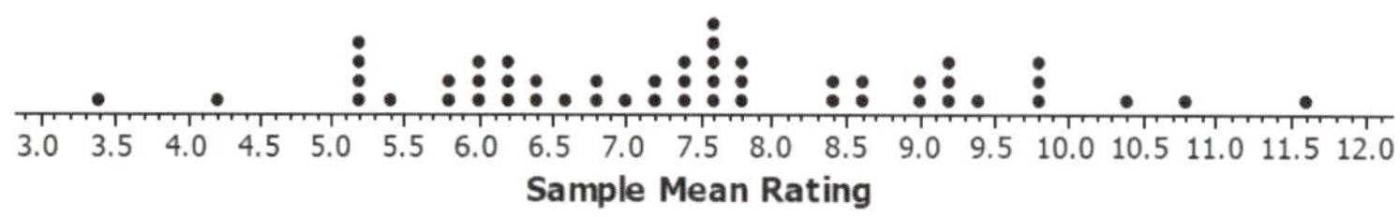

Note that almost all of the sample means are between 4 and 11. That is, almost all are roughly within 3.5 rating points of the population mean 7.5. The value 3.5 is a visual estimate of the margin of error. It is not really an "error" in the sense of a "mistake." Rather, it is how far our estimate for the population mean is likely to be from the actual value of the population mean.

Based on the class distribution of sample means, is the visual estimate of margin of error close to 3.5?

Example 3 (5 minutes): Standard Deviation as a Refinement of Margin of Error

This example refines the concept of margin of error by using the *standard deviation* as the measure of spread. Students need to calculate the standard deviation of the distribution of sample means and should note that twice the standard deviation is close to their visual estimate of margin of error.

Students may ask where the doubling came from. Remind them of their lesson on the normal distribution. The standard deviation of the sample mean variable is called *standard error*. (A formula for standard error follows in the next lesson.) The normal distribution of sample means has 95% of the sample means within *two* standard deviations of the population mean.

Suppose that the margin of error is 3.5. The interpretation of this is that plausible values for the population mean rating are within 3.5 points from their mean estimate of 7.5 points (i.e., from 4 to 11 rating points). Discuss these concepts with students. The following paragraphs summarize these concepts and should provide students with an explanation of margin of error they can use moving forward.

Example 3: Standard Deviation as a Refinement of Margin of Error

Note that the margin of error is measuring how spread out the sample means are relative to the value of the actual population mean. From previous lessons, you know that the standard deviation is a good measure of spread. So, rather than producing a visual estimate for the margin of error from the distribution of sample means, another approach is to use the standard deviation of the sample means as a measure of spread. For example, the standard deviation of the 50 sample means in the example above is 1.7. Note that if you double 1.7, you get a value for margin of error close to the visual estimate of 3.5.

Another way to estimate margin of error is to use two times the standard deviation of a distribution of sample means. For the above example, because $2(1.7) = 3.4$, the refined margin of error (based on the standard deviation of sample means) is 3.4 rating points.

An interpretation of the margin of error is that plausible values for the population mean rating are from $7.5 - 3.4$ to $7.5 + 3.4$ (i.e., 4.1 to 10.9 rating points).

Exploratory Challenge 3/Exercise 7 (8 minutes)

MP.2 & MP.5

It may be somewhat tedious for students to enter approximately 50 numbers into their calculators, but it goes faster if they work in pairs, with one student reading the entries while the other enters them in a calculator. Provide students an opportunity to summarize the standard deviation of the class distribution and the interpretation of it as a margin of error.

> **Exploratory Challenge 3/Exercise 7**
>
> 7. **Calculate and interpret the margin of error for your estimate of the population mean rating of 100 computer games based on the standard deviation of your class distribution of sample means.**
>
> *Answers will vary based on the class distribution of sample means. Sample response: Suppose the sample mean is 7.2 and the standard deviation is 1.9. Because $2(1.9) = 3.8$, the margin of error is 3.8 rating points. This means that the plausible values for the population mean rating are from $7.2 - 3.8$ to $7.2 + 3.8$, or from 3.4 to 11 rating points.*

Closing (2 minutes)

Ask students to summarize the main ideas of the lesson in writing or with a neighbor. Use this opportunity to informally assess comprehension of the lesson. The Lesson Summary below offers some important ideas that should be included.

> **Lesson Summary**
>
> **This lesson revisited margin of error. Previously, you estimated a population proportion of successes and described the accuracy of the estimate by its margin of error. This lesson also focused on margin of error but in the context of estimating the mean of a population of numerical data.**
>
> **Margin of error was estimated in two ways.**
>
> - **The first was through a visual estimation in which you judged the amount of spread in the distribution of sample means.**
> - **The second was more formalized by defining margin of error as twice the standard deviation of the distribution of sample means.**

Exit Ticket (5 minutes)

Name ______________________________ Date ______________

Lesson 20: Margin of Error When Estimating a Population Mean

Exit Ticket

At the beginning of the school year, school districts implemented a new physical fitness program. A student project involves monitoring how long it takes eleventh graders to run a mile. The following data were taken midyear.

a. What is the estimate of the population mean time it currently takes eleventh graders to run a mile based on the following data (minutes) from a random sample of ten students?

$$6.5, 8.4, 8.1, 6.8, 8.4, 7.7, 9.1, 7.1, 9.4, 7.5$$

b. The students doing the project collected 50 random samples of 10 students each and calculated the sample means. The standard deviation of their distribution of 50 sample means was 0.6 minutes. Based on this standard deviation, what is the margin of error for their sample mean estimate? Explain your answer.

c. Interpret the margin of error you found in part (b) in the context of this problem.

Exit Ticket Sample Solutions

At the beginning of the school year, school districts implemented a new physical fitness program. A student project involves monitoring how long it takes eleventh graders to run a mile. The following data were taken midyear.

a. What is the estimate of the population mean time it currently takes eleventh graders to run a mile based on the following data (minutes) from a random sample of ten students?

$$6.5, 8.4, 8.1, 6.8, 8.4, 7.7, 9.1, 7.1, 9.4, 7.5$$

The mean of the ten times is 7.9 minutes.

b. The students doing the project collected 50 random samples of 10 students each and calculated the sample means. The standard deviation of their distribution of 50 sample means was 0.6 minutes. Based on this standard deviation, what is the margin of error for their sample mean estimate? Explain your answer.

The margin of error is twice the standard deviation of the sampling distribution.

Since $2(0.6) = 1.2$, the margin of error is 1.2 minutes.

c. Interpret the margin of error you found in part (b) in the context of this problem.

Because $7.9 - 1.2 = 6.7$ and $7.9 + 1.2 = 9.1$, plausible values for the population mean time it takes eleventh graders to run the mile midyear are between 6.7 to 9.1 minutes.

Problem Set Sample Solutions

1. Suppose you are interested in knowing how many text messages eleventh graders send daily.

   Describe the steps that you would take to estimate the mean number of text messages per day sent by all eleventh graders at a school.

   If I could not get responses from all eleventh graders, I would base my estimate on the responses from a random sample of students. I would need to find a record or list of all eleventh graders. If I had this list, I would number all of the students on it and use random numbers to generate a random selection of students. For example, if there are 450 students, I would number all of the students on the list from 1 to 450 and generate a selection of students using the random number generator on my calculator. If my sample is 10 students, I would generate 10 random numbers from 1 to 450, identify the 10 students based on the random numbers, and ask the 10 students how many text messages they send during a school day. I would then find the mean of those 10 responses. The mean from this sample of students would be my estimate of the mean number of text messages sent by eleventh graders.

2. Suppose that 62 random samples based on ten student responses to the question, "How many text messages do you send per day?" resulted in the 62 sample means (rounded) shown below.

65	68	76	76	78	82	83	83	85	86	87	88	88
88	89	89	89	90	91	91	91	91	92	92	92	92
92	93	93	93	93	93	94	94	94	94	94	94	95
95	95	95	95	95	95	95	96	96	97	97	97	98
98	98	98	98	99	100	100	101	104	106			

a. Draw a dot plot for the distribution of sample means.

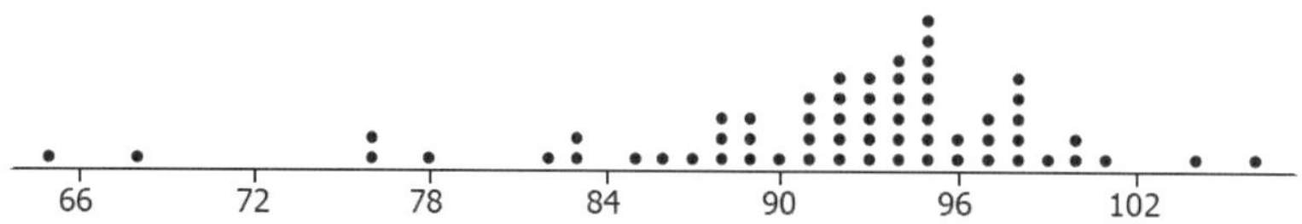

b. Based on your dot plot, would you be surprised if the actual mean number of text messages sent per day for all eleventh graders in the school is 91.7? Why or why not?

No. The distribution appears to be balanced around 92, so 91.7 is plausible.

3. Determine a visual estimate of the margin of error when a random sample of size 10 is used to estimate the population mean number of text messages sent per day.

Almost all sample means are roughly within 10 text messages of the population mean 91.7. So, visually, the margin of error is 10 text messages on average.

4. The standard deviation of the above distribution of sample mean number of text messages sent per day is 7.5. Use this to calculate and interpret the margin of error for an estimate of the population mean number of text messages sent daily by eleventh graders (based on a random sample of size 10 from this population).

Using the standard deviation of the sampling distribution, since $2(7.5) = 15$, the margin of error is 15 text messages. Note that the visual estimate is quite a bit smaller than the one using the standard deviation. However, they are the same if the visual estimate were to include all of the sample means from 77 to 107.

Example 1: Describing a Population of Numerical Data

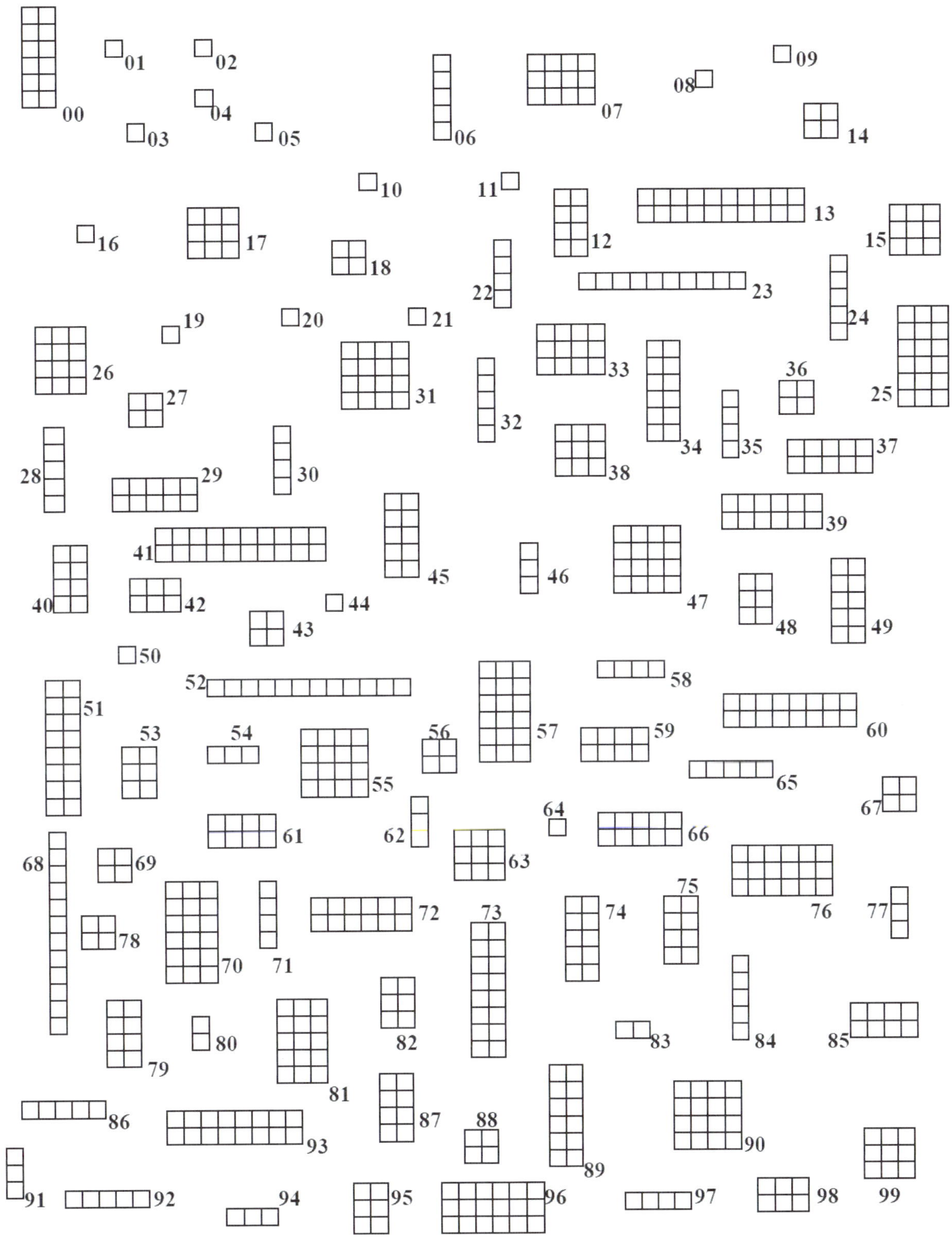

Lesson 21: Margin of Error When Estimating a Population Mean

Student Outcomes

- Students use data from a random sample to estimate a population mean.
- Students calculate and interpret margin of error in context.
- Students know the relationship between sample size and margin of error in the context of estimating a population mean.

Lesson Notes

In the previous lesson, students estimated the population mean using the sample mean based on a random sample of size n. To determine how accurate their estimate was, they had to create a sampling distribution of the sample mean based on computing sample means for a large number of random samples. Finally, they computed the margin of error as twice the standard deviation of the sample means. Although the process was a lot of work, students developed a conceptual understanding of margin of error.

In this lesson, students use a formula for the standard deviation of the sample mean, $\frac{s}{\sqrt{n}}$, where s is the standard deviation of the sample and n is the size of the sample. The margin of error, $2\left(\frac{s}{\sqrt{n}}\right)$, is based on a single random sample, thus making the work much easier.

The formula $\frac{s}{\sqrt{n}}$ is used to calculate the standard deviation of the sample mean when the mean and the standard deviation of the population are stated. Previously, the formula $\sqrt{\frac{p(1-p)}{n}}$ was used to calculate the standard deviation of a sample *proportion* when the number of successes was known. In both formulas, as n gets larger, the standard deviation gets smaller. Both methods are applications of the *central limit theorem*, which says that regardless of the shape of the population from which samples are taken, the distributions of both the sample means and the sample proportions are approximately normal.

Classwork

This lesson continues to discuss using the sample mean as an estimate of the population mean and judging its accuracy based on the concept of margin of error. In the last lesson, the margin of error was defined as twice the standard deviation of the sampling distribution of the sample mean. In this lesson, a formula will be given for the margin of error that allows you to calculate the margin of error from a single random sample rather than having to create a sampling distribution of sample means.

Example 1 (5 minutes): Estimating a Population Mean Using a Random Sample

MP.2 Give students a few minutes to read the introductory material of this example, and remind them of the process they used in the previous lesson to get an estimate of margin of error. Then, write the formula for margin of error on the board, making sure that students understand that this allows them to calculate an estimate of the margin of error using data from a single random sample.

Example 1: Estimating a Population Mean Using a Random Sample

Provide a one-sentence summary of our findings from the previous lesson.

We took lots of random samples of computer game ratings, computed their means, displayed the distribution of their means, and, finally, computed a margin of error.

What were drawbacks of the calculation method?

Many samples are required. If we had increased the sample size or the number of samples, the time required to take all those samples, calculate their means, and analyze the distribution would have increased significantly.

In practice, you do not have to use that process to find the margin of error. Fortunately, just as was the case with estimating a population proportion, there are some general results that lead to a formula that allows you to estimate the margin of error using a single sample. You can then gauge the accuracy of your estimate of the population mean by calculating the margin of error using the sample standard deviation.

> **The standard deviation of the distribution of sample means is approximated by $\frac{s}{\sqrt{n}}$, where s is the standard deviation of the sample and n is the size of the sample.**

Scaffolding:

- For struggling students, the bigger the value of n, the smaller the standard deviation. From a population where $s = 2$ if $n = 36$, the standard deviation is $\frac{1}{3}$; however, if the sample is larger, say 81, the standard deviation would be $\frac{2}{9}$.
- For advanced students, when a sample is taken from a population, the mean of the sample is the same as the mean of the population, but the *variance* (square of the standard deviation) is only $\frac{1}{n}$ as large. Regardless of the shape of the population, the distribution of the sample means approaches normal (central limit theorem). The variance is $\frac{s^2}{n}$; therefore, by applying a square root, the standard deviation is $\frac{s}{\sqrt{n}}$.

Exercises 1–5 (10 minutes)

Have students work independently on the calculations required to answer Exercises 1–3. Then, work through Exercises 4 and 5 as a class.

Exercises 1–5

1. **Suppose a random sample of size ten produced the following ratings in the computer games rating example in the last lesson: $12, 5, 2, 4, 1, 4, 18, 10, 1, 16$. Estimate the population mean rating based on these ten sampled ratings.**

   $\frac{73}{10} = 7.3$

   The sample mean estimate for the population mean rating is 7.3 rating points.

2. **Calculate the sample standard deviation. Round your answer to three decimal places.**

   The sample standard deviation is 6.2725 rating points.

3. **Use the formula given above to calculate the approximate standard deviation of the distribution of sample means. Round your answer to three decimal places.**

$\frac{s}{\sqrt{n}} = \frac{6.273}{\sqrt{10}} \approx 1.984$

The standard deviation of the distribution of sample means is 1.984 *rating points.*

4. **Recall that the margin of error is twice the standard deviation of the distribution of sample means. What is the value of the margin of error based on this sample? Write a sentence interpreting the value of the margin of error in the context of this problem on computer game ratings.**

$2(1.984) = 3.968$

The margin of error is 3.968 *rating points. The population mean rating for the* 100 *computer games is likely to be within* 3.968 *rating points of the sample mean estimate* 7.3*.*

5. **Based on the sample mean and the value of the margin of error, what is an interval of plausible values for the population mean?**

Because $7.3 - 3.968 = 3.332$ *and* $7.3 + 3.968 = 11.268$*, plausible values for the population mean rating are from* 3.332 *to* 11.268 *rating points.*

Exercises 6–13 (20 minutes): The Gettysburg Address

Distribute a copy of the Gettysburg Address to each student in the class. (A copy is provided at the end of this lesson.) Have students work individually or in pairs to answer the questions in this set of exercises. Then, discuss the answers to the last question as a class. Consider challenging students to find the length of a typical word in the Gettysburg Address.

After students do this exercise by hand (using a calculator), consider showing them an applet that displays three different estimates regarding the Gettysburg Address. One is the mean word length. The other two are estimating population proportions; one is the proportion of *long* words, defined as words with more than four letters, and the other is the proportion of nouns. The applet can be found at the following site: http://www.rossmanchance.com/applets/GettysburgSample/GettysburgSample.html.

This applet may require an updated version of an operating system to work correctly. If the applet does not work for all students due to a computer's operating system or network settings, attempt to demonstrate it for the whole class, as it is an effective way to complement how students obtained their answers in the exercises. The applet allows the user to specify a sample size (ten in this exercise) and the number of samples desired. Note that only *one* sample is to be used to answer the questions in this exercise set.

To generate a sampling distribution for the sample mean (or proportion), enter a large number in the Num samples box, such as 500. The Animate box shows the observations for each sample taken and the resulting values of the statistics (mean or proportion) plotted on a histogram. (Unclick the Animate box at any time to see the total results immediately.)

Students should begin work on Exercise 6. Exercises 7–13 are provided as scaffolding if necessary. Students should be able to clearly describe and fully implement a plan on their own. Sample responses are provided but will vary.

Exercises 6–13: The Gettysburg Address

The Gettysburg Address is considered one of history's greatest speeches. Some students noticed that the speech was very short (about 268 words, depending on the version) and wondered if the words were also relatively short. To estimate the mean length of words in the population of words in the Gettysburg Address, work with a partner on the following steps. Your teacher will give you a copy of the Gettysburg Address with words numbered from 001 to 268.

6. Develop and describe a plan for collecting data from the Gettysburg Address and determining the typical length of a word. Then, implement your plan, and report your findings.

Many answers are possible. Every answer should include the following:

- *A description of how a word sample is chosen, making sure to describe how randomization occurs*
- *The actual sample chosen*
- *Calculations of the sample mean, standard deviation, and margin of error*
- *Interpretations in context of the sample mean, standard deviation, and margin of error.*

MP.5

7. Use a random number table or a calculator with a random number generator to obtain ten different random numbers from 001 to 268.

Student answers will vary. Sample solutions to Exercises 8-12 will use the following values.
$219 \;\; 229 \;\; 2 \;\; 113 \;\; 77 \;\; 140 \;\; 185 \;\; 70 \;\; 119 \;\; 54$

8. Use the random numbers found in Exercise 7 as identification numbers for the words that will make up your random sample of words from the Gettysburg Address. Make a list of the ten words in your sample.

219	*the*	140	*The*
229	*resolve*	185	*nobly*
2	*score*	70	*a*
113	*consecrate*	119	*The*
77	*final*	54	*endure*

9. Count the number of letters in each of the ten words in your sample.

$3 \;\; 7 \;\; 5 \;\; 10 \;\; 5 \;\; 3 \;\; 5 \;\; 1 \;\; 3 \;\; 6$

10. Calculate the sample mean number of letters for the ten words in your sample.

$\frac{48}{10} = 4.8$

The mean of the ten word lengths from Exercise 9 is 4.8 letters.

11. Calculate the sample standard deviation of the number of letters for the ten words in your sample. Round your answer to three decimal places.

The standard deviation of the ten word lengths from Exercise 9 is 2.530 letters.

12. Use the sample standard deviation from Exercise 11 to calculate the margin of error associated with using your sample mean as an estimate of the population mean. Round your answer to three decimal places.

$2\left(\frac{2.530}{\sqrt{10}}\right) \approx 1.600$

The margin of error of this estimate is approximately 1.600 letters.

MP.2

13. Write a few sentences describing what you have learned about the mean length of the population of 268 words in the Gettysburg Address. Be sure to include an interpretation of the margin of error.

We estimate the mean word length of words in the Gettysburg Address to be 4.8 letters. The margin of error of this estimate is 1.6 letters. So, plausible values for the population mean word length are from 3.2 to 6.4 letters.

Closing (3 minutes)

Ask students to summarize the main ideas of the lesson in writing or with a neighbor. Use this as an opportunity to informally assess comprehension of the lesson. The Lesson Summary below offers some important ideas that should be included.

Lesson Summary

- **When using the sample mean to estimate a population mean, it is important to know something about how accurate that estimate might be.**
- **Accuracy can be described by the margin of error.**
- **The margin of error can be estimated using data from a single random sample (without the need to create a simulated sampling distribution) by using the formula $2\left(\frac{s}{\sqrt{n}}\right)$, where s is the standard deviation of a single sample and n is the sample size.**

Exit Ticket (7 minutes)

Name ______________________________ Date ______________

Lesson 21: Margin of Error When Estimating a Population Mean

Exit Ticket

A Health Group study recommends that the total weight of a male student's backpack should not be more than 15% of his body weight. For example, if a student weighs 170 pounds, his backpack should not weigh more than 25.5 pounds. Suppose that ten randomly selected eleventh-grade boys produced the following data:

Body Weight	155	136	197	174	165	165	150	142	176	157
Backpack Weight	29.8	27.2	32.5	34.8	31.8	28.8	31.1	26.0	28.3	31.4

a. For each student, calculate backpack weight as a percentage of body weight (round to one decimal place).

b. Based on the data in part (a), estimate the mean percentage of body weight that eleventh-grade boys carry in their backpacks.

c. Find the margin of error for your estimate of part (b). Round your answer to three decimal places. Explain how you determined your answer.

d. Comment on the amount of weight eleventh-grade boys at this school are carrying in their backpacks compared to the recommendation by the Health Group.

Exit Ticket Sample Solutions

A Health Group study recommends that the total weight of a male student's backpack should not be more than 15% of his body weight. For example, if a student weighs 170 pounds, his backpack should not weigh more than 25.5 pounds. Suppose that ten randomly selected eleventh-grade boys produced the following data:

Body Weight	155	136	197	174	165	165	150	142	176	157
Backpack Weight	29.8	27.2	32.5	34.8	31.8	28.8	31.1	26.0	28.3	31.4

a. For each student, calculate backpack weight as a percentage of body weight (round to one decimal place).

Body Weight	155	136	197	174	165	165	150	142	176	157
Backpack Weight	29.8	27.2	32.5	34.8	31.8	28.8	31.1	26.0	28.3	21.4
Percentage	19.2	20.0	16.5	20.0	19.3	17.5	20.7	18.3	16.1	13.6

b. Based on the data in part (a), estimate the mean percentage of body weight that eleventh-grade boys carry in their backpacks.

The sample mean percentage is 18.12 *percentage points.*

c. Find the margin of error for your estimate of part (b). Round your answer to three decimal places. Explain how you determined your answer.

$$2\left(\frac{2.207}{\sqrt{10}}\right) \approx 1.396$$

The standard deviation of the percentages is 2.207 *percent, so the margin of error is approximately* 1.396 *percentage points.*

d. Comment on the amount of weight eleventh-grade boys at this school are carrying in their backpacks compared to the recommendation by the Health Group.

$18.12\% - 1.396\% = 16.724\%$ *and* $18.12\% + 1.396\% = 19.516\%$

Based on the data in this study, plausible percentages of mean body weight percentage that eleventh graders are carrying in their backpacks are 16.724% *to* 19.516%*. The interval* $(16.724\%, 19.516\%)$ *is above the recommended* 15% *maximum. On average, eleventh-grade boys at this school are carrying too much weight in their backpacks.*

Problem Set Sample Solutions

1. A new brand of hot dog claims to have a lower sodium content than the leading brand.

a. A random sample of ten of these new hot dogs results in the following sodium measurements (in milligrams).

$370\ 326\ 322\ 297\ 326\ 289\ 293\ 264\ 327\ 331$

Estimate the population mean sodium content of this new brand of hot dog based on the ten sampled measurements.

Based on the data, an estimate for the population mean sodium content of this new brand of hot dog is 314.5 mg *of sodium.*

b. Calculate the margin of error associated with your estimate of the population mean from part (a). Round your answer to three decimal places.

$2\left(\frac{29.436}{\sqrt{10}}\right) \approx 18.617$

The margin of error is approximately 18.617 mg.

c. The mean sodium content of the leading brand of hot dogs is known to be 350 mg. Based on the sample mean and the value of the margin of error for the new brand, is a mean sodium content of 350 mg a plausible value for the mean sodium content of the new brand? Comment on whether you think the new brand of hot dog has a lower sodium content on average than the leading brand.

$314.5 \text{ mg} - 18.617 \text{ mg} = 295.883 \text{ mg}$ *and* $314.5 \text{ mg} + 18.617 \text{ mg} = 333.117 \text{ mg}$

Plausible values for population mean sodium content are between 295.883 mg *and* 333.117 mg. *This interval is well below* 350 mg, *which is the sodium content for the leading brand. So, the new hot dog brand has lower mean sodium content.*

d. Another random sample of 40 new-brand hot dogs is taken. Should this larger sample of hot dogs produce a more accurate estimate of the population mean sodium content than the sample of size 10? Explain your answer by appealing to the formula for margin of error.

The margin of error will be smaller. Sample size is in the denominator of the formula for margin of error.

2. It is well known that astronauts increase their height in space missions because of the lack of gravity. A question is whether or not we increase height here on Earth when we are put into a situation where the effect of gravity is minimized. In particular, do people grow taller when confined to a bed? A study was done in which the heights of six men were taken before and after they were confined to bed for three full days.

a. The before-after differences in height measurements (in millimeters) for the six men were

$$12.6 \quad 14.4 \quad 14.7 \quad 14.5 \quad 15.2 \quad 13.5.$$

Assuming that the men in this study are representative of the population of all men, what is an estimate of the population mean increase in height after three full days in bed?

Based on the given data, an estimate of the population mean increase in height after three full days in bed is 14.15 mm.

b. Calculate the margin of error associated with your estimate of the population mean from part (a). Round your answer to three decimal places.

$2\left(\frac{0.940}{\sqrt{6}}\right) \approx 0.768$

The margin of error is approximated by 0.768 mm.

c. Based on your sample mean and the margin of error from parts (a) and (b), what are plausible values for the population mean height increase for all men who stay in bed for three full days?

$14.15 \text{ mm} - 0.768 \text{ mm} = 13.382 \text{ mm}$ *and* $14.15 \text{ mm} + 0.768 \text{ mm} = 14.918 \text{ mm}$

Plausible values for the population mean height increase for all men who stay in bed for three full days are those between 13.382 mm *and* 14.918 mm.

Exercises 6–13: The Gettysburg Address

001 Four	045 any	089 nation	133 our	177 they	221 full	265 perish
002 score	046 nation,	090 might	134 poor	178 who	222 measure	266 from
003 and	047 so	091 live.	135 power	179 fought	223 of	267 the
004 seven	048 conceived	092 It	136 to	180 here	224 devotion,	268 earth.
005 years	049 and	093 is	137 add	181 have	225 that	
006 ago,	050 so	094 altogether	138 or	182 thus	226 we	
007 our	051 dedicated,	095 fitting	139 detract.	183 far	227 here	
008 fathers	052 can	096 and	140 The	184 so	228 highly	
009 brought	053 long	097 proper	141 world	185 nobly	229 resolve	
010 forth	054 endure.	098 that	142 will	186 advanced.	230 that	
011 upon	055 We	099 we	143 little	187 It	231 these	
012 this	056 are	100 should	144 note,	188 is	232 dead	
013 continent	057 met	101 do	145 nor	189 rather	233 shall	
014 a	058 on	102 this.	146 long	190 for	234 not	
015 new	059 a	103 But,	147 remember,	191 us	235 have	
016 nation;	060 great	104 in	148 what	192 to	236 died	
017 conceived	061 battlefield	105 a	149 we	193 be	237 in	
018 in	062 of	106 larger	150 say	194 here	238 vain,	
019 liberty,	063 that	107 sense,	151 here,	195 dedicated	239 that	
020 and	064 war.	108 we	152 but	196 to	240 this	
021 dedicated	065 We	109 cannot	153 it	197 the	241 nation,	
022 to	066 have	110 dedicate,	154 can	198 great	242 under	
023 the	067 come	111 we	155 never	199 task	243 God,	
024 proposition	068 to	112 cannot	156 forget	200 remaining	244 shall	
025 that	069 dedicate	113 consecrate,	157 what	201 before	245 have	
026 all	070 a	114 we	158 they	202 us,	246 a	
027 men	071 portion	115 cannot	159 did	203 that	247 new	
028 are	072 of	116 hallow	160 here.	204 from	248 birth	
029 created	073 that	117 this	161 It	205 these	249 of	
030 equal.	074 field	118 ground.	162 is	206 honored	250 freedom,	
031 Now	075 as	119 The	163 for	207 dead	251 and	
032 we	076 a	120 brave	164 us	208 we	252 that	
033 are	077 final	121 men,	165 the	209 take	253 government	
034 engaged	078 resting	122 living	166 living,	210 increased	254 of	
035 in	079 place	123 and	167 rather,	211 devotion	255 the	
036 a	080 for	124 dead,	168 to	212 to	256 people,	
037 great	081 those	125 who	169 be	213 that	257 by	
038 civil	082 who	126 struggled	170 dedicated	214 cause	258 the	
039 war,	083 here	127 here	171 here	215 for	259 people,	
040 testing	084 gave	128 have	172 to	216 which	260 for	
041 whether	085 their	129 consecrated	173 the	217 they	261 the	
042 that	086 lives	130 it,	174 unfinished	218 gave	262 people,	
043 nation,	087 that	131 far	175 work	219 the	263 shall	
044 or	088 that	132 above	176 which	220 last	264 not	

EUREKA MATH™

Lesson 22: Evaluating Reports Based on Data from a Sample

Student Outcomes

- Students interpret margin of error from reports that appear in newspapers and other media.
- Students critique and evaluate statements in published reports that involve estimating a population proportion or a population mean.

Lesson Notes

In this lesson, students read and comment on examples from the media (newspaper and Internet) that involve estimating a population proportion or a population mean. Students calculate the margin of error and compare their calculations with the published results. In addition, students interpret the margin of error in the context of the article and comment on how the survey was conducted.

Classwork

Exercises 1–5 (12 minutes): Election Results

Before starting these exercises, consider discussing the formulas for the margin of error for estimating both a population mean and a population proportion. In addition, review the importance of conducting a survey using a random sample.

Students should work in small groups (two or three students per group). Allow about 10 minutes for students to complete Exercises 1 through 5. Then, discuss answers as a class.

Exercises 1–5: Election Results

The following is part of an article that appeared in a newspaper:

> **With the election for governor still more than a year away, a new poll shows the race is already close. The Republican governor had 47%, and the Democratic challenger had 45% in a poll released Tuesday of 800 registered voters.**
>
> **"That's within the poll's margin of error of 3.5 percentage points, making it essentially a toss-up," said the poll's director.**

1. **Why don't the two percentages add up to 100%?**

   The other 8% might be undecided voters or voters who want to vote for a third-party candidate.

MP.2

2. **What is meant by the margin of error of 3.5 percentage points?**

   It is unlikely that the estimate of 47% of the proportion of all voters who would vote for the Republican candidate will be farther from the actual population proportion than the margin of error of 3.5%, or the proportion of all voters who would vote for the Republican candidate is likely between 43.5% and 50.5%.

Scaffolding:

- As students work to articulate critiques, it may be useful to have sentence frames or sentence stems to organize and begin their writing.
- In particular, if students struggle with interpreting the margin of error, display an example sentence frame on the classroom board for students to refer to during the lesson.
- Use the sample response from Exercise 2 as a guide, with ____ (empty spaces) for percentages.

3. Using the sample size of 800 and the proportion 0.47, calculate the margin of error associated with the estimate of the proportion of all registered voters who would vote for the Republican governor.

   0.035

4. Why did the poll director say that the election is "essentially a toss-up"?

   Even though the sample proportion who preferred the Republican governor is 0.47, this is just an estimate, and it might be too high by as much as 0.035. If this was the case, the Democratic challenger would win.

5. If the sample size had been $2,500$ registered voters, and the results stated 47% would vote for the Republican governor, and 45% said they would vote for the Democratic challenger, what would the margin of error have been? Could the director still say that the election was a toss-up?

   The margin of error would be 0.02. The director could still say the election was a toss-up.

Exercises 6–8 (10 minutes): Chocolate Chip Claim

In these exercises, students estimate a population mean by finding a margin of error. While introducing the scenario of the Nabisco Company claim, point out to students that this claim could be restated as, "Nabisco claims that on average there are at least 1,000 chocolate chips in every 18-ounce bag."

Students should work in small groups (two or three students per group). Allow about 10 minutes for students to complete Exercises 6 through 8. Then, discuss answers as a class.

Exercises 6–8: Chocolate Chip Claim

The Nabisco Company claims that there are at least $1,000$ chocolate chips in every 18-ounce bag of their Chips Ahoy! cookies. An article in a local newspaper reported the efforts of a group of students in their attempt to validate the Nabisco claim. The article reported that the students randomly selected 42 bags of cookies from local grocery stores and counted the number of chocolate chips in the cookies in each bag. The students found the sample mean was $1,261.6$ chips, and the sample standard deviation was 117.6 chips. The article stated that the students' data supported the Nabisco Company claim.

MP.2

6. Using the students' results, calculate the margin of error associated with the estimate of the mean number of chocolate chips in an 18-ounce bag of Chips Ahoy! chocolate chip cookies. Write a sentence interpreting the margin of error.

   The margin of error is 36.3. The population mean number of chocolate chips in an 18-ounce bag is likely between $1,225.3$ and $1,297.9$.

MP.3

7. Do you agree that the student data supported the Nabisco Company claim? Explain.

   Students should agree with the claim. The population mean number of chocolate chips is likely in the range 1261.6 ± 36.3, or between $1,225.3$ and $1,297.9$. This range is consistent with the claim of at least $1,000$ chocolate chips in each 18-ounce bag.

8. Comment on the procedure that the students used to collect their data.

   It was important that the 42 bags were randomly selected. Since they were all collected from local grocery stores, it might not be reasonable to extend the conclusion to beyond the local grocery stores.

Exercises 9–15 (15 minutes): Understanding a Poll

These exercises are similar to Exercises 1–5, estimating a population proportion. The difference is how the data are presented. In this example, students first read a graph and discuss key points in the graph. After this discussion, consider presenting to students how the Gallup organization conducts its polls. Show students the Gallup website (http://www.gallup.com). However, preview the site before suggesting students examine it, as there are topics involved in the polls that on occasion may not be desirable to discuss with the class. The site incorporates polls related to current issues, which are difficult to predict.

Students should work in small groups (two or three students per group). Allow approximately five minutes for students to complete Exercises 9–11. When students have finished, discuss the answers as a class. Then, direct students to read how Gallup conducted the poll, and have them complete Exercises 12–15. When students have finished, discuss answers as a class.

Scaffolding:

- If possible, it might be useful for English language learners to analyze examples that are in their L1 (first) language.
- Many websites have translated versions of reports available.

Exercises 9–15: Understanding a Poll

George Gallup founded the American Institute of Public Opinion (Gallup Poll) in 1935. The company is famous for its public opinion polls, which are conducted in the United States and other countries.

Gallup published a graph in May 2013 titled *Percent in U.S. Who Exercise for at Least 30 Minutes Three or More Days per Week*. Use the graph found on the Gallup website (http://www.gallup.com/poll/162194/americans-exercise-habits-worsen-slightly-2013.aspx) to answer the following questions:

9. **What percent of those surveyed said that they exercise at least 30 minutes three or more days a week at the start of 2013?**

   Approximately 48%

10. **Describe the patterns that you observe in the graph.**

    The percent of adults who say they exercise at least 30 *minutes three or more days a week follows an up-and-down cycle; at the beginning of the year, the percent is low; by midyear, the percent rises to its highest; by the end of the year, the percent goes back down.*

11. **Give some reasons why you think the graph follows the pattern that you described.**

    The pattern seems to be following the seasons of the year. During the winter, the percent is low; during the summer, the percent reaches its peak.

Following are the survey methods that Gallup used to collect the data:

> **Results are based on telephone interviews conducted as part of the Gallup-Healthways Well-Being Index survey June 1–30, 2013, with a random sample of $15,235$ adults, aged 18 and older, living in all 50 U.S. states and the District of Columbia.**
>
> **For results based on the total sample of national adults, one can say with 95% confidence that the maximum margin of sampling error is ± 1 percentage point.**

12. **Using the value of 0.538 for the proportion of those surveyed who said they exercise at least 30 minutes three or more days a week in the most recent poll, calculate the margin of error. How does your margin of error compare to the value reported by Gallup?**

    The margin of error is 0.008*, which rounds up to* 0.01*.*

13. **Interpret the phrase "margin of sampling error is ± 1 percentage point."**

 It is very likely that the proportion of all adults who said they exercise at least 30 minutes three or more days a week is within 0.01 of the sample proportion 0.538.

14. **Why is it important that Gallup selects a random sample of adults?**

 It helps to ensure that the sample taken is representative of the population.

15. **If Gallup had used a random sample of $1,500$, what would happen to the margin of error? Explain your answer.**

 The margin of error would increase to 0.026.

$$ME = 2\sqrt{\frac{0.538(1-0.538)}{1500}}$$

Closing (3 minutes)

- Refer back to the first exercise set about the election results. What do you think would be an appropriate headline for the story?
  - *Race for Governor in a Dead Heat; Gubernatorial Race Too Close to Call*

Ask students to summarize the main ideas of the lesson in writing or with a neighbor. Use this as an opportunity to informally assess comprehension of the lesson. The Lesson Summary below offers some important ideas that should be included.

Lesson Summary

- **The estimated margin of error when a sample proportion from a random sample is used to estimate a population proportion is $ME = 2\sqrt{\frac{\hat{p}(1-\hat{p})}{n}}$, where $\hat{p}$ is the sample proportion.**
- **The estimated margin of error when a sample mean from a random sample is used to estimate a population mean is $ME = 2\left(\frac{s}{\sqrt{n}}\right)$, where s is the sample mean.**
- **It is important to interpret margin of error in context.**
- **It is unlikely that the estimate of a population proportion or mean will be farther from the actual population value than the margin of error.**

Exit Ticket (5 minutes)

Name ______________________________ Date ______________

Lesson 22: Evaluating Reports Based on Data from a Sample

Exit Ticket

The Gallup organization published the following results from a poll that it conducted.

> As health experts increasingly focus on the medical benefits of a healthy lifestyle and preventative healthcare, Americans say their doctor does commonly discuss the benefits of healthy habits with them. Specifically, 71% say their doctor usually discusses the benefits of engaging in regular physical exercise, and 66% say their doctor usually discusses the benefits of eating a healthy diet. Fewer Americans, 50%, say their doctor usually discusses the benefits of not smoking, although that number jumps to 79% among smokers.
>
> **Survey Methods**
>
> Results for this Gallup poll are based on telephone interviews conducted July 10–14, 2013, with a random sample of 2,027 adults, aged 18 and older, living in all 50 U.S. states and the District of Columbia.
>
> For results based on the total sample of national adults, one can say with 95% confidence that the margin of sampling error is ± 3 percentage points.

Source: http://www.gallup.com/poll/163772/americans-say-doctors-advise-health-habits.aspx

1. The headline of the article is "Smokers Much More Likely Than Nonsmokers to Say Doctor Discusses Not Smoking." Do you agree with this headline? Explain your answer.

2. Using the data "71% say their doctor usually discusses the benefits of engaging in regular physical exercise," calculate the margin of error. Show your work.

3. How do your results compare with the margin of error stated in the article?

4. Interpret the margin of error in this context.

Exit Ticket Sample Solutions

The Gallup organization published the following results from a poll that it conducted.

> As health experts increasingly focus on the medical benefits of a healthy lifestyle and preventative healthcare, Americans say their doctor does commonly discuss the benefits of healthy habits with them. Specifically, 71% say their doctor usually discusses the benefits of engaging in regular physical exercise, and 66% say their doctor usually discusses the benefits of eating a healthy diet. Fewer Americans, 50%, say their doctor usually discusses the benefits of not smoking, although that number jumps to 79% among smokers.
>
> Survey Methods
>
> Results for this Gallup poll are based on telephone interviews conducted July 10–14, 2013, with a random sample of $2,027$ adults, aged 18 and older, living in all 50 U.S. states and the District of Columbia.
>
> For results based on the total sample of national adults, one can say with 95% confidence that the margin of sampling error is ± 3 percentage points.

Source: http://www.gallup.com/poll/163772/americans-say-doctors-advise-health-habits.aspx

1. The headline of the article is "Smokers Much More Likely Than Nonsmokers to Say Doctor Discusses Not Smoking." Do you agree with this headline? Explain your answer.

   Yes, I agree with this headline because 79% of smokers say their doctor usually discusses the benefits of nonsmoking, while less than 50% of nonsmokers say their doctor discusses not smoking. The margin of error associated with these estimates is only 3%.

2. Using the data "71% say their doctor usually discusses the benefits of engaging in regular physical exercise," calculate the margin of error. Show your work.

   $$ME = 2\sqrt{\frac{(0.71)(0.29)}{2027}} = 0.02$$

3. How do your results compare with the margin of error stated in the article?

   This value is lower than the stated value of 0.03.

4. Interpret the margin of error in this context.

   It is very likely that the proportion of all adults who would say that their doctor usually discusses the benefits of engaging in physical exercise is within 3% of the sample proportion 0.71.

Problem Set Sample Solutions

1. The *British Medical Journal* published a study whose objective was to investigate estimation of calorie content of meals from fast food restaurants. Below are the published results.

   Participants: $1,877$ adults and 330 school-age children visiting restaurants at dinnertime (evening meal) in 2010 and 2011; $1,178$ adolescents visiting restaurants after school or at lunchtime in 2010 and 2011

   Results: Among adults, adolescents, and school-age children, the mean actual calorie content of meals was 836 calories (SD 465), 756 calories (SD 455), and 733 calories (SD 359), respectively. Compared with the actual figures, participants underestimated calorie content by means of 175 calories, 259 calories, and 175 calories, respectively.

   Source: http://www.bmj.com/content/346/bmj.f2907

   a. Calculate the margin of error associated with the estimate of the mean number of actual calories in the meals eaten by each of the groups: adults, adolescents, and school-age children.

      Margin of error for adults: 21.47

      Margin of error for adolescents: 26.51

      Margin of error for school-age children: 39.52

   b. Write a sentence interpreting the margin of error for the adult group.

      It is very likely that the sample mean of 836 calories is within 21.47 of the actual mean number of calories in meals eaten by all adults.

   c. Explain why the margin of error for the estimate of the mean number of actual calories in meals eaten by adults is smaller than the margin of error of the mean number of actual calories in meals eaten by school-age children.

      The size of the sample for the school-age children is considerably smaller than the sample size for the adults.

   d. Write a conclusion that the researchers could draw from this study.

      The mean number of calories in a meal consumed by adults is greater than the mean number of calories in a meal consumed by adolescents.

2. The Gallup organization published the following results from a poll that it conducted:

   By their own admission, many young Americans, aged 18 to 29, say they spend too much time using the Internet (59%), their cell phones or smartphones (58%), and social media sites such as Facebook (48%). Americans' perceptions that they spend "too much" time using each of these technologies decline with age. Conversely, older Americans are most likely to say they spend too much time watching television, and among all Americans, television is the most overused technology tested.

   Results are based on telephone interviews conducted as part of Gallup Daily tracking April 9–10, 2012, with a random sample of $1,051$ adults, aged 18 and older, living in all 50 U.S. states and the District of Columbia.

   For results based on the total sample of national adults, one can say with 95% confidence that the maximum margin of sampling error is ± 4 percentage points.

   Source: http://www.gallup.com/poll/153863/Young-Adults-Admit-Time-Cell-Phones-Web.aspx

   a. Write a newspaper headline that would capture the main idea from the poll.

      Answers will vary. An example is "U.S. Young Adults Admit Too Much Time on Cell Phones, Web."

b. Use the phrase from the article, "their cell phones or smartphones (58%)," to calculate the margin of error. Show your work.

$$ME = 2\sqrt{\frac{(0.58)(0.42)}{1051}} = 0.03$$

c. How do your results compare with the margin of error stated in the article?

This value is less than the stated value of 0.04.

d. Interpret the statement "the margin of sampling error is ± 3 percentage points."

It is very likely that the sample proportion 0.58 *is within* 0.04 *of the proportion of all adults who would say that they use their cell phones too much.*

e. What would happen to the margin of error if Gallup had surveyed 100 people instead of the $1{,}051$?

The margin of error would increase to 0.10.

3. The Holiday Inn Resort Brand conducted the Kid Classified survey. $1{,}500$ parents and children nationwide were interviewed via an online survey.

The results of the survey state

> While many parents surveyed say they have some financial savings set aside specifically for vacation travel, more than half of parents in the survey (52%) noted that saving enough money was the biggest challenge to planning a family vacation, more so than coordination of family schedules (19%) or taking time off of work (12%).

Source: http://www.lodgingmagazine.com/holiday-inn-resorts-catering-to-kids/

a. Calculate the margin of error associated with the estimate of the proportion of all parents who would say that saving enough money is the biggest challenge to planning a family vacation.

The margin of error is 0.026.

b. Write a sentence interpreting the margin of error.

It is very likely that the sample proportion of 0.52 *is within* 0.026 *of the proportion of all parents who would say that saving enough money is the biggest challenge to planning a family vacation.*

c. Comment on how the survey was conducted.

This was an online survey. It is not known if this was a random sample. Only individuals who have Internet access could take this survey, which would eliminate anyone who did not have Internet access.

EUREKA MATH™

Topic D

Drawing Conclusions Using Data from an Experiment

S-IC.B.3, S-IC.B.5, S-IC.B.6

Focus Standards:	S-IC.B.3	Recognize the purposes of and differences among sample surveys, experiments, and observational studies; explain how randomization relates to each.
	S-IC.B.5	Use data from a randomized experiment to compare two treatments; use simulations to decide if differences between parameters are significant.
	S-IC.B.6	Evaluate reports based on data.
Instructional Days:	8	
Lesson 23:	Experiments and the Role of Random Assignment (P)[1]	
Lesson 24:	Differences Due to Random Assignment Alone (P)	
Lessons 25–27:	Ruling Out Chance (P,P,P)	
Lessons 28–29:	Drawing a Conclusion from an Experiment (E,E)	
Lesson 30:	Evaluating Reports Based on Data from an Experiment (P)	

This topic focuses on drawing conclusions based on data from a statistical experiment. Experiments are introduced as investigations designed to compare the effect of two treatments on a response variable. Students revisit the distinction between random selection and random assignment.

When comparing two treatments using data from a statistical experiment, it is important to assess whether the observed difference in group means indicates a real difference between the treatments in the experiment or whether it is possible that there is no difference and that the observed difference is just a by-product of the random assignment of subjects to treatments (**S-IC.B.5**). To help students understand how this distinction is made, lessons in this topic use simulation to create a randomization distribution as a way of exploring the types of differences they might expect to see by chance when there is no real difference between groups. By understanding these differences, students are able to determine whether an observed difference in means is significant (**S-IC.B.5**).

[1]Lesson Structure Key: **P**-Problem Set Lesson, **M**-Modeling Cycle Lesson, **E**-Exploration Lesson, **S**-Socratic Lesson

Students also critique and evaluate published reports based on statistical experiments that compare two treatments (**S-IC.B.6**). For example, students read a short summary of an article in the online *New England Journal of Medicine* describing an experiment to determine if wearing a brace helps adolescents with scoliosis. Then, they watch an online video report for the *Wall Street Journal* titled "BMW Drivers Really Are Jerks" that describes a study of the relationship between driving behavior and the type of car driven.

Lesson 23: Experiments and the Role of Random Assignment

Student Outcomes

- Given a description of a statistical experiment, students identify the response variable and the treatments.
- Students recognize the different purposes of random selection and of random assignment.
- Students recognize the importance of random assignment in statistical experiments.

Lesson Notes

Experiments are introduced as investigations designed to compare the effect of two treatments on a response variable. This lesson revisits the distinction between random selection and random assignment and also explores the role of random assignment in carrying out a statistical experiment to compare two treatments.

Classwork

Exercises 1–4 (8 minutes): Experiments

Read the information about the two studies to the class. Give students about two to three minutes to work with a partner to answer Exercise 1.

Exercises 1–4: Experiments

Two studies are described below. One is an observational study, while the other is an experiment.

Study A:

A new dog food, specially designed for older dogs, has been developed. A veterinarian wants to test this new food against another dog food currently on the market to see if it improves dogs' health. Thirty older dogs were randomly assigned to either the "new" food group or the "current" food group. After they were fed either the "new" or "current" food for six months, their improvement in health was rated.

Study B:

The administration at a large school wanted to determine if there was a difference in the mean number of text messages sent by ninth-grade students and by eleventh-grade students during a day. Students in a random sample of 30 ninth-grade students were asked how many text messages they sent per day. Students in another random sample of 30 eleventh-grade students were asked how many text messages they sent per day. The difference in the mean number of texts per day was determined.

Scaffolding:

For struggling students, offer a graphic organizer that could be used to remind them of the characteristics of observational studies and experiments.

Also, consider using a visual depiction of the experiments.

1. **Which study is the experiment? Explain. Discuss the answer with your partner.**

   Study A is an experiment because treatments are assigned to the dogs, and a response is measured.

MP.1 Now, let students try to form their own definitions for a *subject*, *response variable*, and *treatment group* by identifying what each one is in the experiment; allow students to work with their partners to answer Exercises 2–4. Convey the following to students before each exercise, and allow two minutes for students to respond to each question:

- We are going to introduce some specific vocabulary used in experiments. In the example above, the dogs are the subjects. Think about what happened in the experiment, and describe what a subject is in an experiment.

2. **In your own words, describe what a subject is in an experiment.**

   A subject is a participant in the experiment.

- The dogs' health was the response variable. Think about the experiment, and describe what a response variable is in an experiment.

3. **In your own words, describe what a response variable is in an experiment.**

   A response variable is not controlled by the experimenter and is measured as part of the experiment.

- "New" food and "old" food were the treatments. Describe what a treatment is in an experiment.

4. **In your own words, describe what a treatment is in an experiment.**

   A treatment is the condition(s) to which subjects are randomly assigned by the experimenter.

Exercises 5–9 (10–13 minutes): Random Selection and Random Assignment

MP.2 Read and discuss the differences between the terms *random selection* and *random assignment*. Have students work with a partner to answer Exercises 5 through 8. Verify that students understand the differences between random selection and random assignment. Allow students about five minutes to complete Exercise 9.

Scaffolding:

- The term *assignment* may be familiar to students (e.g., homework assignments), but in this context it may need explanation.
- In this lesson, *assignment* refers to allocating or assigning a subject to a treatment.
- Consider using a Frayer model diagram for some of the vocabulary in this lesson.

Exercises 5–9: Random Selection and Random Assignment

Take another look at the two studies described above. Study A (the dog food study) is an experiment, while study B (text messages) is an observational study. The term *random sample* implies that a sample was randomly selected from a population. The terms *random selection* and *random assignment* have very different meanings.

***Random selection* refers to randomly selecting a sample from a population. Random selection allows generalization to a population and is used in well-designed observational studies. Sometimes, but not always, the subjects in an experiment are randomly selected.**

***Random assignment* refers to randomly assigning the subjects in an experiment to treatments. Random assignment allows for cause-and-effect conclusions and is used in well-designed experiments.**

EUREKA MATH™

In study B, the data were collected from two random samples of students.

5. Can the results of the survey be generalized to all ninth-grade and all eleventh-grade students at the school? Why or why not? Discuss the answer with your partner.

   Yes, the results of the survey can be generalized to all ninth- and eleventh-grade students at the school because the students were randomly selected.

6. Suppose there really is a difference in the mean number of texts sent by ninth-grade students and by eleventh-grade students. Can we say that the grade level of the students is the cause of the difference in the mean number of texts sent? Why or why not? Discuss the answer with your partner.

   No, we cannot say that the grade level of the students is the cause of the difference in the mean number of texts sent because observational studies do not allow for cause-and-effect conclusions.

In study A, the dogs were randomly assigned to one of the two types of food.

7. Suppose the dogs that were fed the new food showed improved health. Can we say that the new food is the cause of the improvement in the dogs' health? Why or why not? Discuss the answer with your partner.

   Yes, we can say that the new food is the cause of improvement in the dogs' health because the dogs were randomly assigned to treatments. This allows for cause-and-effect conclusions.

8. Can the results of the dog food study be generalized to all dogs? To all older dogs? Why or why not? Discuss the answer with your partner.

   No, the results cannot be generalized to all dogs or to all older dogs because the dogs were not randomly selected from the population of all dogs or from the population of older dogs.

The table below summarizes the differences between the terms *random selection* and *random assignment*.

9. For each statement, put a check mark in the appropriate column(s,) and explain your choices.

	Random Selection	Random Assignment
Used in Experiments	√	√
Used in Observational Studies	√	
Allows Generalization to the Population	√	
Allows a Cause-and-Effect Conclusion		√

Exercises 10–17 (22–25 minutes)

Students should work independently on this set of exercises. Read aloud and discuss the experiment. Be sure to address the concerns that the acres along the river and trees may have a richer nutrient level in the soil and may also get less sunshine than the acres farther away from the river and farther south of the trees. (An acre is $43{,}560\text{ ft}^2$, or approximately 209 ft. by 209 ft. An acre is almost the size of a football field.) Allow about two minutes for students to answer Exercise 10.

Exercise 10

What is the purpose of random assignment in experiments? To answer this, consider the following investigation:

A researcher wants to determine if the yield of corn is different when the soil is treated with one of two different types of fertilizers, fertilizer A and fertilizer B. The researcher has 16 acres of land located beside a river that has several trees along its bank. There are also a few trees to the north of the 16 acres. The land has been divided into 16 one-acre plots. (See the diagram below.) These 16 plots are to be planted with the same type of corn but can be fertilized differently. At the end of the growing season, the corn yield will be measured for each plot, and the mean yields for the plots assigned to each fertilizer will be compared.

10. For the experiment, identify the following, and explain each answer:

a. Subjects (Hint: not always people or animals)

The 16 plots of land are the subjects because that is what the experiment is being carried out on.

b. Treatments

Fertilizer A and fertilizer B are the treatments because that is what is different in the experiment.

c. Response variable

The yield of corn is the response variable because that is what is being measured after treatment.

Next, students need to randomly assign each plot to one of the two treatments, fertilizer A or fertilizer B. Below are instructions for two different methods for the randomization process. Either method is appropriate. Students should develop a plan for randomly assigning the plots, share the responses, discuss, and then choose their preferred method. Consider doing both, starting with the paper method. Allow students 8–10 minutes to carry out the randomization method and answer Exercise 11.

Paper Method

1. Give each student 16 precut slips of paper (approximately a 2-inch square).
2. Have students write a number from 1 to 16 on each square.
3. Turn the squares upside down. Mix well.
4. Separate the squares into 2 piles of 8, identifying one pile as A and the other pile as B.
5. Turn the slips of paper over.
6. The numbers in pile A correspond to the plots that receive the fertilizer A treatment.
7. The numbers in pile B correspond to the plots that receive the fertilizer B treatment.
8. Write the letters *A* and *B* in the corresponding squares in the diagram. (Exercise 11)

Calculator Method

1. When using the random number generator in a graphing calculator, the random generator function on each calculator often needs to be seeded with a unique number.* For the TI-83 or TI-84 calculator, have each student type in a unique number, such as his student ID number. Press the Store, or STO>, key. Go to the Probability, or PRB, menu. This menu is accessed by pressing the MATH key. Select 1: rand. Press Enter to select. Press Enter to perform the command.
2. Next, use the command to generate random integers from 1 to 16. For the TI-83 or TI-84 calculator, go to the Probability, PRB, menu. This menu is accessed by pressing the MATH key. Select 8: randIntNoRep. In the parentheses, type 1,16. Press Enter to do the command.
3. The first 8 unique numbers generated represent the plots of land that are assigned to fertilizer A.
4. The remaining 8 numbers represent the plots of land that are assigned to fertilizer B.
5. Write the letters *A* and *B* in the corresponding squares in the diagram. (Exercise 11)

*The random generator function on a calculator is an algorithm (a program) that produces pseudo random numbers—numbers that behave like random numbers. This program often has a default value that it uses to generate numbers. If this default value is not changed, the generator is not seeded; then, each calculator generates the same set of numbers.

Exercise 11

Next, you need to assign the plots to one of the two treatments. To do this, follow the instructions given by your teacher.

11. **Write A (for fertilizer A) or B (for fertilizer B) in each of the 16 squares in the diagram so that it corresponds to your random assignment of fertilizer to plots.**

    Answers will vary. A sample response is provided.

1 B	2 A	3 A	4 B
5 B	6 A	7 B	8 A
9 A	10 A	11 B	12 A
13 A	14 B	15 B	16 B

After students have completed the table, have them compare their allocations of the two treatments. Be sure to stress that random assignment produces a large variety of possible outcomes. Next, students investigate their results. Students are investigating locations with respect to central vertical (blue) and horizontal (red) lines. Students see that their chosen random distributions gave a good spread of fertilizer A and B above and below the red line and to the right and left of the blue line. Allow five to eight minutes for students to answer Exercises 12–15.

Exercises 12–15

Let's investigate the results of the random assignment of the fertilizer types to the plots.

A sample response is provided.

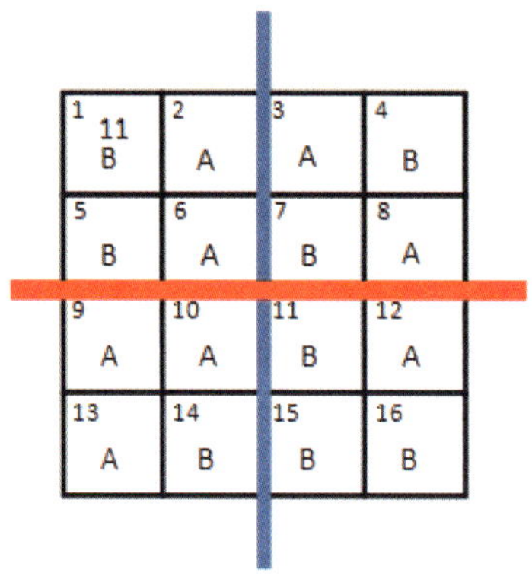

12. **On the diagram above, draw a vertical line down the center of the 16 plots of land.**

    See the chart above.

13. **Count the number of plots on the left side of the vertical line that will receive fertilizer A. Count the number of plots on the right side of the vertical line that will receive fertilizer A.**

    There are 5 plots on the left and 3 plots on the right. Most students get results that have an approximately equal split. A few students may get 6 for one side of the line and 2 for the other side. Point out that this is possible with random assignment but not very likely.

    Left _________ Right _________

14. **On the diagram above, draw a horizontal line through the center of the 16 plots of land.**

    See the chart above.

15. **Count the number of plots above the horizontal line that will receive fertilizer A. Count the number of plots below the horizontal line that will receive fertilizer A.**

    There are 4 plots above the line and 4 plots below the line. Most students get results that have an approximately equal split. A few students may get 6 for one side of the line and 2 for the other side. Point out that this is possible with random assignment but not very likely.

    Above _________ Below _________

When students have completed Exercises 12–15, ask them to raise their hands if their randomization method produced an approximately even split left of the vertical line versus right of the vertical line. Comment on the fact that most students raised their hands.

Next, ask students to raise their hands if their randomization methods produced an approximately even split above the horizontal line versus below the horizontal line. Comment on the fact that most students raised their hands.

Then, hold a class discussion about the proximity of the river and of the northern trees. Stress the following ideas:

- About half of the plots of land close to the river received fertilizer A, while the other half received fertilizer B.
- About half of the plots of land close to the northern trees received fertilizer A, while the other half received fertilizer B.

Allow students five to eight minutes to answer Exercises 16 and 17.

Exercises 16–17

In experiments, random assignment is used as a way of ensuring that the groups that receive each treatment are as much alike as possible with respect to other factors that might affect the response.

16. **Explain what this means in the context of this experiment.**

    This means that approximately half of the plots that are close to the river were assigned fertilizer A, and the remaining half were assigned fertilizer B. Also, approximately half of the plots that are close to the northern trees were assigned fertilizer A, and the remaining half were assigned fertilizer B.

MP.3

17. **Suppose that, at the end of the experiment, the mean yield for one of the fertilizers is quite a bit higher than the mean yield for the other fertilizer. Explain why it would be reasonable to say that the type of fertilizer is the cause of the difference in yield and not the proximity to the river or to the northern trees.**

    Because about half the plots of land close to the river receive fertilizer A and the other half receive fertilizer B, we can see how the fertilizer works regardless of its proximity to the river. Also, because about half of the plots of land close to the northern trees receive fertilizer A and the other half receive fertilizer B, we can see how the fertilizer works regardless of its proximity to the northern trees.

Closing (2 minutes)

Ask students to summarize the main ideas of the lesson in writing or with a neighbor. Use this as an opportunity to informally assess comprehension of the lesson. The Lesson Summary below offers some important ideas that should be included.

Lesson Summary

- **An *experiment* is an investigation designed to compare the effect of two or more treatments on a response variable.**
- **A *subject* is a participant in the experiment.**
- **The *response variable* is a variable that is not controlled by the experimenter and that is measured as part of the experiment.**
- **The *treatments* are the conditions to which subjects are randomly assigned by the experimenter.**
- ***Random selection* refers to randomly selecting a sample from a population.**
  - **Random selection allows for generalization to a population.**
- ***Random assignment* refers to randomly assigning subjects to treatment groups.**
  - **Random assignment allows for cause-and-effect conclusions.**
  - **The purpose of random assignment in an experiment is to create similar groups of subjects for each of the treatments in the experiment.**

Exit Ticket (2 minutes)

Name ______________________________ Date ______________

Lesson 23: Experiments and the Role of Random Assignment

Exit Ticket

Runners who suffered from shin splints were randomly assigned to one of two stretching routines. One of the routines involved a series of pre-run and post-run dynamic stretches that last approximately 5 minutes before and after the run. The other routine involved a 1-minute hamstring stretch pre-run and no stretching post-run. After a 45-minute run, each runner will be assessed for shin splints.

a. Explain why this is an experiment.

b. Identify the subjects.

c. Identify the treatments.

d. Identify the response variable.

e. Why are the runners randomly assigned to one of two stretching routines?

Exit Ticket Sample Solutions

Runners who suffered from shin splints were randomly assigned to one of two stretching routines. One of the routines involved a series of pre-run and post-run dynamic stretches that last approximately 5 minutes before and after the run. The other routine involved a 1-minute hamstring stretch pre-run and no stretching post-run. After a 45-minute run, each runner will be assessed for shin splints.

a. Explain why this is an experiment.

It is an investigation that is designed to compare the effect of two treatments to a response variable.

b. Identify the subjects.

The subjects are runners.

c. Identify the treatments.

The treatments are dynamic pre-run and post-run stretching routine and the 1-minute hamstring stretch.

d. Identify the response variable.

The response variable is the incidence of shin splints.

e. Why are the runners randomly assigned to one of two stretching routines?

The experimenter wants to create similar groups of runners for each of the treatments (stretching routines) in the experiment, which would allow for a cause-and-effect conclusion.

Problem Set Sample Solutions

For Problems 1 through 5, identify (i) the subjects, (ii) the treatments, and (iii) the response variable for each experiment.

1. A botanist was interested in determining the effects of watering (three days a week or daily) on the heat rating of jalapeño peppers. The botanist wanted to know which watering schedule would produce the highest heat rating in the peppers. He conducted an experiment, randomly assigning each watering schedule to half of 12 plots that had similar soil and full sun. The average final heat rating for the peppers grown in each plot was recorded at the end of the growing season.

   i. *Plots of peppers*

   ii. *Watering three days a week and watering daily*

   iii. *Heat rating*

2. A manufacturer advertises that its new plastic cake pan bakes cakes more evenly. A consumer group wants to carry out an experiment to see if the plastic cake pans do bake more evenly than standard metal cake pans. Twenty cake mixes (same brand and type) are randomly assigned to either the plastic pan or the metal pan. All of the cakes are baked in the same oven. The rating scale was then used to rate the evenness of each cake.

   i. *Cake mixes*

   ii. *Plastic pan and metal pan*

   iii. *Evenness rating*

3. The city council of a large city is considering a new law that prohibits talking on a cell phone while driving. A consumer rights organization wants to know if talking on a cell phone while driving distracts a person's attention, causing that person to make errors while driving. An experiment is designed that uses a driving simulator to compare the two treatments: driving while talking on a cell phone and driving while not talking on a cell phone. The number of errors made while driving on an obstacle course will be recorded for each driver. Each person in a random sample of 200 licensed drivers in the city was asked to participate in the experiment. All of the drivers agreed to participate in the experiment. Half of the drivers were randomly assigned to drive an obstacle course while talking on a phone. The remaining half were assigned to drive the obstacle course while not talking on a phone.

   i. The 200 randomly selected drivers who took part in the experiment

   ii. Driving while talking on a cell phone and driving while not talking on a cell phone

   iii. Number of errors made while driving on an obstacle course

4. Researchers studied 208 infants whose brains were temporarily deprived of oxygen as a result of complications at birth (*The New England Journal of Medicine*, October 13, 2005). An experiment was performed to determine if reducing body temperature for three days after birth improved their chances of surviving without brain damage. Infants were randomly assigned to usual care or whole-body cooling. The amount of brain damage was measured for each infant.

   i. Infants born with a temporary deprivation of oxygen at birth

   ii. Usual care or whole-body cooling

   iii. Amount of brain damage

5. The head of the quality control department at a printing company would like to carry out an experiment to determine which of three different glues results in the greatest binding strength. Copies of a book were randomly assigned to one of the three different glues.

   i. Copies of a book

   ii. Three different glues

   iii. Binding strength

6. In Problem 3, suppose that drivers who talked on the phone while driving committed more errors on the obstacle course than drivers who did not talk on the phone while driving. Can we say that talking on the cell phone while driving is the cause of the increased errors on the obstacle course? Why or why not?

   Yes, we can say that talking on the cell phone while driving is the cause of the increased errors on the obstacle course because the drivers were randomly assigned to one of two treatments.

7. Can the results of the experiment in Problem 3 be generalized to all licensed drivers in the city? Why or why not?

   Yes, the results can be generalized to all licensed drivers in the city because the drivers were randomly selected from licensed drivers in the city.

8. In Problem 4, one of the treatment groups was to use usual care for the infants. Why was this treatment group included in the experiment?

   The usual care treatment group is used as a comparison for how well the cooling treatment worked.

9. In Problem 5, why were copies of only one book used in the experiment?

   This prevents other factors, such as number of pages, from interfering with the results of the experiment.

Lesson 24: Differences Due to Random Assignment Alone

Student Outcomes

- Students understand that when one group is randomly divided into two groups, the two groups' means differ just by chance (a consequence of the random division).
- Students understand that when one group is randomly divided into two groups, the distribution of the difference in the two groups' means can be described in terms of shape, center, and spread.

Lesson Notes

This lesson investigates differences in group means when a single group is randomly divided into two groups. The goal of this lesson is for students to understand that when a single group is randomly divided into two groups, the two group means tend to differ just by chance. Students are given 20 values, which they randomly divide into two groups. The mean is then calculated for each group. The process is repeated two more times, and all group means are used to create a class dot plot, which confirms that the distribution of the random groups' means are centered at the single set's mean. This idea is fundamental to the lessons that follow, which involve distinguishing meaningful differences in means from differences that might be due only to chance.

Classwork

This lesson is designed for students to work individually. However, students can discuss some answers with their neighbors before a class discussion of the answers. Prior to class, make a copy of Appendix A for each student. Scissors may also be needed to cut the table into pieces.

Exercises 1–17 (40–45 minutes)

Read the scenario regarding the fastest speeds driven by twenty adult drivers. Consider talking about the data before beginning the lesson.

Optional questions:

- One driver answered that the fastest speed he had driven was 40 mph. How is this possible?
  - *This person may be learning how to drive and is still unsure of his driving abilities.*
- Another driver answered that the fastest speed she had driven was 110 mph. How is this possible?
  - *This person may have driven on a racetrack.*
- What would it mean if a driver's response was 0?
  - *This person may have never driven a car.*

Allow about five minutes for students to answer Exercises 1 and 2.

Exercises 1–17

Twenty adult drivers were asked the following question:

"What speed is the fastest that you have driven?"

The table below summarizes the fastest speeds driven in miles per hour (mph).

70	60	70	95	50	60	80	75	55	90
110	65	65	65	55	70	75	70	65	40

1. What is the mean fastest speed driven?

   69.25 mph

2. What is the range of fastest speed driven?

   70 mph

Scaffolding:

- Include a visual of a car, and discuss the meaning of the question.
- Review the concepts of *mean* and *range*, model the process for determining these in the example, and discuss their meaning in context.

Next, students investigate what happens to the mean fastest speed driven when the original 20 values are randomly divided into two groups. Give each student a copy of Appendix A, and let them independently answer Exercise 3.

MP.3

3. Imagine that the fastest speeds were randomly divided into two groups. How would the means and ranges compare to one another? To the means and ranges of the whole group? Explain your thinking.

   Answers will vary. Sample response: I think the mean of each group will be fairly close because most of the values appear to be between 60 and 80. I do not think the ranges of each group will be the same. Depending on the values in the group, one possible range could be as large as 70 mph (from 40 to 110) or as small as 10 mph (from 60 to 70). I think the means of the two groups should be close to the mean of the whole group, again because most values appear to be between 60 and 80. The range of the whole group is 70 mph, so the range of the two groups may be smaller than that.

Scaffolding:

Advanced students may be encouraged to develop their own plans for investigating the answer to the question of how the means of randomly selected groups are related to one another. Consider allowing them time to write, carry out, and evaluate plans for exploring this question.

Allow about five minutes for students to answer Exercises 4 and 5. Discuss students' answers to Exercise 5.

Instructions for randomly dividing the 20 fastest speeds driven into two groups:

1. Cut (or tear) along the lines in the table so that 20 equal slips of paper are obtained.
2. Turn the squares upside down. Mix well.
3. Separate the squares into 2 piles of 10, identifying one pile as Group 1 and the other pile as Group 2.
4. Turn the slips of paper over.
5. Record the numbers for Group 1 and Group 2 in the table (Exercise 4).

Let's investigate what happens when the fastest speeds driven are randomly divided into two equal-sized groups.

4. **Following the instructions from your teacher, randomly divide the 20 values in the table above into two groups of 10 values each.**

   Sample student responses; answers will vary. One example follows.

											Mean
Group 1	70	40	65	75	65	70	65	110	60	55	
Group 2	75	70	65	60	70	95	50	55	90	80	

5. **Do you expect the means of these two groups to be equal? Why or why not?**

   The values of the means of the two groups will probably not be exactly equal, but I don't expect them to be very different.

Allow students about five minutes to answer Exercises 6–8. Then, discuss answers as a class.

Answers will vary. Sample responses are based on the sample answer provided for Exercise 4.

6. **Compute the means of these two groups. Write the means in the chart above.**

   Group 1 mean: 67.5

   Group 2 mean: 71

7. **How do these two means compare to each other?**

   The values of these two means are not very different from each other.

8. **How do these two means compare to the mean fastest speed driven for the entire group (Exercise 1)?**

   The values of these two means are close to the original mean. The value of one mean is larger than the original mean, and the value of the other mean is smaller than the original mean. (Note: The two group mean values are equidistant from the original mean value.)

Allow about 15 minutes for students to answer Exercises 9 and 10. For the class dot plot in Exercise 10, create a number line on a white board or paper. The horizontal scale should range from 61 mph to 78 mph, using tick marks of 0.5. The horizontal label should be "Mean Fastest Speed Driven." Provide two or three markers so that multiple students can place their means on the graph at the same time.

Sample student responses are provided below; answers will vary.

9. **Use the instructions provided for Exercise 4 to repeat the random division process two more times. Compute the mean of each group for each of the random divisions into two groups. Record your results in the table below.**

											Mean
Group 3	55	60	95	70	75	50	65	55	70	75	67
Group 4	65	110	70	65	70	65	40	90	80	60	71.5
Group 5	65	70	55	65	75	70	75	70	80	70	69.5
Group 6	60	65	50	65	95	60	110	40	90	55	69

10. **Plot the means of all six groups on a class dot plot.**

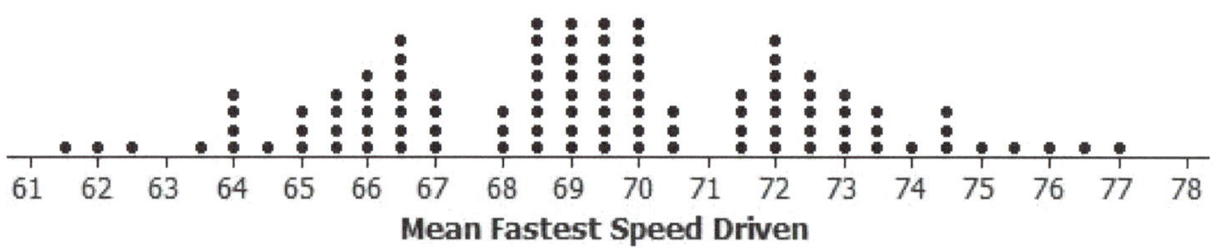

When the class dot plot is finished, allow students about three minutes to answer Exercise 11. Discuss the answer as a class.

11. **Based on the class dot plot, what can you say about the possible values of the group means?**

Answers will vary.

The group sample means are centered at the original mean of 69.25 mph.

There is variability in the group means. Some groups' means vary more from the original mean of 69.25 *than others.*

The group means that are closer to the original mean of 69.25 mph *occur more often than the group means that are farther away from* 69.25 mph.

Allow about five minutes for students to answer Exercises 12–14. Then, discuss answers as a class.

12. **What is the smallest possible value for a group mean? Largest possible value?**

Smallest possible mean (the 10 smallest values) $= \dfrac{40+50+55+55+60+60+65+65+65+65}{10} = 58$

Largest possible mean (the 10 largest values) $= \dfrac{70+70+70+70+75+75+80+90+95+110}{10} = 80.5$

The smallest possible mean is 58 mph, *and the largest possible mean is* 80.5 mph.

13. **What is the largest possible range for the distribution of group means?**

$80.5\text{ mph} - 58\text{ mph} = 22.5\text{ mph}$

From Exercise 12, the largest possible range is 22.5 mph.

14. **How does the largest possible range in the group means compare to the range of the original data set (Exercise 2)? Why is this so?**

The range of the original data is 70 mph*. The largest possible range for the distribution of group means is* 22.5 mph*, which is much smaller. This difference is due to the use of means. The means of the two groups of* 10 *do not vary as much as the individual observations in the data set.*

Allow students about three minutes to answer Exercises 15–16. Discuss these answers as a class. Anticipate that students might need help with Exercise 16.

15. **What is the shape of the distribution of group means?**

It is symmetrical.

16. **Will your answer to the above question always be true? Explain.**

Yes. When a single set of values is divided into two equal groups, the two group means will be equidistant from the single set's mean. Thus, it will always produce a symmetrical distribution.

Allow students about three minutes to answer Exercise 17 independently or with a partner. Discuss this answer.

MP.2

17. **When a single set of values is randomly divided into two equal groups, explain how the means of these two groups may be very different from each other and may be very different from the mean of the single set of values.**

It is possible that the random division could result in most of the smaller values being in one group and most of the larger values in the other group. This would produce group means that were very different from each other and from the single set's mean. But with a random division, this is not very likely to occur.

Closing (2 minutes)

MP.3

- Compare your conjecture about the randomized groups (Exercise 3) with what you have learned.

Ask students to summarize the main ideas of the lesson in writing or with a neighbor. Use this as an opportunity to informally assess comprehension of the lesson. The Lesson Summary below offers some important ideas that should be included.

Lesson Summary

When a single set of values is randomly divided into two groups:

- **The two group means will tend to differ just by chance.**
- **The distribution of random groups' means will be centered at the single set's mean.**
- **The range of the distribution of the random groups' means will be smaller than the range of the data set.**
- **The shape of the distribution of the random groups' means will be symmetrical.**

Exit Ticket (3 minutes)

Name ______________________________ Date ______________

Lesson 24: Differences Due to Random Assignment Alone

Exit Ticket

When a single group is randomly divided into two groups, why do the two group means tend to be different?

Exit Ticket Sample Solutions

When a single group is randomly divided into two groups, why do the two group means tend to be different?

The two group means tend to be different because of random chance involved in the process of dividing the original group into two subgroups.

Problem Set Sample Solutions

In one high school, there are eight math classes during second period. The number of students in each second-period math class is recorded below.

$32 \quad 27 \quad 26 \quad 23 \quad 25 \quad 22 \quad 30 \quad 19$

This data set is randomly divided into two equal-sized groups, and the group means are computed.

1. Will the two group means be the same? Why or why not?

   No. The two group means tend to not be the same just due to chance.

The random division into two groups process is repeated many times to create a distribution of group mean class size.

2. What is the center of the distribution of group mean class size?

   The distribution of the group mean class size is centered at the mean of the original set of 8 *class sizes.*

   $$\frac{19+22+23+25+26+27+30+32}{8} = 25.5$$

   The center of the distribution of group mean class size is 25.5.

3. What is the largest possible range of the distribution of group mean class size?

   The range of the distribution of mean class sizes would be the difference of the largest possible group mean and the smallest possible group mean.

   Largest possible group mean: $\dfrac{26+27+30+32}{4} = 28.75$

   Smallest possible group mean: $\dfrac{19+22+23+25}{4} = 22.25$

   Largest possible group mean range: $28.75 - 22.25 = 6.5$*; the range is* 6.5 *students.*

4. What possible values for the mean class size are more likely to happen than others? Explain why you chose these values.

   Below is a sample partial distribution of the group mean class sizes.

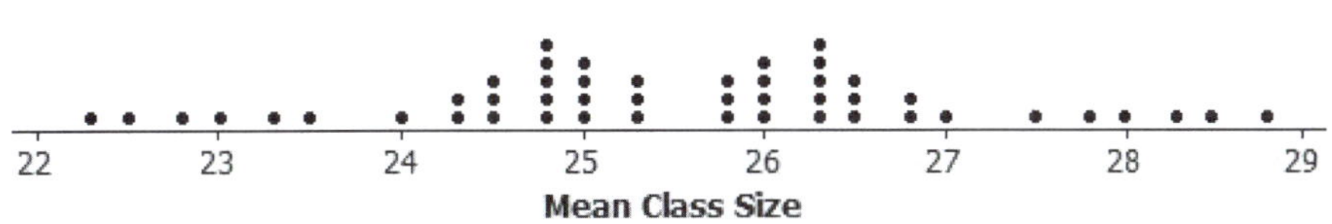

There are 3 different sets of numbers: Set A, Set B, and Set C. Each set contains 10 numbers. In two of the sets, the 10 numbers were randomly divided into two groups of 5 numbers each, and the mean for each group was calculated. These two means are plotted on a dot plot. This procedure was repeated many times, and the dot plots of the group means are shown below.

The third set did not use the above procedure to compute the means.

For each set, the smallest possible group mean and the largest possible group mean were calculated, and these two means are shown in the dot plots below.

Use the dot plots below to answer Problems 5–8.

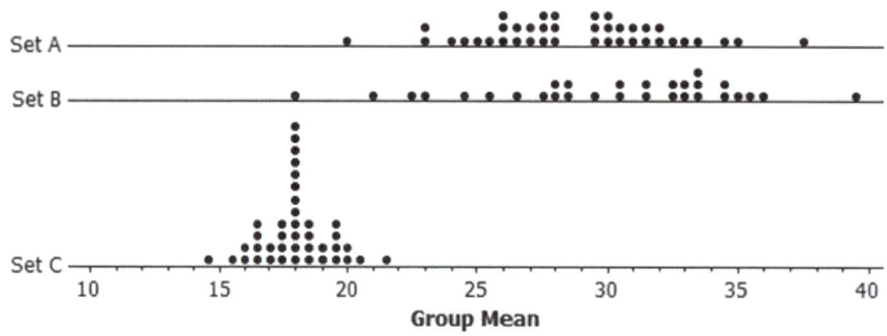

5. Which set is *not* one of the two sets that were randomly divided into two groups of 5 numbers? Explain.

   Set B is not one of the two sets that were randomly divided into two subsets. When a set of numbers is divided into two equal groups, the resulting dot plot of group means will be symmetrical.

6. Estimate the mean of the original values in Set A. Show your work.

   Estimated mean of Set A: $\frac{27 + 29.5}{2} = 28.25$

7. Estimate the range of the group means shown in the dot plot for Set C. Show your work.

   $21.5 - 14.5 = 7$

   The estimated range of the distribution of sample means for Set C is 7.

8. Is the range of the original values in Set C smaller or larger than your answer in Problem 7? Explain.

   The range of Set C is larger than the range of the distribution of sample means from Set C. The range of the group means will always be smaller than the range of the original set of values.

Appendix A

65	40	55	90
70	75	80	75
55	70	50	60
65	65	95	70
110	65	60	70

Lesson 25: Ruling Out Chance

Student Outcomes

- Given data from a statistical experiment with two treatments, students create a randomization distribution.
- Students use a randomization distribution to determine if there is a significant difference between two treatments.

Lesson Notes

MP.4

In Lessons 25–27, students are introduced to *randomization testing*. By the end of the three lessons, they will have experienced and executed all steps of this procedure, and they will have engaged in a "start to finish" example with provided data. Once that is accomplished, students participate in a capstone experience in Lessons 28 and 29 where they collect their own data, execute their own randomization tests, and present their findings in a poster.

Note that randomization testing is somewhat different from more traditional methods for comparing the means of two groups such as the two-sample t-test and other procedures found in many college-level statistics courses or AP statistics. While a randomization test still requires developing competing hypotheses (null and alternative) and coming to a conclusion based on the probability of obtaining results as extreme as those one obtains in one's sample or experiment, the randomization testing method develops the probability distribution of the statistic of interest via simulation and using the data at hand. As a result, the randomization testing method does not carry with it the usual concerns regarding the characteristics of a population distribution and other assumptions required of traditional methods. The technique is thus also referred to as a *nonparametric* technique.

This lesson starts by reminding students of previous lessons' work regarding randomization distributions. Specifically, students review that when a single group of observations with any variability is randomly divided into two groups, the means of these two groups tend to differ just by chance. In some cases, the difference in the means of these two groups may be very small (or 0), but in other cases, this difference may be quite large. As the difference in the two groups' means is the statistic of interest, students consider that difference value in context with an experiment regarding the use of a nutrient supplement to encourage tomato growth.

In subsequent lessons, students estimate a probability distribution for the possible values of this "difference value," and students assess whether or not the actual difference in means obtained from the experiment (in this case, the difference observed between the treatment group's mean weight and the control group's mean weight) is consistent with usual chance behavior. When the observed difference is not typical of chance behavior, it may indicate a significant difference. Specifically, it may indicate that the nutrient treatment was effective.

Throughout these lessons, experimental design principles are presented, and some questions prompt students to consider the reasoning behind the principles.

Lesson 27 may require slightly more time than Lessons 25 and 26.

Classwork

Explain that the next three lessons explore some examples of the method called *randomization testing.* Then, after completing these three lessons, students conduct an experiment and use this randomization testing method to assess if the result observed in the experiment is statistically significant.

Opening Exercise (3 minutes)

Let students answer part (a) in writing and share their responses with a neighbor. Pose the question in part (b) to the class, and allow multiple responses.

Opening Exercise

a. **Explain in writing what you learned about randomly divided groups from the last lesson. Share your thoughts with a neighbor.**

In the previous lesson, I saw that when a single group of observations with any variability is randomly divided into two groups, the means of these two groups will tend to differ just by chance. In some cases, the difference in the means of these two groups may be very small (or 0), but in other cases, this difference may be quite large.

b. **How could simulation be used to understand the typical differences between randomized groups?**

Through simulation, you can learn about the values of the difference that would be expected just by chance. You can then use this information to decide if a given set of data from an experiment supports a claim of a statistically significant difference between two treatment means.

Explain that this is done by assessing whether or not the actual difference in means obtained from the experiment (such as the difference observed between a control group mean and a treatment group mean or between two treatment groups' means) is consistent with usual chance behavior. When the observed difference is not typical of chance behavior, it may indicate a significant difference in treatment means.

Scaffolding:

- Consider using a visual or physical model to explain the experiment clearly.
- Consider asking students to restate, write, or draw a picture of the experiment.
- For advanced learners, consider posing the question: Suppose the nutrient group weighed, on average, 0.25 pounds more than the non-nutrient group. Would that be sufficient evidence that the nutrients caused a difference in weight? Explain why or why not.

Exercises 1–3 (10 minutes): Random Assignments and Computing the Difference of the Group Means

Read through the example. Let students work independently on the exercises and confirm answers as a class.

Exercises 1–3: Random Assignments and Computing the Difference of the Group Means

Imagine that 10 tomatoes of varying shapes and sizes have been placed in front of you. These 10 tomatoes (all of the same variety) have been part of a nutrient experiment where the application of the nutrient is expected to yield larger tomatoes that weigh more. All 10 tomatoes have been grown under similar conditions regarding soil, water, and sunlight, but 5 of the tomatoes received the additional nutrient supplement. Using the weight data of these 10 tomatoes, you wish to examine the claim that the nutrient yields larger tomatoes on average.

1. Why would it be important in this experiment for the 10 tomatoes to all be of the same variety and grown under the same conditions (except for the treatment applied to 5 of the tomatoes)?

Our variable is the weight of the tomatoes, and our objective is to see if the nutrient treatment yields heavier tomatoes. We do not want any other factors in our experiment that might cause some tomatoes to grow better than others; their presence might distort our findings. So, we prefer that the tomatoes all have similar growing conditions such as the same soil, the same amount of water, and the same amount of sunlight.

Note: When data collected from an experiment is examined, it is assumed that the experiment was conducted in a reasonable way with respect to important experimental design principles such as proper control measures. In most cases, if these principles are not incorporated, inaccurate conclusions may be reached.

Here are the 10 tomatoes with their weights shown. They have been ordered from largest to smallest based on weight.

9.1	8.4	8.0	7.7	7.3	6.4	5.9	5.2	4.4	3.8
#1	#2	#3	#4	#5	#6	#7	#8	#9	#10

For now, do not be concerned about which tomatoes received the additional nutrients. The object here is to randomly assign the tomatoes to two groups.

Imagine that someone assisting you uses a random number generator or some other impartial selection device and randomly selects tomatoes 1, 4, 5, 7, and 10 to be in Group A. By default, tomatoes 2, 3, 6, 8, and 9 will be in Group B. The result is illustrated below.

Group A	Group B
9.1	8.4
7.7	8.0
7.3	6.4
5.9	5.2
3.8	4.4

2. Confirm that the mean for Group A is 6.76 ounces, and calculate the mean for Group B.

The mean for Group B is 6.48 *ounces.*

3. Calculate the difference between the mean of Group A and the mean of Group B (that is, calculate $\bar{x}_A - \bar{x}_B$).

$\bar{x}_A - \bar{x}_B = 6.76 - 6.48 = 0.28$

The difference between the means of Group A and Group B is 0.28 *ounces.*

Exercises 4–6 (5–7 minutes): Interpreting the Value of a Difference

In these exercises, students interpret the difference between the group means. Let students work independently on the exercises and confirm answers as a class.

Exercises 4–6: Interpreting the Value of a Difference

The statistic of interest that you care about is the difference between the mean of the 5 tomatoes in Group A and the mean of the 5 tomatoes in Group B. For now, call that difference *Diff*. $\text{Diff} = \overline{x}_A - \overline{x}_B$

4. **Explain what a Diff value of 1.64 ounces would mean in terms of which group has the larger mean weight and the number of ounces by which that group's mean weight exceeds the other group's mean weight.**

   Group A has a mean weight that is 1.64 ounces higher than Group B's mean weight.

5. **Explain what a Diff value of -0.4 ounces would mean in terms of which group has the larger mean weight and the number of ounces by which that group's mean weight exceeds the other group's mean weight.**

   Group A has a mean weight that is 0.4 ounce lower than Group B's mean weight.

6. **Explain what a Diff value of 0 ounces would mean regarding the difference between the mean weight of the 5 tomatoes in Group A and the mean weight of the 5 tomatoes in Group B.**

   Group A's mean weight is the same as Group B's mean weight.

Exercises 7–8 (10 minutes): Additional Random Assignments

Let students work independently on the exercises and confirm answers as a class.

Exercises 7–8: Additional Random Assignments

7. **Below is a second random assignment of the 10 tomatoes to two groups. Calculate the mean of each group, and then calculate the Diff value for this second case. Also, interpret the Diff value in context using your responses to the previous questions as a guide.**

Group A	Group B
9.1	7.7
8.4	5.9
8.0	5.2
7.3	4.4
6.4	3.8

   Group A has a mean weight of 7.84 ounces, and that is 2.44 ounces higher than Group B's mean weight (5.4 ounces).

8. **Here is a third random assignment of the 10 tomatoes. Calculate the mean of each group, and then calculate the value of Diff for this case. Interpret the Diff value in context using your responses to the previous questions as a guide.**

Group A	Group B
9.1	8.4
7.7	8.0
7.3	6.4
5.2	5.9
3.8	4.4

There is no difference between Group A's mean weight and Group B's mean weight; each mean is 6.62 ounces.

Closing (10 minutes)

MP.3

- Using the 10 tomatoes, what arrangement of tomatoes would yield the largest value of Diff? How would you explain your reasoning without performing calculations to verify your answer? How big is the largest value of Diff?
  - *The largest Diff value would occur when Group A had the 5 largest tomatoes. That would create a situation where the mean of Group A has the largest possible mean for any grouping of the 5 tomatoes, and, consequently, Group B would have the smallest possible mean for any grouping of the 5 tomatoes.* $\text{Diff} = 8.10 - 5.14 = 2.96$
- Using your logic above, what arrangement of tomatoes would yield the most negative value of Diff? How would you explain your reasoning without performing calculations to verify your answer? What is this Diff value?
  - *The most negative Diff value would occur in the exact opposite case from above—that is when Group A had the 5 smallest tomatoes. That would create a situation where the mean of Group A has the smallest possible mean for any grouping of the 5 tomatoes, and, consequently, Group B would have the largest possible mean for any grouping of the 5 tomatoes.* $\text{Diff} = 5.14 - 8.10 = -2.96$
- What would it mean for the mean of all Diff values to be close to zero?
  - *A mean close to zero would indicate that the tomatoes in each group were mixed, or neither group had only the largest or smallest values. The resulting differences were both above and below the mean of the differences. Use Exercise 8 as a way to highlight the answer to this question.*

Ask students to summarize the main ideas of the lesson in writing or with a neighbor. Use this as an opportunity to informally assess comprehension of the lesson. The Lesson Summary below offers some important ideas that should be included.

> Lesson Summary
>
> **In this lesson, when the single group of observations was randomly divided into two groups, the means of these two groups differed by chance. These differences have a context based on the purpose of the experiment and the units of the original observations.**
>
> **The differences varied. In some cases, the difference in the means of these two groups was very small (or 0), but in other cases, this difference was larger. However, in order to determine which differences were typical and ordinary versus unusual and rare, a sense of the center, spread, and shape of the distribution of possible differences is needed. In the following lessons, you will develop this distribution by executing repeated random assignments similar to the ones you saw in this lesson.**

Exit Ticket (7 minutes)

Name ______________________________ Date ______________

Lesson 25: Ruling Out Chance

Exit Ticket

Six ping-pong balls are labeled as follows: $0, 3, 6, 9, 12, 18$. Three ping-pong balls will be randomly assigned to Group A; the rest will be assigned to Group B. $\text{Diff} = \bar{x}_A - \bar{x}_B$

1. Calculate Diff, the difference between the mean of the numbers, on the balls assigned to Group A and the mean of the numbers on the balls assigned to Group B (i.e., $\bar{x}_A - \bar{x}_B$) when the 3 ping-pong balls selected for Group A are 3, 6, and 12.

2. Calculate Diff, the difference between the mean of the numbers, on the balls assigned to Group A and the mean of the numbers on the balls assigned to Group B (i.e., $\bar{x}_A - \bar{x}_B$) when the 3 ping-pong balls selected for Group A are 3, 12, and 18.

3. What is the greatest possible value of Diff, and what selection of ping-pong balls for Group A corresponds to that value?

4. What is the smallest (most negative) possible value of Diff, and what selection of ping-pong balls for Group A corresponds to that value?

5. If these 6 observations represent the burn times of 6 candles (in minutes), explain what a Diff value of 6 means in terms of (a) which group (A or B) has the longer average burn time and (b) the amount of time by which that group's mean exceeds the other group's mean.

Exit Ticket Sample Solutions

Six ping-pong balls are labeled as follows: $0, 3, 6, 9, 12, 18$. Three ping-pong balls will be randomly assigned to Group A; the rest will be assigned to Group B. $\text{Diff} = \bar{x}_A - \bar{x}_B$

1. **Calculate Diff, the difference between the mean of the numbers, on the balls assigned to Group A and the mean of the numbers on the balls assigned to Group B (i.e., $\bar{x}_A - \bar{x}_B$) when the 3 ping-pong balls selected for Group A are 3, 6, and 12.**

   $7 - 9 = -2$

2. **Calculate Diff, the difference between the mean of the numbers, on the balls assigned to Group A and the mean of the numbers on the balls assigned to Group B (i.e., $\bar{x}_A - \bar{x}_B$) when the 3 ping-pong balls selected for Group A are 3, 12, and 18.**

   $11 - 5 = 6$

3. **What is the greatest possible value of Diff, and what selection of ping-pong balls for Group A corresponds to that value?**

   If 9, 12, and 18 (the three highest numbers) are selected for Group A, $\text{Diff} = 13 - 3 = 10$.

4. **What is the smallest (most negative) possible value of Diff and what selection of ping-pong balls for Group A corresponds to that value?**

   If 0, 3, and 6 (the three lowest numbers) are selected for Group A, $\text{Diff} = 3 - 13 = -10$.

5. **If these 6 observations represent the burn times of 6 candles (in minutes), explain what a Diff value of 6 means in terms of (a) which group (A or B) has the longer average burn time and (b) the amount of time by which that group's mean exceeds the other group's mean.**

   Group A has a mean burn time that is 6 minutes higher (longer) than Group B's mean burn time.

Problem Set Sample Solutions

Six ping-pong balls are labeled as follows: $0, 3, 6, 9, 12, 18$. Three ping-pong balls will be randomly assigned to Group A; the rest will be assigned to Group B. Diff $= \overline{x}_A - \overline{x}_B$

In the Exit Ticket problem, 4 of the 20 possible randomizations have been addressed.

1. **Develop the remaining 16 possible random assignments to two groups, and calculate the Diff value for each.**

   (Note: Avoid redundant cases; selecting 0, 3, and 6 for Group A is *not* a distinct random assignment from selecting 6, 0, and 3, so do not record both.)

   Note: Reference notes on the far right of the table below refer to the corresponding question in the Exit Ticket.

Group A Selection			$\overline{x}_A$	$\overline{x}_B$	Diff	
3	6	12	7	9	-2	Question #1
3	12	18	11	5	6	Question #2
9	12	18	13	3	10	Question #3
0	3	6	3	13	-10	Question #4
0	3	9	4	12	-8	
0	3	12	5	11	-6	
0	3	18	7	9	-2	
0	6	9	5	11	-6	
0	6	12	6	10	-4	
0	6	18	8	8	0	
0	9	12	7	9	-2	
0	9	18	9	7	2	
0	12	18	10	6	4	
3	6	9	6	10	-4	
3	6	18	9	7	2	
3	9	12	8	8	0	
3	9	18	10	6	4	
6	9	12	9	7	2	
6	9	18	11	5	6	
6	12	18	12	4	8	

2. **Create a dot plot that shows the 20 Diff values obtained from the 20 possible randomizations. By visual inspection, what is the mean and median value of the distribution?**

   The mean and median appear to be 0 due to the center and symmetry.

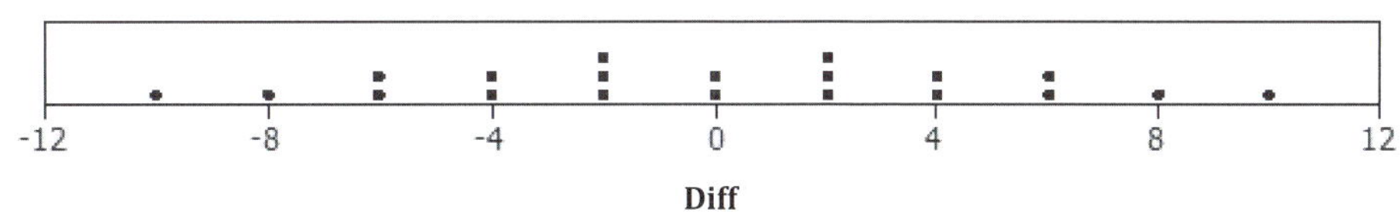

3. **Based on your dot plot, what is the probability of obtaining a Diff value of 8 or higher?**

   Two of the 20 values are 8 or higher.

   $\frac{2}{20} = 0.10 = 10\%$

4. **Would a Diff value of 8 or higher be considered a difference that is likely to happen or one that is unlikely to happen? Explain.**

   A Diff value of 8 or higher would be considered unlikely to happen since a difference this large would happen only 10% of the time.

5. **Based on your dot plot, what is the probability of obtaining a Diff value of -2 or smaller?**

   Nine of the 20 values are -2 or smaller.

   $\frac{9}{20} = 0.45 = 45\%$

6. **Would a Diff value of -2 or smaller be considered a difference that is likely to happen or one that is unlikely to happen? Explain.**

   A Diff value of -2 or smaller would be considered likely to happen since a difference this small would happen 45% of the time.

Lesson 26: Ruling Out Chance

Student Outcomes

- Given data from a statistical experiment with two treatments, students create a randomization distribution.
- Students use a randomization distribution to determine if there is a significant difference between two treatments.

Lesson Notes

In the previous lesson, students investigated examples of the random assignment of 10 tomatoes to two groups of 5 each. In each case, they calculated $\text{Diff} = \bar{x}_A - \bar{x}_B$, the difference between the mean weight of the 5 tomatoes in Group A and the mean weight of the 5 tomatoes in Group B. The Diff values varied.

This lesson asks students to consider if certain Diff values may be unusual or extreme. To do this, students need to develop a sense of the center, spread, and shape of the distribution of possible Diff values. Developing a distribution of *all* possible values based on this random assignment approach would be very difficult, particularly if there were a greater number of tomatoes involved. Thus, repeated simulation is employed to develop something called a *randomization distribution* to adequately approximate the true probability distribution of Diff. The randomization distributions are predeveloped when presented in this lesson; in the next lesson, students create their own randomization distributions.

Classwork

Opening Exercise (3 minutes)

Recall the scenario that students worked through in the previous lesson. Let students answer the questions independently and share their responses with a neighbor. Then, discuss the following as a class:

- Even a distribution that is centered at 0 could have some variability. This means it is important to examine the entire distribution of the Diff statistic to establish just how unusual or extreme a specific Diff value is. To do this, you need a sense of the center, spread, and shape of the distribution in order to determine if a specific value is unusual or extreme.
- Developing a distribution of *all* possible differences based on this random assignment approach could be very difficult—particularly if there were a greater number of tomatoes involved. Fortunately, you can use simulation to develop something called a *randomization distribution* to adequately approximate the true probability distribution of Diff.

Scaffolding:

Use the following as a concrete example that illustrates the point made in this opening:

If the treatment was not effective, which of the following mean weights would you expect?

A	B
4.2	4.0
2.9	2.3
3.0	3.4
2.0	1.2
5.0	4.0

This highlights the difficulty in quantifying what is a meaningful difference in means and illustrates the point of the lesson.

Opening Exercise

Previously, you considered the random assignment of 10 tomatoes into two distinct groups of 5 tomatoes, Group A and Group B. With each random assignment, you calculated $\text{Diff} = \bar{x}_A - \bar{x}_B$, the difference between the mean weight of the 5 tomatoes in Group A and the mean weight of the 5 tomatoes in Group B.

a. **Summarize in writing what you learned in the last lesson. Share your thoughts with a neighbor.**

In the last lesson, when the single group of observations was randomly divided into two groups, the means of these two groups differed by chance. In some cases, the difference in the means of these two groups was very small (or 0), but in other cases, this difference was larger. However, in order to determine which differences were typical and ordinary versus unusual and rare, a sense of the center, spread, and shape of the distribution of possible differences is needed.

b. **Recall that 5 of these 10 tomatoes are from plants that received a nutrient treatment in the hope of growing bigger tomatoes. But what if the treatment was *not* effective? What difference would you expect to find between the group means?**

I would expect there to be no difference between the mean of Group A and the mean of Group B when performing these randomization assignments; in other words, I would expect a value of Diff equal to 0.

Exercises 1–2 (7 minutes): The Distribution of Diff and Why 0 Is Important

Read through the beginning of the exercise, and verify that students understand the information displayed in the dot plot. Let students work with a partner on Exercises 1 and 2. Then, confirm answers as a class.

Exercises 1–2: The Distribution of Diff and Why 0 Is Important

In the previous lesson, 3 instances of the tomato randomization were considered. Imagine that the random assignment was conducted an additional 247 times, and 250 Diff values were computed from these 250 random assignments. The results are shown graphically below in a dot plot where each dot represents the Diff value that results from a random assignment:

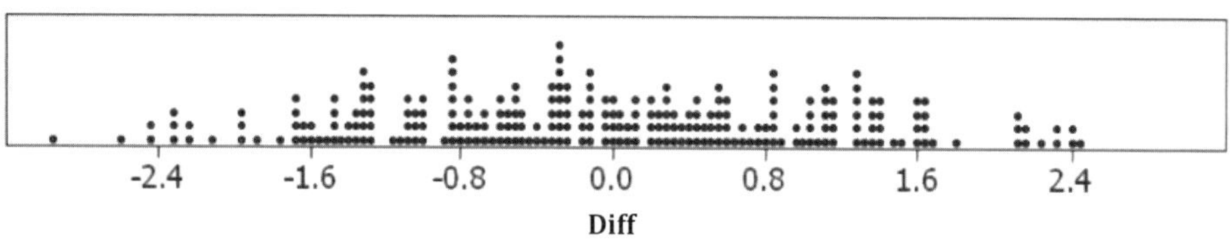

This dot plot will serve as your *randomization distribution* for the Diff statistic in this tomato randomization example. The dots are placed at increments of 0.04 ounces.

1. **Given the distribution picture above, what is the *approximate* value of the median and mean of the distribution? Specifically, do you think this distribution is centered near a value that implies "No Difference" between Group A and Group B?**

   The Diff value that implies "No Difference" between Group A and Group B would be 0. From visual inspection, the approximate value of the median and mean of the distribution appears to be near 0, based on the near symmetry and the center of the distribution. (Note: The actual mean in this case is -0.053 ounces, and the median is -0.08 ounces—both just slightly below 0.)

2. Given the distribution pictured above and based on the simulation results, determine the approximate probability of obtaining a Diff value in the cases described in (a), (b), and (c).

a. Of 1.64 ounces or more

17 out of 250 are 1.64 or more.

$$\frac{17}{250} = 0.068 = 6.8\%$$

b. Of -0.80 ounces or less

69 out of 250 are -0.80 or less.

$$\frac{69}{250} = 0.276 = 27.6\%$$

c. Within 0.80 ounces of 0 ounces

121 out of 250 are between -0.80 and 0.80.

$$\frac{121}{250} = 0.484 = 48.4\%$$

d. How do you think these probabilities could be useful to people who are designing experiments?

The probabilities could be used to help determine if the differences occurred by chance or not.

Exercises 3–5 (15 minutes): Statistically Significant Diff Values

Determining how unusual or extreme a Diff value is allows students to then consider if their experiments' results are statistically significant. The reasoning is as follows:

- 5 of these 10 tomatoes are from plants that received a nutrient treatment in the hope of growing bigger tomatoes, and the other 5 received no such treatment. If the treatment was *not* effective, then one would generally expect there to be no difference between the mean of Group A and the mean of Group B when performing these randomization assignments; in other words, we would expect a value of Diff equal to 0.
- However, as seen in the previous lesson, the value of Diff varies due to chance behavior.
- After establishing a sense of the full distribution of Diff, if the observed difference from an experiment is "extreme" (far from 0) and not typical of chance behavior, it may be considered "statistically significant" and possibly not the result of chance behavior.
- If the difference is not the result of chance behavior, then maybe the difference did not just happen by chance alone.
- If the difference didn't just happen by chance alone, maybe the difference observed in the experiment is caused by the treatment in question, which, in this case, is the nutrient.

Keep in mind that saying *statistically significant* in this case means that the observed difference between two groups is not likely due to chance.

In Exercises 3–5 and 6–8, students are asked to speculate if certain values of the Diff statistic are statistically significant. Specifically, students are told that the Diff statistic value should be considered statistically significant if there is a *low* probability of obtaining a result as extreme as or more extreme than the value in question. However, a "cutoff value" as to what is considered a low probability has been deliberately omitted. Ideally, students should give some thought as to what probability values might be associated with unusual events. In real-world situations, the cutoff probability value used for determining statistical significance (called a *significance level*) varies from situation to situation based on context; however, many introductory statistics references use a value of 0.05, or 5%, as a benchmark.

Exercises 3–5: Statistically Significant Diff Values

In the context of a randomization distribution that is based upon the assumption that there is no real difference between the groups, consider a Diff value of X to be statistically significant if there is a low probability of obtaining a result that is as extreme as or more extreme than X.

3. **Using that definition and your work above, would you consider any of the Diff values below to be statistically significant? Explain.**

   a. **1.64 ounces**

   Possibly statistically significant; an event with a 6.8% probability of occurring is not a very frequent occurrence.

   b. **-0.80 ounces**

   Not statistically significant; an event with a 27.6% probability of occurring is a fairly common occurrence.

   c. **Values within 0.80 ounces of 0 ounces**

   Values within 0.80 ounces of 0 ounces are not statistically significant; these values are not very far from 0, and they are fairly common. Also, given the symmetry, if -0.80 is not considered statistically significant (in part (b), above), then values that are closer to 0 would also not be considered statistically significant.

4. **In the previous lessons, you obtained Diff values of 0.28 ounces, 2.44 ounces, and 0 ounces for 3 different tomato randomizations. Would you consider any of those values to be statistically significant for this distribution? Explain.**

   The values of 0 and 0.28 ounces would not be statistically significant based on the work above and the fact that neither value is very far from 0 in the distribution. However, the value of 2.44 would be statistically significant because it is very far from 0 (maximum observation), and there is only a 1 in 250 chance (0.004, or 0.4% chance) of obtaining a value that extreme in this distribution.

5. **Recalling that Diff is the mean weight of the 5 Group A tomatoes minus the mean weight of the 5 Group B tomatoes, how would you explain the meaning of a Diff value of 1.64 ounces in this case?**

   The 5 tomatoes of Group A have a mean weight that is 1.64 ounces higher than the mean weight of Group B's 5 tomatoes.

Exercises 6–8 (10 minutes): The Implication of Statistically Significant Diff Values

Read through the exercise as a class. Work through one or two of the Diff values in Exercise 6 as a class. Let students continue to work with their partners on the exercises. Then, confirm answers as a class.

Exercises 6–8: The Implication of Statistically Significant Diff Values

Keep in mind that for reasons mentioned earlier, the randomization distribution above is demonstrating what is likely to happen *by chance alone* if the treatment was *not* effective. As stated in the previous lesson, you can use this randomization distribution to assess whether or not the *actual* difference in means *obtained from your experiment* (the difference between the mean weight of the 5 actual control group tomatoes and the mean weight of the 5 actual treatment group tomatoes) is consistent with usual chance behavior. The logic is as follows:

- **If the observed difference is "extreme" and not typical of chance behavior, it may be considered statistically significant and possibly not the result of chance behavior.**
- **If the difference is not the result of chance behavior, then maybe the difference did not just happen by chance alone.**
- **If the difference did not just happen by chance alone, maybe the difference you observed is caused by the treatment in question, which, in this case, is the nutrient. In the context of our example, a statistically significant Diff value provides evidence that the nutrient treatment did in fact yield heavier tomatoes on average.**

6. **For reasons that will be explained in the next lesson, for your tomato example, Diff values that are *positive* and statistically significant will be considered as good evidence that your nutrient treatment did in fact yield heavier tomatoes on average. Again, using the randomization distribution shown earlier in the lesson, which (if any) of the following Diff values would you consider to be statistically significant and lead you to think that the nutrient treatment did in fact yield heavier tomatoes on average? Explain for each case.**

$$\text{Diff} = 0.4,\ \text{Diff} = 0.8,\ \text{Diff} = 1.2,\ \text{Diff} = 1.6,\ \text{Diff} = 2.0,\ \text{Diff} = 2.4$$

$\text{Diff} = 0.4$: *not statistically significant; not very far from* 0*;* 91 *of* 250 *values* (36.4%) *are greater than or equal to* 0.4.

$\text{Diff} = 0.8$: *not statistically significant; not very far from* 0*;* 64 *of* 250 *values* (25.6%) *are greater than or equal to* 0.8.

$\text{Diff} = 1.2$: *not statistically significant; not too far from* 0*;* 38 *of* 250 *values* (15.2%) *are greater than or equal to* 1.2.

$\text{Diff} = 1.6$: *possibly statistically significant; somewhat far from* 0*;* 21 *of* 250 *values* (8.4%) *are greater than or equal to* 1.6.

$\text{Diff} = 2.0$: *statistically significant; very far from* 0*; only* 11 *of* 250 *values* (4.4%) *are greater than or equal to* 2.0.

$\text{Diff} = 2.4$: *statistically significant; very far from* 0*; only* 3 *of* 250 *values* (1.2%) *are greater than or equal to* 2.4.

MP.2

7. **In the first random assignment in the previous lesson, you obtained a Diff value of 0.28 ounce. Earlier in this lesson, you were asked to consider if this might be a statistically significant value. Given the distribution shown in this lesson, if you had obtained a Diff value of 0.28 ounces *in your experiment* and the 5 Group A tomatoes had been the "treatment" tomatoes that received the nutrient, would you say that the Diff value was extreme enough to support a conclusion that the nutrient treatment yielded heavier tomatoes on average? Or do you think such a Diff value may just occur by chance when the treatment is ineffective? Explain.**

I would say that the Diff *value of* 0.28 *was NOT extreme enough to support a conclusion that the nutrient treatment yielded heavier tomatoes on average. Such a* Diff *value may just occur by chance in this case. See earlier work in Exercise 4. Also, referencing the question above,* 0.28 *is even closer to* 0 *than other not statistically significant values.*

MP.2

8. **In the second random assignment in the previous lesson, you obtained a Diff value of 2.44 ounces. Earlier in this lesson, you were asked to consider if this might be a statistically significant value. Given the distribution shown in this lesson, if you had obtained a Diff value of 2.44 ounces *in your experiment* and the 5 Group A tomatoes had been the "treatment" tomatoes that received the nutrient, would you say that the Diff value was extreme enough to support a conclusion that the nutrient treatment yielded heavier tomatoes on average? Or do you think such a Diff value may just occur by chance when the treatment is ineffective? Explain.**

   I would say that the Diff value of 2.44 was extreme enough to support a conclusion that the nutrient treatment yielded heavier tomatoes on average. This most likely did NOT just occur by chance. See earlier work in Exercise 4. Also, referencing the question above, 2.44 is even farther away from 0 than other statistically significant values.

Closing (5 minutes)

- If you were about to prepare for a debate, a court case, or some other situation where you were challenging an existing claim such as "the treatment is not effective" or "the process is not harming anyone," what kind of probability value would you be looking for in your results before you would claim statistical significance and feel comfortable challenging the existing claim? What characteristics of the situation might influence your decision?
  - *In order to be confident in your challenge, your probability value needs to be fairly small. In the tomato example, consider that you are saying "If the nutrient was not effective, the chances of our obtaining a value this extreme in our experiment is ____; therefore, we think the nutrient must be working." You are acknowledging that your experiment's results could still occur a certain percent of the time even if the nutrient really did not work. If you were trying to convince an audience, a constituency, or a jury, you probably would not want to say, "There's a 1 in 3 (33%) chance of obtaining a value this extreme." Rather, you would want to say that the chances are "1 in 20 (5%)," or "1 in 100 (1%)," or "1 in $1{,}000$ (0.1%)." Considerations include the amount of money and time required for sampling, the severity of the claim that is being examined, and the risks of falsely rejecting the claim of "no difference" when it was true all along, and so on.*
- Calculate and interpret $\text{Diff} = \bar{x}_A - \bar{x}_B$ in the following instance: Group A: 10 similar homes with insulated windows have an average monthly electric bill of $\$123$. Group B: 10 similar homes with noninsulated windows have an average monthly electric bill of $\$157$.
  - *The homes with insulated windows have an average monthly electric bill that is $\$34$ less than the homes with noninsulated windows.*

Ask students to summarize the main ideas of the lesson in writing or with a neighbor. Use this as an opportunity to informally assess comprehension of the lesson. The Lesson Summary below offers some important ideas that should be included.

Lesson Summary

In the previous lesson, the concept of randomly separating 10 tomatoes into 2 groups and comparing the means of each group was introduced. The randomization distribution of the difference in means that is created from multiple occurrences of these random assignments demonstrates what is likely to happen *by chance alone* if the nutrient treatment is *not* effective. When the results of your tomato growth experiment are compared to that distribution, you can then determine if the tomato growth experiment's results were typical of chance behavior.

If the results appear typical of chance behavior and near the center of the distribution (that is, not relatively very far from a Diff value of 0), then there is little evidence that the treatment was effective. However, if it appears that the experiment's results are not typical of chance behavior, then maybe the difference you are observing didn't just happen by chance alone. It may indicate a statistically significant difference between the treatment group and the control group, and the source of that difference might be (in this case) the nutrient treatment.

Exit Ticket (5 minutes)

Name ______________________________ Date ______________

Lesson 26: Ruling Out Chance

Exit Ticket

Medical patients who are in physical pain are often asked to communicate their level of pain on a scale of 0 to 10 where 0 means no pain and 10 means worst pain. (Note: Sometimes a visual device with pain faces is used to accommodate the reporting of the pain score.) Due to the structure of the scale, a patient would desire a lower value on this scale after treatment for pain.

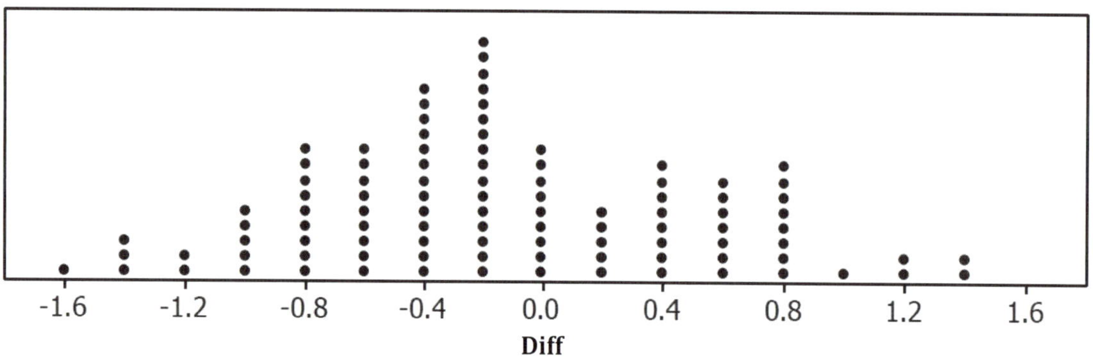

Imagine that 20 subjects participate in a clinical experiment and that a variable of "ChangeinScore" is recorded for each subject as the subject's pain score after treatment minus the subject's pain score before treatment. Since the expectation is that the treatment would lower a patient's pain score, you would desire a *negative* value for "ChangeinScore." For example, a "ChangeinScore" value of -2 would mean that the patient was in less pain (for example, now at a 6, formerly at an 8).

Recall that $\text{Diff} = \bar{x}_A - \bar{x}_B$. Although the 20 "ChangeinScore" values for the 20 patients are not shown here, below is a randomization distribution of the value Diff based on 100 random assignments of these 20 observations into two groups of 10 (Group A and Group B).

1. From the distribution above, what is the probability of obtaining a Diff value of -1 or less?

2. With regard to this distribution, would you consider a Diff value of -0.4 to be statistically significant? Explain.

3.

a. With regard to how Diff is calculated, if Group A represented a group of patients in your experiment who received a new pain relief treatment and Group B received a pill with no medicine (called a *placebo*), how would you interpret a Diff value of -1.4 pain scale units in context?

b. Given the distribution above, if you obtained a Diff value such as -1.4 from your experiment, would you consider that to be significant evidence of the new treatment being effective on average in relieving pain? Explain.

Exit Ticket Sample Solutions

A graphic of the Wong-Baker FACES Pain Rating Scale (as seen in many physicians' offices) or a similar visual reference may assist students with the Exit Ticket questions. One can be found at http://pain.about.com/od/testingdiagnosis/ig/pain-scales/Wong-Baker.htm.

Medical patients who are in physical pain are often asked to communicate their level of pain on a scale of 0 to 10 where 0 means no pain and 10 means worst pain. (Note: Sometimes a visual device with pain faces is used to accommodate the reporting of the pain score.) Due to the structure of the scale, a patient would desire a lower value on this scale after treatment for pain.

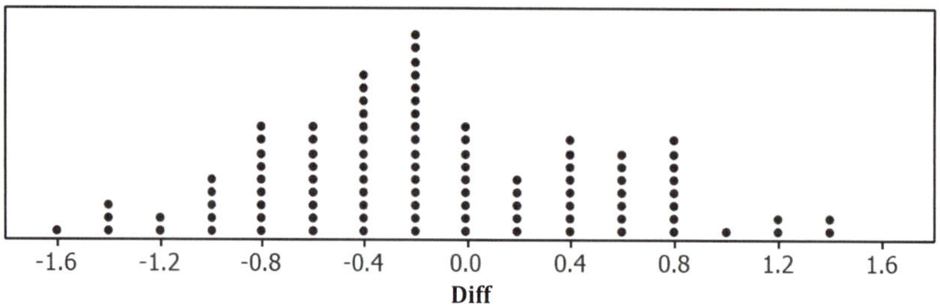

Imagine that 20 subjects participate in a clinical experiment and that a variable of "ChangeinScore" is recorded for each subject as the subject's pain score after treatment minus the subject's pain score before treatment. Since the expectation is that the treatment would lower a patient's pain score, you would desire a negative value for "ChangeinScore." For example, a "ChangeinScore" value of -2 would mean that the patient was in less pain (for example, now at a 6, formerly at an 8).

Recall that $\text{Diff} = \overline{x}_A - \overline{x}_B$. Although the 20 "ChangeinScore" values for the 20 patients are not shown here, below is a randomization distribution of the value Diff based on 100 random assignments of these 20 observations into two groups of 10 (Group A and Group B).

1. **From the distribution above, what is the probability of obtaining a Diff value of -1 or less?**

   $\frac{11}{100} = 0.11$, *which is* 11%

2. **With regard to this distribution, would you consider a Diff value of -0.4 to be statistically significant? Explain.**

   No. The value is not far from 0*, and* 42 *of the* 100 *values in the distribution are at or below* -0.4.

3.
   a. **With regard to how Diff is calculated, if Group A represented a group of patients in your experiment who received a new pain relief treatment and Group B received a pill with no medicine (called a *placebo*), how would you interpret a Diff value of -1.4 pain scale units in context?**

      The group that received the pain relief treatment had an average reduction in pain that was 1.4 *units better (lower) than the group that received the placebo.*

   b. **Given the distribution above, if you obtained Diff value such as -1.4 from your experiment, would you consider that to be significant evidence of the new treatment being effective on average in relieving pain? Explain.**

      Yes. -1.4 *is far from* 0*, and the probability of obtaining a* Diff *value of* -1.4 *or less is only* 4%. *The value provides evidence that the new treatment may be effective in relieving pain.*

Problem Set Sample Solutions

In each of the 3 cases below, calculate the Diff value as directed, and write a sentence explaining what the Diff value means in context. Write the sentence for a general audience.

1. Group A: 8 dieters lost an average of 8 pounds, so $\overline{x}_A = -8$.

   Group B: 8 nondieters lost an average of 2 pounds over the same time period, so $\overline{x}_B = -2$.

   Calculate and interpret $\text{Diff} = \overline{x}_A - \overline{x}_B$.

   $\text{Diff} = \overline{x}_A - \overline{x}_B = -8 - (-2) = -6$. *The 8 dieters lost an average of 6 pounds more than the 8 nondieters.*

2. Group A: 11 students were on average 0.4 seconds faster in their 100-meter run times after following a new training regimen.

   Group B: 11 students were on average 0.2 seconds slower in their 100-meter run times after not following any new training regimens.

   Calculate and interpret $\text{Diff} = \overline{x}_A - \overline{x}_B$.

   $\text{Diff} = -0.4 - 0.2 = -0.6$. *The 11 students following the new training regimen were on average 0.6 seconds faster in their 100-meter run times than the 11 students not following any new training regimens.*

3. Group A: 20 squash that have been grown in an irrigated field have an average weight of 1.3 pounds.

   Group B: 20 squash that have been grown in a nonirrigated field have an average weight of 1.2 pounds.

   Calculate and interpret $\text{Diff} = \overline{x}_A - \overline{x}_B$.

   $\text{Diff} = 1.3 - 1.2 = 0.1$. *The 20 squash grown in an irrigated field have an average weight that is 0.1 pound higher than the 20 squash grown in a nonirrigated field.*

4. Using the randomization distribution shown below, what is the probability of obtaining a Diff value of -0.6 or less?

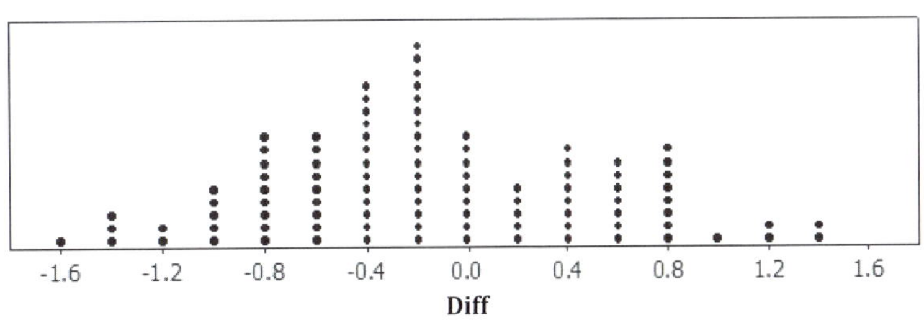

   $\frac{29}{100} = 0.29$, *which is* 29%

5. Would a Diff value of -0.6 or less be considered a statistically significant difference? Why or why not?

   No. -0.6 *is not far from* 0, *and the probability of obtaining a* Diff *value of* -0.6 *or less is* 29%.

6. Using the randomization distribution shown in Problem 4, what is the probability of obtaining a Diff value of -1.2 or less?

   The probability of obtaining a Diff *value of* -1.2 *or less is* 6 *out of* 100*, or* 6%.

7. Would a Diff value of -1.2 or less be considered a statistically significant difference? Why or why not?

   Possibly statistically significant. -1.2 *is far from* 0*, and the probability of obtaining a* Diff *value of* -1.2 *or less is only* 6%*. The value provides evidence that the new treatment may be effective in relieving pain.*

Lesson 27: Ruling Out Chance

Student Outcomes

- Given data from a statistical experiment with two treatments, students create a randomization distribution.
- Students use a randomization distribution to determine if there is a significant difference between two treatments.

Lesson Notes

MP.4 In this lesson, students perform the five steps of a randomization test to assess if the tomato data provide evidence that the nutrient was effective. All of the steps have been practiced or indirectly discussed in the previous two lessons; however, now everything is detailed and performed together in context.

In the second half of the lesson, students begin the important task of developing their own randomization distributions—first manually and then using technology. The use of technology is strongly encouraged to assist in the steps of a randomization test, specifically a Web-based randomization testing applet/calculator available at http://www.rossmanchance.com/applets/AnovaShuffle.htm.

This lesson may require slightly more time than the two previous lessons, but it is important that students have a solid understanding of these steps (and some familiarity with the technology) prior to moving on to the capstone experiences in the subsequent lessons.

Classwork

Opening (5 minutes)

Before the start of the lesson, ask students to recall what they have learned from earlier lessons, either in writing or by speaking to a neighbor, and then share with the class.

- To recap some highlights of the past few lessons:
  - *You have learned how to carry out repeated random assignments to develop a randomization distribution for the difference in means of two groups.*
  - *You have also established a method for determining if a specific difference value in this distribution is considered extreme or significant.*
  - *When a specific difference is significant, it is evidence that the specific difference did not occur by chance alone and may have in fact occurred due to something other than chance; for example, a treatment may have been imposed in one of an experiment's groups.*

Exercises 1–4 (15 minutes): Carrying Out a Randomization Test

Before students begin the exercises, use the following question to encourage discussion about what is needed to carry out a randomization test. Allow for multiple student responses.

- Think back to the work done with the tomato experiment. Statisticians agree there are five important parts to the process of using randomization to reach a conclusion in an experiment. Using what we've learned, see how many you can come up with on your own.

Work through Exercises 1–4 as a class. Pose one question at a time, and allow sufficient time for discussion of each step for carrying out a randomization test.

Exercises 1–4: Carrying Out a Randomization Test

The following are the general steps for carrying out a randomization test to analyze the results of an experiment. The steps are also presented in the context of the tomato example of the previous lessons.

***Step 1*—Develop competing claims: no difference versus difference.**

One claim corresponds to no difference between the two groups in the experiment. This claim is called the *null hypothesis*.

- **For the tomato example, the null hypothesis is that the nutrient treatment is not effective in increasing tomato weight. This is equivalent to saying that the average weight of treated tomatoes may be the same as the average weight of nontreated (control) tomatoes.**

The competing claim corresponds to a difference between the two groups. This claim could take the form of a *different from*, *greater than*, or *less than* statement. This claim is called the *alternative hypothesis*.

- **For the tomato example, the alternative hypothesis is that the nutrient treatment is effective in increasing tomato weight. This is equivalent to saying that the average weight of treated tomatoes *is greater than* the average weight of nontreated (control) tomatoes.**

Scaffolding:

- Students may want to create a Frayer diagram and/or rehearse with *null hypothesis* and *alternative hypothesis*.
- After the 5 steps are introduced, consider asking students to restate, write, or draw their own representations of them.
- For advanced students, ask them to use the 5 steps to design their own randomized experiment to answer a question of interest.

1. **Previously, the statistic of interest that you used was the difference between the mean weight of the 5 tomatoes in Group A and the mean weight of the 5 tomatoes in Group B. That difference was called *Diff*. $\text{Diff} = \overline{x}_A - \overline{x}_B$ If the treatment tomatoes are represented by Group A and the control tomatoes are represented by Group B, what type of statistically significant values of Diff would support the claim that the average weight of treated tomatoes *is greater than* the average weight of nontreated (control) tomatoes: negative values of Diff positive values of Diff or both? Explain.**

   For the tomato example, since the treatment group is Group A, the Diff value of $\overline{x}_A - \overline{x}_B$ is $\overline{x}_{\text{Treatment}} - \overline{x}_{\text{Control}}$. Since the alternative claim is supported by $\overline{x}_{\text{Treatment}} > \overline{x}_{\text{Control}}$, we are seeking statistically significant Diff values that are positive since if $\overline{x}_{\text{Treatment}} > \overline{x}_{\text{Control}}$, then $\overline{x}_{\text{Treatment}} - \overline{x}_{\text{Control}} > 0$. Statistically significant values of Diff that are negative in this case would imply that the treatment made the tomatoes smaller on average.

   Note: The answer/information above appears in Step 4 of the randomization test of the student material.

***Step 2*—Take measurements from each group, and calculate the value of the Diff statistic from the experiment.**

For the tomato example, first, measure the weights of the 5 tomatoes from the treatment group (Group A); next, measure the weights of the 5 tomatoes from the control group (Group B); finally, compute $\text{Diff} = \overline{x}_A - \overline{x}_B$, which will serve as the result from your experiment.

2. Assume that the following represents the two groups of tomatoes from the *actual* experiment. Calculate the value of $\text{Diff} = \overline{x}_A - \overline{x}_B$. This will serve as the result from your experiment.

These are the same 10 tomatoes used in previous lessons; the identification of which tomatoes are treatment versus control is now revealed.

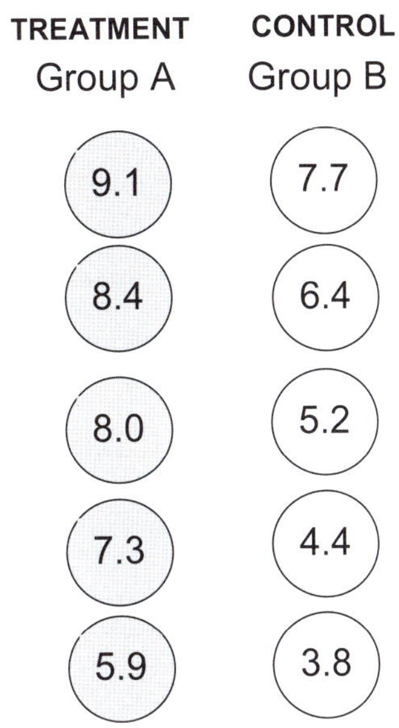

$\text{Diff} = 7.74 - 5.5 = 2.24$

Again, these tomatoes represent the actual result from your experiment. You will now create the randomization distribution by making repeated random assignments of these 10 tomatoes into 2 groups and recording the observed difference in means for each random assignment. This develops a randomization distribution of the many possible difference values that could occur under the assumption that there is no difference between the mean weights of tomatoes that receive the treatment and tomatoes that don't receive the treatment.

Step 3—Randomly assign the observations to two groups, and calculate the difference between the group means. Repeat this several times, recording each difference. This will create the randomization distribution for the Diff *statistic.*

Examples of this technique were presented in a previous lesson. For the tomato example, the randomization distribution has already been presented in a previous lesson and is shown again here. The dots are placed at increments of 0.04 ounces.

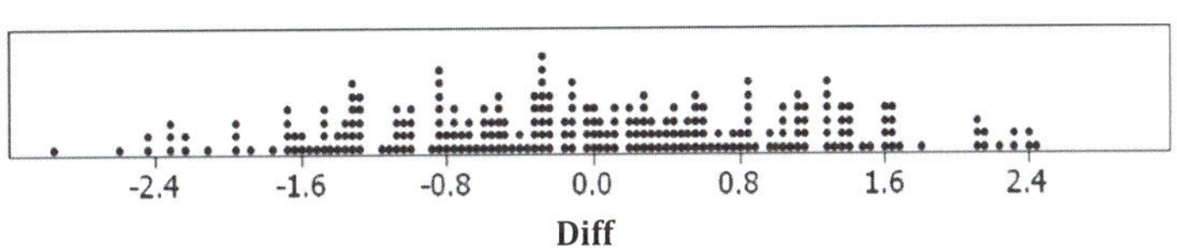

Step 4—With reference to the randomization distribution (Step 3) and the inequality in your alternative hypothesis (Step 1), compute the probability of getting a Diff *value as extreme as or more extreme than the* Diff *value you obtained in your experiment (Step 2).*

For the tomato example, since the treatment group is Group A, the Diff value of $\bar{x}_A - \bar{x}_B$ is $\bar{x}_{\text{Treatment}} - \bar{x}_{\text{Control}}$. Since the alternative claim is supported by $\bar{x}_{\text{Treatment}} > \bar{x}_{\text{Control}}$, you are seeking statistically significant Diff values that are positive since if $\bar{x}_{\text{Treatment}} > \bar{x}_{\text{Control}}$, then $\bar{x}_{\text{Treatment}} - \bar{x}_{\text{Control}} > 0$.

Statistically significant values of Diff that are negative, in this case, would imply that the treatment made the tomatoes smaller on average.

3. Using your calculation from Exercise 2, determine the probability of getting a Diff value as extreme as or more extreme than the Diff value you obtained for this experiment (in Step 2).

   6 out of 250 is 0.024, which is 2.4%

Step 5—Make a conclusion in context based on the probability calculation (Step 4).

If there is a *small probability* of obtaining a Diff value as extreme as or more extreme than the Diff value you obtained in your experiment, then *the Diff value from the experiment is unusual* and not typical of chance behavior. Your experiment's results probably did not happen by chance, and the results probably occurred because of a statistically significant difference in the two groups.

- In the tomato experiment, if you think there is a statistically significant difference in the two groups, *you have evidence that the treatment may in fact be yielding heavier tomatoes on average.*

If there is *not a small probability* of obtaining a Diff value as extreme as or more extreme than the Diff value you obtained in your experiment, then your Diff *value from the experiment is NOT considered unusual* and could be typical of chance behavior. The experiment's results may have just happened by chance and not because of a statistically significant difference in the two groups.

- In the tomato experiment, if you don't think that there is a statistically significant difference in the two groups, then you do *not* have evidence that the treatment results in larger tomatoes on average.

In some cases, a specific cutoff value called a *significance level* might be employed to assist in determining how small this probability must be in order to consider results statistically significant.

4. Based on your probability calculation in Exercise 3, do the data from the tomato experiment support the claim that the treatment yields heavier tomatoes on average? Explain.

   The data from the tomato experiment support the claim that the treatment yields heavier tomatoes on average. The probability of obtaining a Diff value as extreme as or more extreme than 2.24 is very small if the treatment is ineffective. This means that there is a statistically significant difference between the average weights of the treatment and control tomatoes, with the treatment tomatoes being heavier on average.

Exercises 5–10 (10 minutes): Developing the Randomization Distribution

MP.5 In this set of exercises, students begin the important task of developing their own randomization distributions—first manually and then using technology. Students' work will vary, so there are no specific single answers for certain tasks in this teacher material.

For Exercises 5–7 under Manually Generated, it is strongly recommended that the performing of random assignments (and computing of the Diff statistic) be distributed among the class rather than having any one individual perform too many of these randomizations by hand. While the task may be repetitive, it is important for students to grasp what a randomization distribution is, how it is developed, and how technology can greatly assist.

Exercises 5–10: Developing the Randomization Distribution

Although you are familiar with how a randomization distribution is created in the tomato example, the randomization distribution was provided for you. In this exercise, you will develop two randomization distributions based on the same group of 10 tomatoes. One distribution will be developed by hand and will contain the results of at least 250 random assignments. The second distribution will be developed using technology and will contain the results of at least 250 random assignments. Once the two distributions have been developed, you will be asked to compare the distributions.

Manually Generated

Your instructor will provide you with specific guidance regarding how many random assignments you need to carry out. Ultimately, your class should generate at least 250 random assignments, compute the Diff value for each, and record these 250 or more Diff values on a class or an individual dot plot.

5. **To begin, write the 10 tomato weights on 10 equally sized slips of paper, one weight on each slip. Place the slips in a container, and shake the container well. Remove 5 slips, and assign those 5 tomatoes as Group A. The remaining tomatoes will serve as Group B.**

   For the manually generated randomization distribution, it states that students can use 10 equally sized pieces of paper to be selected from a container. However, manipulatives such as small chips or checkers would be fine as well.

6. **Calculate the mean weight for Group A and the mean weight for Group B. Then, calculate $\text{Diff} = \overline{x}_A - \overline{x}_B$ for this random assignment.**

   See above.

7. **Record your Diff value, and add this value to the dot plot. Repeat as needed per your instructor's request until a manually generated randomization distribution of at least 250 differences has been achieved.**

   (Note: This distribution will most likely be slightly different from the tomato randomization distribution given earlier in this lesson.)

   Students' work will vary. Overall, the manually generated distribution for the Diff statistic in this tomato example should not differ too much from the distribution presented previously in this and other lessons.

As stated earlier, when it is time to move on to Exercises 8–10, students' computers (or at least one station in the classroom) require Internet access in order to use the Anova Shuffle applet. Instructor practice with the applet beforehand is recommended. Specific instructions, comments, and recommendations regarding its use appear in the student lesson material.

This computer-generated distribution is most likely slightly different from both the randomization distribution that appeared earlier in this lesson and the randomization distribution that was manually generated.

Computer Generated

At this stage, you will be encouraged to use a Web-based randomization testing applet/calculator to perform the steps above. The applet is located at http://www.rossmanchance.com/applets/AnovaShuffle.htm. To supplement the instructions below, a screenshot of the applet appears as the final page of this lesson.

Upon reaching the applet, do the following:

- Press the Clear button to clear the data under Sample Data.
- Enter the tomato data exactly as shown below. When finished, press the Use Data button.

Group	Ounces
Treatment	9.1
Treatment	8.4
Treatment	8
Control	7.7
Treatment	7.3
Control	6.4
Treatment	5.9
Control	5.2
Control	4.4
Control	3.8

Once the data are entered, notice that dot plots of the two groups appear. Also, the statistic window below the data now says "difference in means," and an Observed Diff value of 2.24 is computed for the experiment's data (just as you computed in Exercise 2).

By design, the applet will determine the difference of means based on the first group name it encounters in the data set—specifically, it will use the first group name it encounters as the first value in the difference of means calculation. In other words, to compute the difference in means as $\bar{x}_{\text{Treatment}} - \bar{x}_{\text{Control}}$, a Treatment observation needs to appear prior to any Control observations in the data set as entered.

- Select the check box next to Show Shuffle Options, and a dot plot template will appear.
- Enter 250 in the box next to Number of Shuffles, and press the Shuffle Responses button. A randomization distribution based on 250 randomizations (in the form of a histogram) is created.

This distribution will most likely be *slightly* different from both the tomato randomization distribution that appeared earlier in this lesson and the randomization distribution that was manually generated in Exercise 7.

8. Write a few comments comparing the manually generated distribution and the computer-generated distribution. Specifically, did they appear to have roughly the same shape, center, and spread?

Students should compare and contrast the shape, center, and spread characteristics of the manually generated and computer-generated distributions. There should be a great deal of similarity, but outliers and slightly different clustering patterns may be present.

The applet also allows you to compute probabilities. For this case:

- Under Count Samples, select Greater Than. Then, in the box next to Greater Than, enter 2.2399.

Since the applet computes the count value as *strictly* greater than and not greater than or equal to, in order to obtain the probability of obtaining a value as extreme as or more extreme than the Observed Diff value of 2.24, you will need to enter a value just slightly below 2.24 to ensure that Diff observations of 2.24 are included in the count.

- Select the Count button. The probability of obtaining a Diff value of 2.24 or more in this distribution will be computed for you.

  The applet displays the randomization distribution in the form of a histogram, and it shades in red *all* histogram classes that contain *any* difference values that meet your Count Samples criteria. Due to the grouping and binning of the classes, some of the red shaded classes (bars) may also contain difference values that do not fit your Count Samples criteria. Just keep in mind that the Count value stated in red below the histogram will be exact; the red shading in the histogram may be approximate.

9. How did the probability of obtaining a Diff value of 2.24 or more using your computer-generated distribution compare with the probability of obtaining a Diff value of 2.24 or more using your manually generated distribution?

   It is expected that the event of finding a Diff value of 2.24 or more will have a similarly rare probability of occurrence in both the manually generated and the computer-generated distributions.

10. Would you come to the same conclusion regarding the experiment using either the computer-generated or manually generated distribution? Explain. Is this the same conclusion you came to using the distribution shown earlier in this lesson back in Step 3?

    Based on the assumptions regarding the similarity of the distributions as described in Exercise 9 above, it is expected that students will come to the same conclusion (that the treatment is effective) from both the manually generated and the computer-generated distributions as was seen in Exercise 4.

Closing (5 minutes)

Realizing that this lesson may require more time than others, if time permits, consider posing the following questions; allow a few student responses for each, or consider investigation/research outside of class.

- Although the idea of randomization testing has been around since the 1930s, it was not used much in practice until the later part of the 20th century. Why might that be?
  - *The computing power necessary to develop a randomization distribution based on numerous random assignments—particularly from data sets with far more observations than those seen in these lessons—was just not as easily or cheaply available in the early 20th century. With technology today, we can develop a randomization distribution from a large data set and from a very high number of random assignments in seconds.*
- In the previous lesson's Exit Ticket problems, a pain reliever was tested. Subjects who did not receive the pain reliever still received a pill, but the pill had no medicine. Why would this step be taken?
  - *Many experiments have shown that human beings tend to exhibit a positive result such as an improvement in pain or a loss of weight even when they are unknowingly taking medicine that has no real medicine or benefit, known as a placebo. Since some people in the experiment might exhibit this "placebo effect," rather than compare the treatment group to a group that receives nothing at all, it is important to establish a baseline of comparison that accounts for this power of suggestion in the subjects' minds; thus, the control group is given a placebo without their knowledge.*

> *Scaffolding:*
>
> Some students may not be familiar with the word *placebo*. Have students look up the definition and put it in their own words to ensure they understand the meaning.

Ask students to summarize the main ideas of the lesson in writing or with a neighbor. Use this as an opportunity to informally assess comprehension of the lesson. The Lesson Summary below offers some important ideas that should be included.

Lesson Summary

The following are the general steps for carrying out a randomization test to analyze the results of an experiment:

Step 1—Develop competing claims: no difference versus difference.

***Develop the null hypothesis*: This claim is that there is no difference between the two groups in the experiment.**

***Develop the alternative hypothesis*: The competing claim is that there *is* a difference between the two groups. This difference could take the form of a *different from*, *greater than*, or *less than* statement depending on the purpose of the experiment and the claim being assessed.**

Step 2—Take measurements from each group, and calculate the value of the Diff statistic from the experiment.

This is the observed Diff value from the experiment.

Step 3—Randomly assign the observations to two groups, and calculate the difference between the group means. Repeat this several times, recording each difference.

This will create the *randomization distribution* for the Diff statistic under the assumption that there is no statistically significant difference between the two groups.

Step 4—With reference to the randomization distribution (from Step 3) and the inequality in your alternative hypothesis (from Step 1), compute the probability of getting a Diff value as extreme as or more extreme than the Diff value you obtained in your experiment (from Step 2).

Step 5—Make a conclusion in context based on the probability calculation (from Step 4).

Small probability: If the Diff value from the experiment is unusual and not typical of chance behavior, your experiment's results probably did not happen by chance. The results probably occurred because of a statistically significant difference in the two groups.

Not a small probability: If the Diff value from the experiment is *not* considered unusual and could be typical of chance behavior, your experiment's results may have just happened by chance and *not* because of a statistically significant difference in the two groups.

Note: The use of technology is strongly encouraged to assist in Steps 2–4.

Exit Ticket (10 minutes)

Name ______________________________ Date ______________

Lesson 27: Ruling Out Chance

Exit Ticket

In the Exit Ticket of a previous lesson, an experiment with 20 subjects investigating a new pain reliever was introduced. The subjects were asked to communicate their level of pain on a scale of 0 to 10 where 0 means no pain and 10 means worst pain. Due to the structure of the scale, a patient would desire a lower value on this scale after treatment for pain. The value "ChangeinScore" was recorded as the subject's pain score after treatment minus the subject's pain score before treatment. Since the expectation is that the treatment would lower a patient's pain score, a *negative* value would be desired for "ChangeinScore." For example, a "ChangeinScore" value of -2 would mean that the patient was in less pain (for example, now at a 6, formerly at an 8).

In the experiment, the null hypothesis would be that the treatment had no effect. The average change in pain score for the treatment group would be the same as the average change in pain score for the placebo (control) group.

1. The alternative hypothesis would be that the treatment was effective. Using this context, which mathematical relationship below would match this alternative hypothesis? Choose one.

    a. The average change in pain score (the average "ChangeinScore") for the treatment group would be less than the average change in pain score for the placebo group (supported by $\bar{x}_{\text{Treatment}} < \bar{x}_{\text{Control}}$, or $\bar{x}_{\text{Treatment}} - \bar{x}_{\text{Control}} < 0$).

    b. The average change in pain score (the average "ChangeinScore") for the treatment group would be greater than the average change in pain score for the placebo group (supported by $\bar{x}_{\text{Treatment}} > \bar{x}_{\text{Control}}$, or $\bar{x}_{\text{Treatment}} - \bar{x}_{\text{Control}} > 0$).

2. Imagine that the 20 "ChangeinScore" observations below represent the change in pain levels of the 20 subjects (chronic pain sufferers) who participated in the clinical experiment. The 10 individuals in Group A (the treatment group) received a new medicine for their pain while the 10 individuals in Group B received the pill with no medicine (placebo). Assume for now that the 20 individuals have similar initial pain levels and medical conditions. Calculate the value of $\text{Diff} = \bar{x}_A - \bar{x}_B = \bar{x}_{\text{Treatment}} - \bar{x}_{\text{Control}}$. This is the result from the experiment.

Group	ChangeinScore
A	0
A	0
A	-1
A	-1
A	-2
A	-2
A	-3
A	-3
A	-3
A	-4
B	0
B	0
B	0
B	0
B	0
B	0
B	-1
B	-1
B	-1
B	-2

3. Below is a randomization distribution of the value $\text{Diff} = \bar{x}_A - \bar{x}_B$ based on 100 random assignments of these 20 observations into two groups of 10 (shown in a previous lesson).

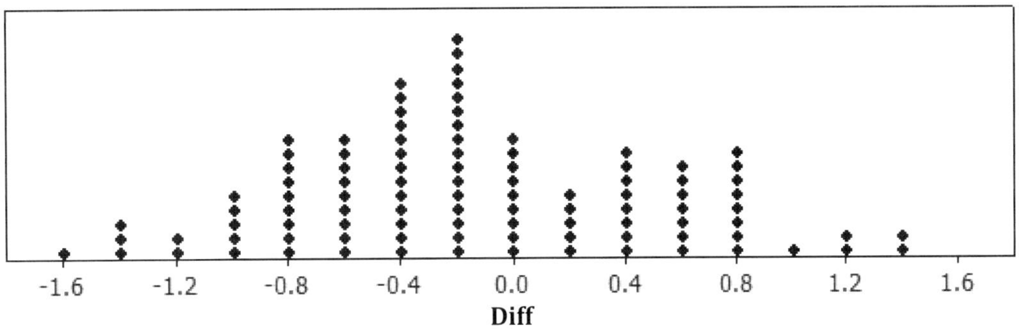

With reference to the randomization distribution above and the inequality in your alternative hypothesis, compute the probability of getting a Diff value as extreme as or more extreme than the Diff value you obtained in the experiment.

4. Based on your probability value from Problem 3 and the randomization distribution above, *choose one* of the following conclusions:

   a. Due to the small chance of obtaining a Diff value as extreme as or more extreme than the Diff value obtained in the experiment, we believe that the observed difference did not happen by chance alone, and we support the claim that the treatment is effective.

   b. Because the chance of obtaining a Diff value as extreme as or more extreme than the Diff value obtained in the experiment is not small, it is possible that the observed difference may have happened by chance alone, and we cannot support the claim that the treatment is effective.

Exit Ticket Sample Solutions

In the Exit Ticket of a previous lesson, an experiment with 20 subjects investigating a new pain reliever was introduced. The subjects were asked to communicate their level of pain on a scale of 0 to 10 where 0 means no pain and 10 means worst pain. Due to the structure of the scale, a patient would desire a lower value on this scale after treatment for pain. The value "ChangeinScore" was recorded as the subject's pain score after treatment minus the subject's pain score before treatment. Since the expectation is that the treatment would lower a patient's pain score, a *negative* value would be desired for "ChangeinScore." For example, a "ChangeinScore" value of -2 would mean that the patient was in less pain (for example, now at a 6, formerly at an 8).

In the experiment, the null hypothesis would be that the treatment had no effect. The average change in pain score for the treatment group would be the same as the average change in pain score for the placebo (control) group.

1. The alternative hypothesis would be that the treatment was effective. Using this context, which mathematical relationship below would match this alternative hypothesis? Choose one.

   a. The average change in pain score (the average "ChangeinScore") for the treatment group would be less than the average change in pain score for the placebo group (supported by $\bar{x}_{\text{Treatment}} < \bar{x}_{\text{Control}}$, or $\bar{x}_{\text{Treatment}} - \bar{x}_{\text{Control}} < 0$).

   b. The average change in pain score (the average "ChangeinScore") for the treatment group would be greater than the average change in pain score for the placebo group (supported by $\bar{x}_{\text{Treatment}} > \bar{x}_{\text{Control}}$, or $\bar{x}_{\text{Treatment}} - \bar{x}_{\text{Control}} > 0$).

   Choice (a): The average change in pain score (the average "ChangeinScore") for the treatment group would be less than the average change in pain score for the placebo group.

2. Imagine that the 20 "ChangeinScore" observations below represent the change in pain levels of the 20 subjects (chronic pain sufferers) who participated in the clinical experiment. The 10 individuals in Group A (the treatment group) received a new medicine for their pain while the 10 individuals in Group B received the pill with no medicine (*placebo*). Assume for now that the 20 individuals have similar initial pain levels and medical conditions. Calculate the value of $\text{Diff} = \bar{x}_A - \bar{x}_B = \bar{x}_{\text{Treatment}} - \bar{x}_{\text{Control}}$. This is the result from the experiment.

   $\text{Diff} = -1.4$ *(Group A mean* $= -1.9$; *Group B mean* $= -0.5$*)*

Group	ChangeinScore
A	0
A	0
A	-1
A	-1
A	-2
A	-2
A	-3
A	-3
A	-3
A	-4
B	0
B	0
B	0
B	0
B	0
B	0
B	-1
B	-1
B	-1
B	-2

EUREKA MATH™

3. Below is a randomization distribution of the value $\text{Diff} = \bar{x}_A - \bar{x}_B$ based on 100 random assignments of these 20 observations into two groups of 10 (shown in a previous lesson).

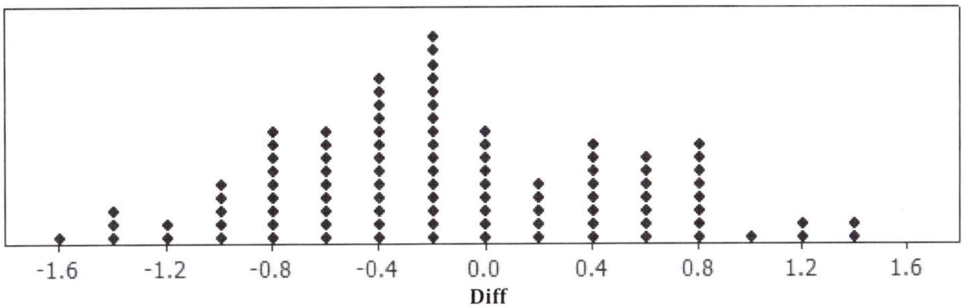

With reference to the randomization distribution above and the inequality in your alternative hypothesis, compute the probability of getting a Diff value as extreme as or more extreme than the Diff value you obtained in the experiment.

4% (4 out of 100)

4. Based on your probability value from Problem 3 and the randomization distribution above, *choose one* of the following conclusions:

   a. Due to the small chance of obtaining a Diff value as extreme as or more extreme than the Diff value obtained in the experiment, we believe that the observed difference did not happen by chance alone, and we support the claim that the treatment is effective.

   b. Because the chance of obtaining a Diff value as extreme as or more extreme than the Diff value obtained in the experiment is not small, it is possible that the observed difference may have happened by chance alone, and we cannot support the claim that the treatment is effective.

   Choice (a): Due to the small chance of obtaining a Diff value as extreme as or more extreme than the Diff value obtained in the experiment, we believe that the observed difference did not happen by chance alone, and we support the claim that the treatment is effective.

Problem Set Sample Solutions

1. Using the 20 observations that appear in the table below for the changes in pain scores of 20 individuals, use the Anova Shuffle applet to develop a randomization distribution of the value $\text{Diff} = \bar{x}_A - \bar{x}_B$ based on 100 random assignments of these 20 observations into two groups of 10. Enter the data exactly as shown below. Describe similarities and differences between this new randomization distribution and the distribution shown.

Group	ChangeinScore
A	0
A	0
A	-1
A	-1
A	-2
A	-2
A	-3
A	-3
A	-3
A	-4
B	0
B	0
B	0
B	0
B	0
B	0
B	-1
B	-1
B	-1
B	-2

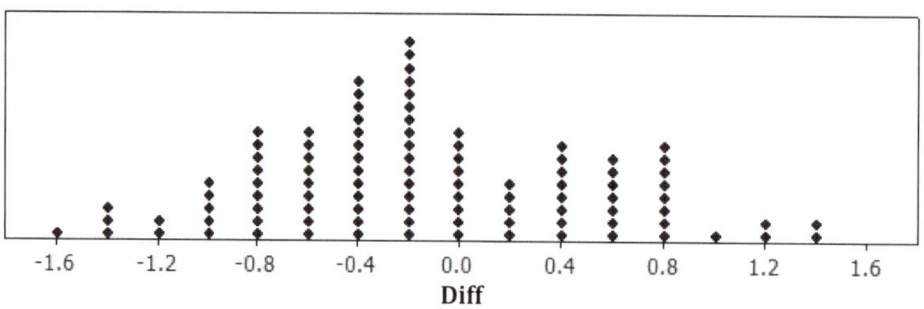

Answers will vary, but the distribution should not differ too much from the distribution presented previously in this lesson (Exit Ticket Problem 3) and other lessons.

2. In a previous lesson, the burn times of 6 candles were presented. It is believed that candles from Group A will burn longer on average than candles from Group B. The data from the experiment (now shown with group identifiers) are provided below.

Group	Burntime
A	18
A	12
B	9
A	6
B	3
B	0

Perform a randomization test of this claim. Carry out all 5 steps, and use the Anova Shuffle applet to perform Steps 2–4. Enter the data exactly as presented above, and in Step 3, develop the randomization distribution based on 200 random assignments.

Step 1—Null hypothesis: Candles from Group A will burn for the same amount of time on average as candles from Group B (no difference in average burn time).

Alternative hypothesis: Candles from Group A will burn longer on average than candles from Group B.

Step 2—$\text{Diff} = 12 - 4 = 8$

Step 3—Randomization distribution of Diff *developed by student*

Step 4—Compute probability of Diff *greater than or equal to* 8 *(value should be close to* 10%*).*

Step 5—While there are no specific criteria stated in the question for what is a "small probability," students should consider probability values from previous work in determining "small" versus "not small." Again, student values will vary. The important point is that students' conclusions should be consistent with the probability value and students' assessment of that value as follows:

If students deem the probability to be "small," then they should state a conclusion based on a statistically significant result. More specifically, due to the small chance of obtaining a Diff *value as extreme as or more extreme than the* Diff *value obtained in the experiment, it is believed that the observed difference did not happen by chance alone, and we support the claim that the Group A candles burn longer on average.*

If students deem the probability to be NOT "small," then they should state a conclusion based on a result that is NOT statistically significant. More specifically, it is believed that the observed difference may have occurred by chance, and we do NOT have evidence to support the claim that the Group A candles burn longer on average.

Appendix: Screenshot of Applet

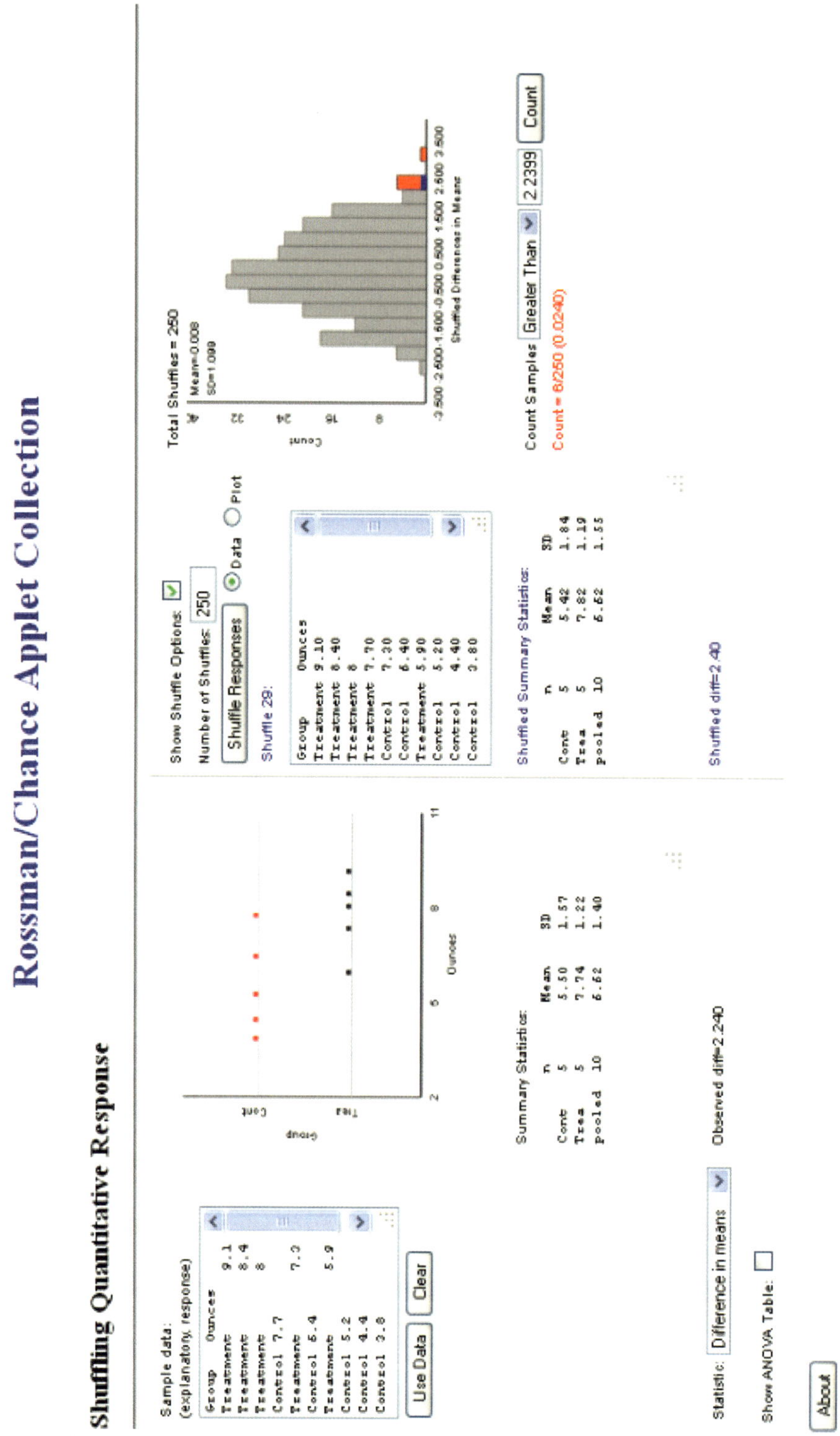

Lesson 28: Drawing a Conclusion from an Experiment

Student Outcomes

- Students carry out a statistical experiment to compare two treatments.
- Given data from a statistical experiment with two treatments, students create a randomization distribution.
- Students use a randomization distribution to determine if there is a significant difference between two treatments.

Lesson Notes

In these next two lessons, students participate in the capstone experience of conducting all phases of an experiment: collecting data, creating a randomization distribution based on these data, determining if there is a significant difference in treatment effects, and reporting their findings. The first of these two lessons deals with executing the experiment, collecting the data, and coming to a conclusion via a randomization test. The subsequent lesson asks students to develop a comprehensive report.

As mentioned in the student material, the following experiments are in homage to George E. P. Box, a statistician who worked extensively in the areas of quality control, design of experiments, and other topics. He earned the honor of Fellow of the Royal Society during his career and is a former president of the American Statistical Association. Several resources are available regarding his work and life including the book *Statistics for Experimenters: Design, Innovation, and Discovery* by Box, Hunter, and Hunter. There is other material written by and about him that would be accessible to secondary school students interested in learning more.

Box would use paper helicopters in some of his classes to provide students with a tangible (and low-cost) experience with experimental design and analysis. The experiments in this lesson investigate if modifications in certain dimensions of a paper helicopter affect its flight time. A blueprint and construction notes appear at the end of the student lesson. Consider practicing making and flying a few of these helicopters in the days leading up to the lesson. Students are advised that they may want to use a piece of tape to secure the two folded body panels to the body of the helicopter for greater stability. By design, there is some overlap from this folding in some helicopters.

Classwork

At the start of the lesson, ask students to restate to a partner or in writing the important parts of experimental design they have learned.

First, read through the opening text with students, and briefly summarize the experiment. Then, allow students to work on the exercises with a partner or in a small group.

While students' data will vary, generally speaking:

- The helicopters with longer wings should stay in flight longer, on average, than the helicopters with shorter wings.
- The helicopters with wider bodies (this variable is explored in the Problem Set) should have an average flight time that is significantly less than the average flight time of the helicopters with narrower bodies.

> *Scaffolding:*
> - Consider showing a visual, such as a poster with all directions, to help describe the experiment.
> - Ask students to restate the directions either verbally or by writing them.

Two key points before starting:

- If students discover a significant difference in any of the experiments, they should be careful *not* to overgeneralize or extrapolate their findings. For example, observed differences may or may not occur with larger helicopters or helicopters made of different material. Moreover, students should not assume that the significant difference in average flight time observed by a longer wing length of 1 inch would necessarily be doubled (or even occur at all) if the wing length was increased by 2 inches, for example.
- Encourage students to minimize potentially confounding factors that could distort the data collection. For example, find an environment that is as windless as possible, use the same type of paper for all helicopters, drop them all from the same height, and so on. Also, have students randomize the order in which they drop the first 30 helicopters so that any peculiar, unforeseen, or unknown factors that may affect flight times are distributed among the groups and not necessarily localized to any one group.

This lesson does require the use of scissors, tape, a stopwatch, a ruler, and possibly a measuring tape.

In this lesson, you will be conducting all phases of an experiment: collecting data, creating a randomization distribution based on these data, and determining if there is a significant difference in treatment effects. In the next lesson, you will develop a report of your findings.

The following experiments are in homage to George E. P. Box, a famous statistician who worked extensively in the areas of quality control, design of experiments, and other topics. He earned the honor of Fellow of the Royal Society during his career and is a former president of the American Statistical Association. Several resources are available regarding his work and life including the book *Statistics for Experimenters: Design, Innovation, and Discovery* by Box, Hunter, and Hunter.

The experiments will investigate whether modifications in certain dimensions of a paper helicopter will affect its flight time.

Exercise (5 minutes): Build the Helicopters

This task may be performed in advance of the day of data collection. Also, it is strongly recommended that multiple copies of the original helicopter blueprint be available. Consider posing the following question to the class, and allow for multiple responses:

- What are some statistical questions you would want to answer about helicopters?

Scaffolding:

According to the needs of students, consider making the experimental design and execution as unstructured as possible. Give students an opportunity to design an effective experiment as independently as possible. Consider using each exercise as a task that students should complete and use to assess their progress in the experiment.

Exercise: Build the Helicopters

In preparation for your data collection, you will need to construct 20 paper helicopters following the blueprint given at the end of this lesson. For consistency, use the same type of paper for each helicopter. For greater stability, you may want to use a piece of tape to secure the two folded body panels to the body of the helicopter. By design, there will be some overlap from this folding in some helicopters.

You will carry out an experiment to investigate the effect of wing length on flight time.

a. **Construct 20 helicopters with wing length of 4 inches and body length of 3 inches. Label 10 each of these helicopters with the word *long*.**

b. **Take the other 10 helicopters, and cut 1 inch off each of the wings so that you have 10 helicopters with 3-inch wings. Label each of these helicopters with the word *short*.**

c. **How do you think wing length will affect flight time? Explain your answer.**

Answers will vary. I think that a longer wing length will produce more resistance to the air, which will result in a longer flight time.

Exploratory Challenge 1 (15 minutes): Data Collection

From a consistent height (preferably 12 feet or higher), drop each of the 20 helicopters one at a time in a random order. Record the time it takes for each helicopter to reach the ground. Encourage students to record any relevant comments such as "this group's helicopters had a smoother flight," and so on.

Exploratory Challenge 1: Data Collection

Once you have built the 20 helicopters, each of them will be flown by dropping the helicopter from a fixed distance above the ground (preferably 12 feet or higher—record this height for use when presenting your findings later). For consistency, drop all helicopters from the same height each time, and try to perform this exercise in a space where possible confounding factors such as wind gusts and drafts from heating and air conditioning are eliminated.

a. Place the 20 helicopters in a bag, shake the bag, and randomly pull out one helicopter. Drop the helicopter from the starting height and, using a stopwatch, record the amount of time it takes until the helicopter reaches the ground. Write down this flight time in the appropriate column in the table below. Repeat for the remaining 19 helicopters.

Some helicopters might fly more smoothly than others; you may want to record relevant comments in your report.

Answers will vary.

Flight Time (seconds)

Long Wings (Group A)	Short Wings (Group B)

b. Why might it be important to randomize (impartially select) the order in which the helicopters were dropped?

(This is *different from* the randomization you perform later when you are allocating observations to groups to develop the randomization distribution.)

Randomizing the order in which the 20 helicopters are dropped will allow any peculiar, unforeseen, or unknown factors that may affect flight times to be distributed among the groups and not necessarily be localized to any one group.

Exploratory Challenge 2 (15 minutes): Developing Claims and Using Technology

Exploratory Challenge 2: Developing Claims and Using Technology

With the data in hand, you will now perform your analysis regarding the effect of wing length.

Experiment: Wing Length

MP.3

In this experiment, you will examine whether wing length makes a difference in flight time. You will compare the helicopters with long wings (wing length of 4 inches, Group A) to the helicopters with short wings (wing length of 3 inches, Group B). Since you are dropping the helicopters from the same height in the same location, using the same type of paper, the only difference in the two groups will be the different wing lengths.

MP.5

Questions: Does a 1-inch addition in wing length appear to result in a change in average flight time? If so, do helicopters with longer wing length or shorter wing length tend to have longer flight times on average?

Carry out a complete randomization test to answer these questions. Show all 5 steps, and use the Anova Shuffle applet described in the previous lessons to assist both in creating the distribution and with your computations. Be sure to write a final conclusion that clearly answers the questions in context.

Step 1—Null hypothesis: A 1-inch increase in wing length does not change average flight time.

Alternative hypothesis: A 1-inch increase in wing length changes average flight time.

Step 2—Students compute Diff from the experiment's data. The value WILL most likely be statistically significant.

Step 3—Randomization distribution of Diff developed by student using applet

Step 4—Compute the probability of obtaining a Diff more extreme than the value from the experiment. Since the original question is asking about a change in flight times (as opposed to a strict increase or decrease), the alternative hypothesis is of the form "different from," and students should select the beyond choice from the applet under Count Samples.

Step 5—While there are no specific criteria stated in the question for what is a small probability, students should consider probability values from previous work in determining "small" versus "not small." Again, student values will vary. The important point is that students' conclusions should be consistent with the probability value and their assessments of that value as follows:

- *If students deem the probability to be "small," then they should state a conclusion based on a statistically significant result. More specifically, we support the claim that a 1-inch increase in wing length changes average flight time. Given the sign of the observed difference and the method students have chosen for computing Diff (e.g., was it Group A's mean minus Group B's mean?), they should state as to whether the 1-inch increase in wing length appears to increase or decrease the average flight time.*
- *If students deem the probability to be NOT "small," then they should state a conclusion based on a result that is NOT statistically significant. More specifically, we DO NOT have evidence to support the claim that a 1-inch increase in wing length changes average flight time.*

The expectation is that the helicopters with longer wings should stay in flight longer on average than the helicopters with shorter wings.

Closing (5 minutes)

- What other characteristics of the helicopter could you modify to perform further experiments in flight times?
  - *Body length, wing width, "middle body" width, material used, etc.*

Ask students to summarize the main ideas of the lesson in writing or with a neighbor. Use this as an opportunity to informally assess comprehension of the lesson. The Lesson Summary below offers some important ideas that should be included.

Lesson Summary

In previous lessons, you learned how to carry out a randomization test to decide if there was a statistically significant difference between two groups in an experiment. Throughout these previous lessons, certain aspects of proper experimental design were discussed. In this lesson, you were able to carry out a complete experiment and collect your own data. When an experiment is developed, you must be careful to minimize confounding effects that may compromise or invalidate findings. When possible, the treatment groups should be created so that the only distinction between the groups in the experiment is the treatment imposed.

Exit Ticket (5 minutes)

Name ______________________________ Date ______________

Lesson 28: Drawing a Conclusion from an Experiment

Exit Ticket

Explain why you constructed a randomization distribution in order to decide if wing length has an effect on flight time.

Exit Ticket Sample Solutions

Explain why you constructed a randomization distribution in order to decide if wing length has an effect on flight time.

There is variability in flight times even for helicopters that have the same wing length. This means that two groups of helicopters with the same wing lengths will still have different mean flight times. So, when we see a difference in the mean flight time for short-wing helicopters and long-wing helicopters, we need to know if that difference is bigger than the kind of differences we would see just by chance when there is no difference in wing length. This is how we can tell if our observed difference between the flight times of long- and short-winged helicopters is significant enough that we don't think it is just due to chance.

Problem Set Sample Solutions

One other variable that can be adjusted in the paper helicopters is body width. See the blueprints for details.

1. **Construct 10 helicopters using the blueprint from the lesson. Label each helicopter with the word *narrow*.**
2. **Develop a blueprint for a helicopter that is identical to the blueprint used in class except for the fact that the body width will now be 1.75 inches.**
3. **Use the blueprint to construct 10 of these new helicopters, and label each of these helicopters with the word *wide*.**
4. **Place the 20 helicopters in a bag, shake the bag, and randomly pull out one helicopter. Drop the helicopter from the starting height and, using a stopwatch, record the amount of time it takes until the helicopter reaches the ground. Write down this flight time in the appropriate column in the table below. Repeat for the remaining 19 helicopters.**

Flight Time (seconds)

Narrow Body (Group C)	Wide Body (Group D)

5. **Questions: Does a 0.5-inch addition in body width appear to result in a change in average flight time? If so, do helicopters with wider body width (Group D) or narrower body width (Group C) tend to have longer flight times on average? Carry out a complete randomization test to answer these questions. Show all 5 steps, and use the Anova Shuffle applet described in previous lessons to assist both in creating the distribution and with your computations. Be sure to write a final conclusion that clearly answers the questions in context.**

Once the 20 helicopter flight time data are collected…

Step 1—Null hypothesis: A 0.5-inch addition in body width does not change average flight time.

Alternative hypothesis: A 0.5-inch addition in body width changes average flight time.

Step 2—Students compute Diff from the experiment's data. The value WILL most likely be statistically significant.

Step 3—Randomization distribution of Diff developed by student using applet

Step 4—Compute the probability of obtaining a Diff more extreme than the value from the experiment. Since the original question is asking about a change in flight times (as opposed to a strict increase or decrease), the alternative hypothesis is of the form "different from," and students should select the beyond choice from the applet under Count Samples.

Step 5—While there are no specific criteria stated in the question for what is a small probability, students should consider probability values from previous work in determining "small" versus "not small." Again, student values will vary. The important point is that students' conclusions should be consistent with the probability value and their assessments of that value as follows:

- *If students deem the probability to be "small," then they should state a conclusion based on a statistically significant result. More specifically, we support the claim that a 0.5-inch addition in body width changes average flight time. Given the sign of the observed difference and the method students have chosen for computing Diff (e.g., was it Group D's mean minus Group C's mean?), they should state as to whether the 0.5-inch increase in body width appears to increase or decrease the average flight time.*
- *If students deem the probability to be NOT "small," then they should state a conclusion based on a result that is NOT statistically significant. More specifically, we DO NOT have evidence to support the claim that a 0.5-inch addition in body width changes average flight time.*

The expectation is that the Group D helicopters (with wider body width) should have an average flight time that is significantly less than the average flight time of the control helicopters of Group C.

Appendix: Blueprint

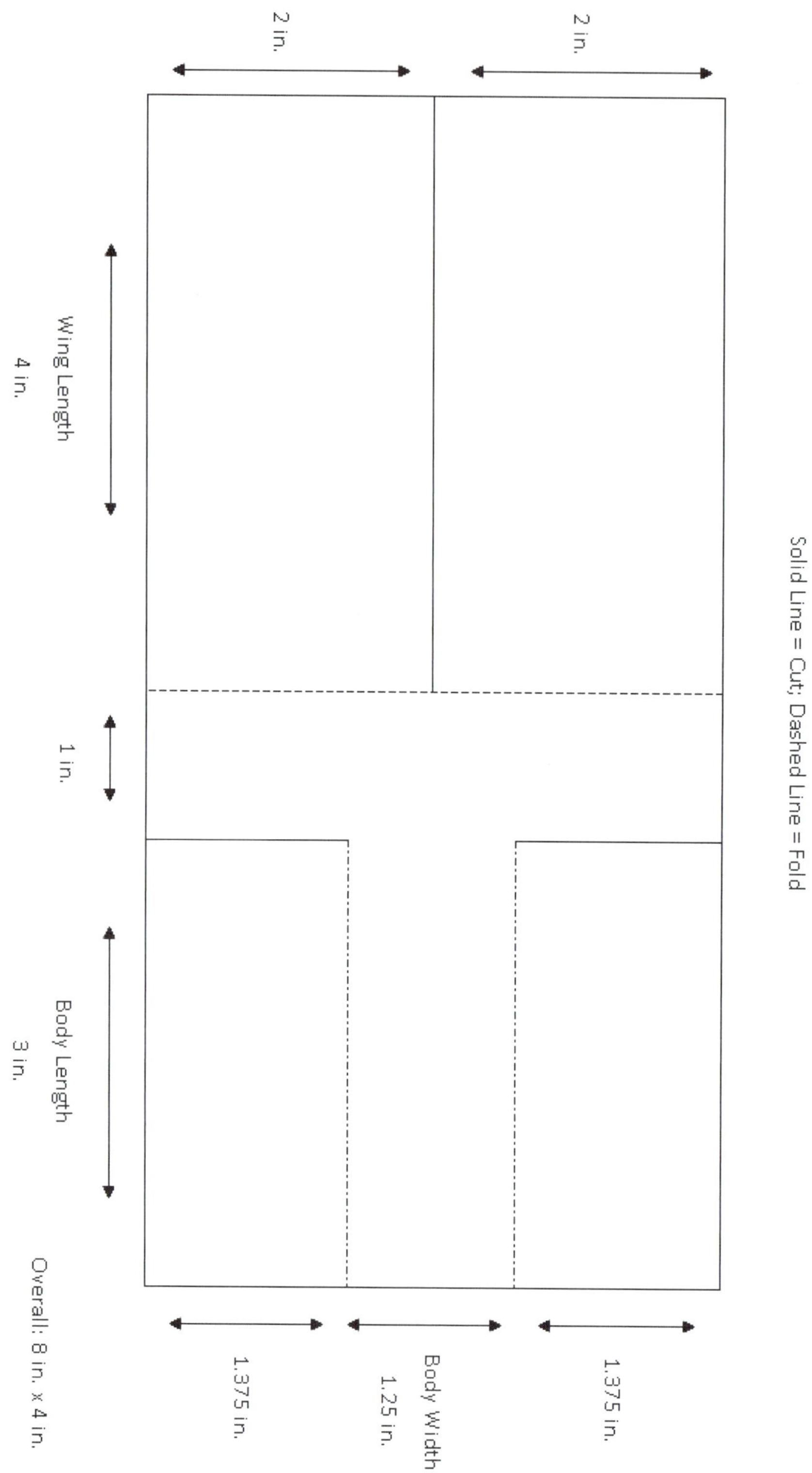

Lesson 29: Drawing a Conclusion from an Experiment

Student Outcomes

- Students carry out a statistical experiment to compare two treatments.
- Given data from a statistical experiment with two treatments, students create a randomization distribution.
- Students use a randomization distribution to determine if there is a significant difference between two treatments.

Classwork

MP.3 In this lesson, students develop a comprehensive poster summarizing the experiments involving helicopter Groups A, B, and C. The poster should address the results of both Experiments 1 and 2 regarding the effects of both body width and wing length.

As part of the lesson, students should be provided with an instructor-customized rubric for assessing a poster. A sample rubric from the American Statistical Association (ASA) is attached at the end of the teacher notes and was taken from "Poster Judging Rubric" at the "Poster Competition and Project Competition" page of the American Statistical Association, www.amstat.org/education/posterprojects/pdfs/PosterJudgingRubric.pdf. It is stated in the student lesson that the instructor provides more specific instructions and possibly a more defined (or otherwise modified) rubric. Specifically, it also says, "Your instructor will provide guidance as to groups, amount of time to spend, the rubric to be used for evaluation, etc. Your poster should address the results of both Experiments 1 and 2 regarding the effects of both body width and wing length." Please consult this ASA rubric, and modify as needed.

Read through the lesson with students. Convey to students that in addition to making their posters visually pleasing, they need to include answers to the focus questions presented in the lesson. Allow students to work with their partners or groups to prepare the posters.

In this lesson, you will develop a comprehensive poster summarizing your experiments.

Characteristics of a Good Poster

Your instructor will provide you with specific instructions and a rubric for assessing your poster (taken from "Poster Judging Rubric" at the "Poster Competition and Project Competition" page of the American Statistical Association, www.amstat.org/education/posterprojects/pdfs/PosterJudgingRubric.pdf).

Generally speaking, the presentation of a statistical analysis and/or an experiment should clearly state the question or purpose. The presentation should lead to the conclusion on a path that is easy to follow. The results of the study should be immediately obvious to the viewer. Any graphs included should be relevant to the question of interest and appropriate for the type of data collected.

Exploratory Challenge (45 minutes): Explaining the Experiment and Results

Exploratory Challenge: Explaining the Experiment and Results

Your classwork will involve developing your poster. Your instructor will provide guidance as to groups, amount of time to spend, the rubric to be used for evaluation, etc. Your poster should address the results of both Experiments 1 and 2 regarding the effects of both body width and wing length.

In addition to the general concerns of colors, fonts to use, etc., in preparation for creating your poster, consider (and answer) these classwork questions:

- **What was the objective of the experiment?**
- **How did you collect your data?**
- **What summary values and graphs should you present?**
- **How will you develop and present a summary of the experiment in a way that it is easy to follow and effortlessly leads the viewer to the conclusion?**
- **How will you explain statistical significance?**

Note: There is no specific Exit Ticket or Problem Set for this lesson. The finished poster represents these lesson components.

To Be Customized by the Instructor:

Note: The rubric was taken from "Poster Judging Rubric" at the "Poster Competition and Project Competition" page of the American Statistical Association, www.amstat.org/education/posterprojects/pdfs/PosterJudgingRubric.pdf.

Rubric for the Judging of Statistics Posters

Score	**Overall Impact of the Display** (Use of space, dimensions of question, readability, neatness, poster design aspects)	**Technical Aspects** (Spelling, Grammar, Consistency of colors or patterns)	**Clarity of the Message** (How well is a story told?)	**Appropriateness of the Graphs for the Data** (Statistical Appropriateness)	**Creativity** (Data collection methods, sample size issues, who cares factor)
5	Poster is neatly constructed, including good use fonts, pictures, and extras. The overall display is eye-catching but retains statistical substance. Good use of space for graphical presentation. Addresses multiple dimensions of the question or problem.	Poster uses colors and patterns well. Correct grammar and spelling are used.	Question or purpose is clearly stated, and the presentation leads to the conclusion on a path that is easy to follow. The results of the study are immediately obvious to the viewer.	Graphs are appropriate for the question and data, and they are correctly constructed.	Overall question is interesting, phrasing of titles, captions, and question are creative. Shows creative thought in topic, graph design, or data collection. Collects data appropriately. Answers an important topic.
4	Addresses multiple dimensions of a question. Good use of space. Fonts could be larger but do not really detract from the message. Could be a little neater but really does not detract from the message.	Better use of color or patterns would help the presentation, but in general the poster grabs the attention of the viewer. Correct grammar and spelling are used.	At least one link in the chain from the question through the results to the conclusion is difficult to follow.	Errors or inaccuracies are present in at least one graph. More appropriate display(s) would improve the presentation.	Overall question is interesting. Some creativity in design or data collection. Collects appropriate data.
3	Good use of space. Addresses multiple dimensions of a question. Readability or neatness detract from the overall appeal of the poster.	Use of more or different colors would vastly improve the appeal of the poster. Minor grammar and/or spelling mistakes.	The progression from question to conclusion can be followed, at least in part, but only with considerable effort, and the information on the back may be needed to confirm.	Significant gap exists in the demonstration of understanding of the graphics, or how the graphics relate to the purpose of the poster.	Some creativity. Data could be better but it doesn't distract.
2	Serious problems with neatness or organization prevent the poster from being eye-catching and understandable. Multiple dimensions of the question addressed. Could use space better.	Serious problems colors or patterns prevent the poster from being eye-catching and understandable. OR Multiple mistakes in grammar or spelling prevent the poster from being eye-catching and understandable.	The information on the back is required in order for any relationships in the poster to be understood.	Although some part of the graphs is correct, substantial errors lead to invalid or inappropriate conclusions.	Creativity and topic are of some interest. Data collection could be improved with larger samples.
1	The poster is unidimensional. Poor use of space for graphics. Major neatness or readability issues.	The poster is has multiple spelling or grammar/spelling errors AND isn't consistent with colors or patterns so much so that it severely distracts from the poster.	The poster is virtually incomprehensible.	The displays are inappropriate and incorrect for the research question and data types. The question is badly misunderstood and the results are nonsensical.	The poster appears to have been constructed with very little or no creativity or with improper data collection methods.

Lesson 30: Evaluating Reports Based on Data from an Experiment

Student Outcomes

- Students critique and evaluate statements in published reports that involve determining if there is a significant difference between two treatments in a statistical experiment.

Lesson Notes

In this lesson, students read and comment on examples from the media that involve statistical experiments that compare two treatments. For this lesson, students should work in small groups (two or three students per group). This allows students to discuss and evaluate the articles together. Students need to have access to the printed or online articles. This may be done by one of the following methods:

- Print a copy of the article for each student prior to class. Review the article before discussing with students. If using an online version, screen the links before distributing to students. In a few cases, the instructor is required to provide some background information in order to download the article.
- If there is access to a computer lab, students can read the articles online.
- Students may also use their personal electronic devices (laptop computer, smartphone, iPad, or tablet) to read the articles. Again, preview the article or the link before using this option.

Identifying relevant reports for students to review and evaluate poses challenges; however, the Common Core Standards explicitly states the importance of evaluating statistical studies by students at this grade level. The reports and examples in this lesson are identified by a link to a reliable source. These links, however, may change over time. Consider the source of the article and the general title to identify possible changes in the links.

Classwork

Exercises 1–7 (15–20 minutes)

Before students begin working on the exercises, pose the following question to the class. Have students share and discuss ideas with their neighbors:

- Frequently, we read reports of studies and experiments, but it's important to know if we can trust the results. Based on what we've learned so far, what would be some good questions to keep in mind when evaluating a report on a statistical study?

Allow 10–15 minutes for students to read the article "A Randomized Trial of Colchicine for Acute Pericarditis" in *The New England Journal of Medicine* (October 2013) and answer Exercises 1–6. (*The New England Journal of Medicine* website requires some background information before the article can be downloaded. Only the teacher should provide this information.)

> *Scaffolding:*
> It may be useful to allow students to draw from resources that are in their first languages. Providing structured sentence frames and exemplars of solid commentaries may be useful as students work to create their own comments.

When students have finished answering Exercises 1–6, discuss the answers to the questions. Then, allow two to three minutes for students to answer Exercise 7.

Exercises 1–7

Pericarditis is an inflammation (irritation and swelling) of the pericardium, the thin sac that surrounds the heart. When extra fluid builds up between the two layers of the pericardium, the heart's actions are restricted. An experiment reported in the article "A Randomized Trial of Colchicine for Acute Pericarditis" in *The New England Journal of Medicine* (October 2013) tested the effects of the drug colchicine on acute pericarditis.

Read the abstract of the article, and answer the following questions:

Website: http://www.nejm.org/doi/full/10.1056/NEJMoa1208536

1. How many treatment groups are there?

   There are 2 *treatment groups.*

2. What treatments are being compared?

   The two treatment groups are colchicine and a placebo with the standard anti-inflammatory treatment.

3. Is there a placebo group? Explain.

   Yes, the second treatment group is a placebo in addition to an aspirin or an ibuprofen.

4. How many subjects are in each treatment group?

   There are 120 *subjects in each treatment group.*

5. Do you think that the number of subjects in each treatment group is enough? Explain.

   Yes; 120 *subjects should be large enough to allow the researchers to see the variation in treatment effects.*

6. What method was used to assign the subjects to the treatment groups? Explain why this is important.

   The subjects were randomly assigned to the treatment groups. This is important because the random assignment evenly disperses the extraneous variables into both treatment groups.

Suppose newspaper reporters brainstormed some headlines for an article on this experiment. These are their suggested headlines:

A. "New Treatment Helps Pericarditis Patients"

B. "Colchicine Tends to Improve Treatment for Pericarditis"

C. "Pericarditis Patients May Get Help"

7. Which of the headlines above would be best to use for the article? Explain why.

   Headline A would be the best because this is a well-designed experiment. Therefore, a cause-and-effect relationship has been established. Headlines B and C talk about a tendency relationship, not a cause-and-effect relationship.

Exercises 8–10 (25–30 minutes)

MP.3 In this set of exercises, students critique and evaluate statements in published reports that involve determining if there is a significant difference between two treatments in a statistical experiment. Below is a discussion for each criterion so that the class can have a discussion on the meaning of each criterion listed for evaluating experiments.

- Were the subjects randomly assigned to treatment groups?

Students should look to see if the article explicitly states that the subjects were randomly assigned to each treatment group. This is important because random assignment negates the effects of extraneous variables that may have an effect on the response by evenly distributing these variables into both treatment groups. (Refer back to Lesson 23.)

- Was there a *control group* or a *comparison group*?
  - *A control group is a group that receives either no treatment or a placebo. A comparison group is a group that receives the "standard treatment." A comparison group is used when the context of the experiment requires a treatment (used especially in medical experiments when not giving a treatment would be unethical). The use of a control group or comparison group provides a baseline for the response in the experiment. This allows researchers to see how much of the response is due to the actual variable being tested.*
- Were the sample sizes reasonably large?
  - *The number of subjects in each treatment group should be large enough so that the variation in the response from one subject to another is evident. A small size ($n < 50$) may not show all the variation that might exist in the subjects' responses.*
- Do the results show a cause-and-effect relationship?
  - *If the first three criteria are met, then the results of an experiment can be interpreted as showing a cause-and-effect relationship.*

Allow students 5–10 minutes to read the abstract of the article about exercise intervention with mild cognitive impairment (MCI) patients and to answer Exercise 8. When students have finished, discuss their answers. Note that students' work should contain the underlined explanations in the sample responses.

Exercises 8–10

What you should look for when evaluating an experiment:

- **Were the subjects randomly assigned to treatment groups?**
- **Was there a control group or a comparison group?**
- **Were the sample sizes reasonably large?**
- **Do the results show a cause-and-effect relationship?**

Read the abstracts of the two articles below. Write a few sentences evaluating these articles using the guidelines above.

8. **The study "Semantic Memory Functional MRI and Cognitive Function After Exercise Intervention in Mild Cognitive Impairment" (*Journal of Alzheimer's Disease*, August 2013) was performed to see if exercise would increase memory retrieval in older adults with mild cognitive impairment (associated with early memory loss).**

   Website: http://iospress.metapress.com/content/xm8t241628h37h7t/

   While there <u>was a control group</u> of nonimpaired adults, the <u>sample sizes are too small</u> to demonstrate the variation in the responses. Also, the subjects were <u>not randomly assigned</u> to treatment groups. Therefore, <u>no cause-and-effect relationship</u> can be shown with this study. The authors correctly stated, "These findings suggest exercise may improve … semantic memory retrieval in MCI…"

Allow students 5–10 minutes to read the summary of the article about adolescent scoliosis patients wearing a brace and to answer Exercise 9. When students have finished, discuss their answers.

9. **The article "Effects of Bracing in Adolescents with Idiopathic Scoliosis" (*New England Journal of Medicine*, October 2013) reports on the role of bracing patients with adolescent idiopathic scoliosis (curvature of the spine) for prevention of back surgery.**

Website: www.nejm.org/doi/full/10.1056/NEJMoa1307337

Researchers conducted a study of one hundred sixteen (reasonably large enough) adolescent patients with scoliosis in which each patient was randomly assigned to either a treatment where the patient wore a brace or a treatment where the patient's progress was observed. The treatment where the patients' progress was observed is a control group. Therefore, the conclusion that wearing a brace significantly decreased the progression of high-risk curves toward the need for surgery is accurate. The researchers also examined what the treatment effects were if the patients were allowed to choose their treatment (which is not an experiment).

Have the class watch the video of the report by Tom Bemis. Allow students three to five minutes to answer Exercise 10. When students have finished, discuss their answers. Preview this website before showing it to students as there is a commercial preceding the video. Also, copy and paste the website into a browser to view the video.

10. **View the report by Tom Bemis (Market Watch, *Wall Street Journal*, August 13, 2013) about the type of car driven by a person and the person's driving behavior.**

Website: http://live.wsj.com/video/bmw-drivers-really-are-jerks-studies-find/29285015-BB1A-4E41-B0C0-0A41CB990F60.html

Is the title "BMW Drivers Really Are Jerks" an accurate title for these reported studies? Why or why not? If not, suggest a better title.

The title may imply that because a person drives a BMW, this causes the person to use bad driving habits. Since these studies are observational studies and not experiments, this is inaccurate. A more appropriate title would be "BMW Drivers Tend to Have Bad Driving Behaviors."

Closing (2 minutes)

Ask students to summarize the main ideas of the lesson in writing or with a neighbor. Use this as an opportunity to informally assess comprehension of the lesson. The Lesson Summary below offers some important ideas that should be included.

Lesson Summary

- **A cause-and-effect relationship can only be shown by a well-designed experiment.**
- **Randomly assigning the subjects to treatment groups evens out the effects of extraneous variables to create comparable treatment groups.**
- **A control group (which may be a placebo group) or a comparison group (a standard treatment) is sometimes included in an experiment so that you can evaluate the effect of the treatment.**
- **The number of subjects in each treatment group (sample size) should be large enough for the random assignment to experimental groups to create groups with comparable variability between the subjects.**

Exit Ticket (2–3 minutes)

Name ______________________________ Date ______________

Lesson 30: Evaluating Reports Based on Data from an Experiment

Exit Ticket

What are the aspects of a well-designed experiment that show a causal relationship?

Exit Ticket Sample Solutions

What are the aspects of a well-designed experiment that show a causal relationship?

The use of a control group or a comparison group sets a baseline for the treatment effect. The use of random assignment evenly spreads the effects of extraneous variables into the treatment groups. Thus, any significant difference between the treatment groups can be attributed to the treatment, and you can conclude that the treatment is the cause of the observed difference.

Problem Set Sample Solutions

Read the following articles and summaries. Write a few sentences evaluating each one using the guidelines given in the lesson.

1. The article "Emerging Technology" (*Discover Magazine*, November 2005) reports a study on the effect of "infomania" on IQ scores.

   Website: discovermagazine.com/2005/nov/emerging-technology

   This article discusses two groups of people: one who had to check email and respond to IM while testing and the other who was not distracted by email or IM. It does not specify that subjects were randomly assigned to the 2 groups. Also, the sample sizes were not stated. Therefore, based on this article, it would not be reasonable to state a cause-and-effect relationship.

2. In *The New England Journal of Medicine*, October 2013, the article "Increased Survival in Pancreatic Cancer with nab-Paclitaxel Plus Gemcitabine" reports on an experiment to test which treatment, nab-paclitaxel plus gemcitabine or gemcitabine alone, is the most effective in treating advanced pancreatic cancer.

   Website: www.nejm.org/doi/full/10.1056/NEJMoa1304369

   The article states that pancreatic cancer patients were randomly assigned to the 2 groups. The sample sizes (431 and 430) were large enough to observe the variation in responses. The standard treatment of gemcitabine alone was the comparison group. The conclusion, that the treatment group of nab-paclitaxel plus gemcitabine significantly improved the overall survival rate, is accurate.

3. Doctors conducted a randomized trial of hypothermia in infants with a gestational age of at least 36 weeks who were admitted to the hospital at or before six hours of age with either severe acidosis or perinatal complications and resuscitation at birth and who had moderate or severe encephalopathy. The trial "Whole-Body Hypothermia for Neonates with Hypoxic–Ischemic Encephalopathy" tested two treatments, standard care and whole-body cooling for 72 hours.

   Website: www.nejm.org/doi/full/10.1056/NEJMcps050929

   The article states that the infants were randomly assigned to usual care (the control group—106 infants) or whole-body cooling (102 infants). Thus, the sample sizes were large enough to observe the variation in responses. The conclusion, that whole-body cooling reduces the risk of death or disability in infants with moderate or severe encephalopathy, is accurate.

Name ______________________________ Date ______________

1. Suppose you wanted to determine whether students who close their eyes are better able to estimate when 30 seconds have passed than students who do not close their eyes. (You ask students to tell you when to stop a stopwatch after they think 30 seconds have passed.) You find the first 50 students arriving at school one day. For those 50, you flip a coin to decide whether or not they will close their eyes during the test. Then, you compare the amounts by which each group overestimated or underestimated.

   a. Did this study use *random sampling*? Explain your answer by describing what purpose random sampling serves in such a study.

   b. Did this study make use of *random assignment*? Explain your answer by describing what purpose random assignment serves in such a study.

   c. Would the study described above be an *observational study* or an *experimental study*? Explain how you are deciding.

2. A Gallup poll conducted July 10–14, 2013, asked a random sample of U.S. adults: "How much attention do you pay to the nutritional information that is printed on restaurant menus or posted in restaurants, including calories and sugar and fat content?" The sample results were that 43% of the respondents said they pay a "fair amount" or a "great deal" of attention. Suppose there had been 500 people in the study.

   The following graph displays the results from $1{,}000$ random samples (each with sample size 500) from a very large population where 50% of respondents "pay some attention" and 50% "pay little or no attention."

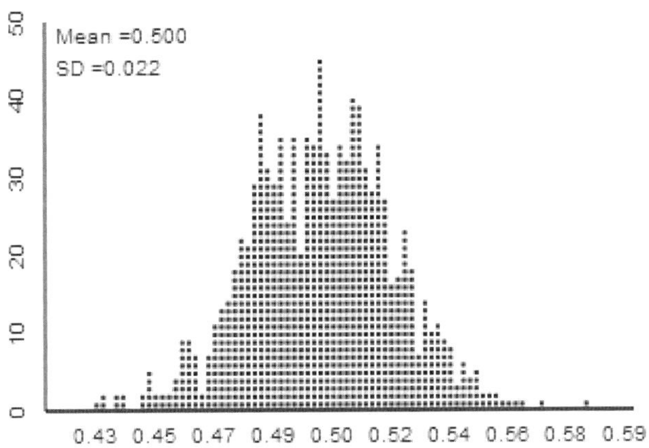

   a. Based on the simulation results above, are the sample data (43% responding "pay some attention") consistent with the simulation? In other words, do these results cause you to question whether the population is $50/50$ on this issue? Explain.

   Do you believe it is reasonable to generalize the results from this study to all U.S. adults? Explain.

Suppose Gallup plans to conduct a new poll of a random sample of 1,000 U.S. adults on an issue where the population is evenly split between two responses. The following graph displays the results from 2,000 random samples (each with sample size 1,000) from such a population.

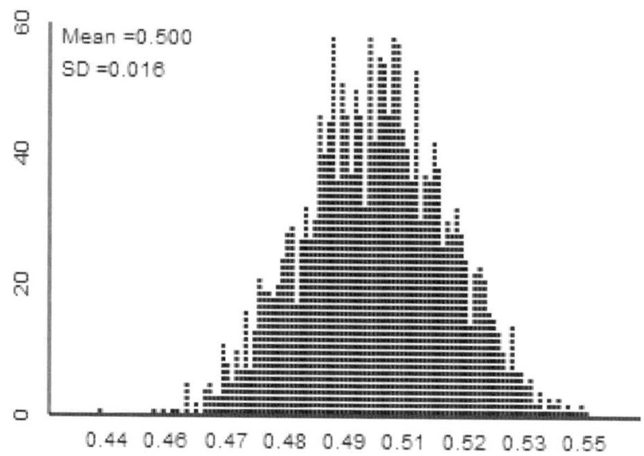

b. Based on these simulation results, estimate the expected margin of error for the Gallup poll. Explain how you developed your estimate.

c. Suppose the study used a sample size of 2,000 instead of 1,000. Would you expect the margin of error to be larger or smaller?

3. A randomized experiment compared the reaction time (in milliseconds) for subjects who had been sleep deprived (group 1) and subjects who had not (group 2).

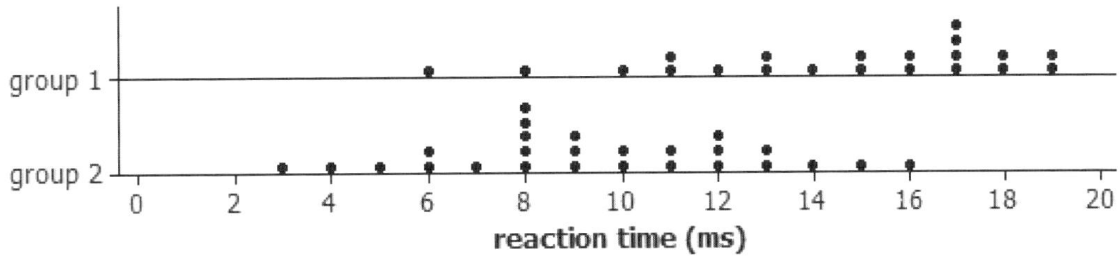

Variable	N	Mean	StDev
group 1	21	14.38	3.64
group 2	26	9.50	3.33

a. Based on the above output, for which group would it be more reasonable to use a normal curve to model the reaction time distribution? Justify your choice.

b. The difference in means is $14.38 - 9.50 = 4.88$. One of the researchers claims that the reaction time if you are sleep deprived is 5 ms greater than the reaction time if you are not sleep deprived. Explain one reason why this claim is potentially misleading.

c. Describe how to carry out a simulation analysis to determine whether the mean reaction time for group 1 is significantly larger than the mean reaction time for group 2.

d. The graph below displays the results of 100 repetitions of a simulation to investigate the difference in sample means when there is no real difference in the treatment means. Use this graph to determine whether the observed mean reaction time for group 1 is significantly larger than the observed mean reaction time for group 2. Explain your reasoning.

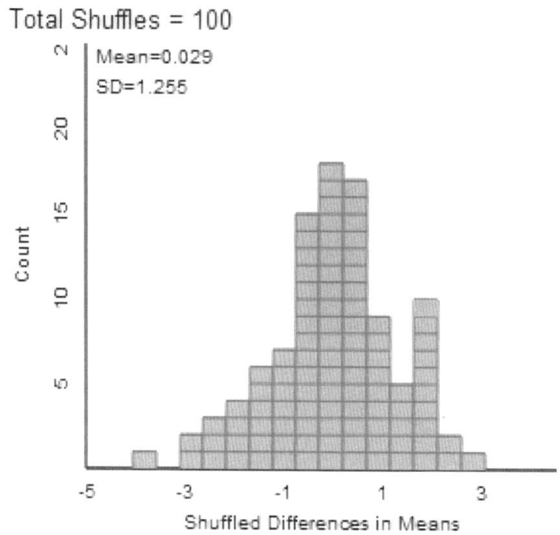

A Progression Toward Mastery

Assessment Task Item		STEP 1 Missing or incorrect answer and little evidence of reasoning or application of mathematics to solve the problem.	STEP 2 Missing or incorrect answer but evidence of some reasoning or application of mathematics to solve the problem.	STEP 3 A correct answer with some evidence of reasoning or application of mathematics to solve the problem. OR An incorrect answer with substantial evidence of solid reasoning or application of mathematics to solve the problem.	STEP 4 A correct answer supported by substantial evidence of solid reasoning or application of mathematics to solve the problem.
1	a S-IC.B.3	Student provides an answer that is not germane to the design of the study.	Student confuses random sampling with other random forms such as random assignment.	Student indicates that the study did not use random sampling, but the description of the purpose of random sampling is inadequate (e.g., to draw cause-and-effect conclusions).	Student indicates that the study did not use random sampling and describes the purpose as obtaining a representative sample.
	b S-IC.B.3	Student provides an answer that is not germane to the design of the study.	Student confuses random assignment with other random methods such as random sampling.	Student indicates that the study did use random assignment, but the description of the purpose is inadequate (e.g., only to draw cause-and-effect conclusions but with no explanation why that works).	Student indicates that the study did use random assignment and describes the purpose in terms of balancing out confounding variables.
	c S-IC.B.3	Student is not consistent with a choice between the two types of studies.	Student identifies the study as observational, and/or the justification contradicts student's choice.	Student identifies the study as experimental, but the justification is inadequate (e.g., "They controlled variables").	Student identifies the study as experimental and justifies based on the imposition of the eyes being open or closed and/or use in random assignment.

2	a S-IC.A.2	Student makes no use of the simulated distribution.	Student answers no because the center of the simulated distribution is close to 0.50.	Student attempts to integrate information from the graph and the observed statistic, but the response is inconsistent (e.g., "0.43 is on the graph, so I believe the population is $50/50$"), or the response is too definitive (e.g., "We have proven that the population cannot be $50/50$").	Student uses the observed statistic (0.43) and the simulated distribution to address the question of whether the population is $50/50$ (e.g., "Yes, because 0.43 is far in the tail of this distribution").
	b S-IC.A.1	Student does not include statistical justification (e.g., "I was not surveyed," or anecdotal evidence about who pays attention).	Student says no based on the sample size being too small or that only adults were surveyed or says the task of representing the entire population is too difficult.	Student attempts to use the random sampling as justification but is concerned by the 43% being too different from 50%.	Student appeals to the use of a random sample of U.S. adults to argue that it would be appropriate to generalize the results to the larger population of U.S. adults.
	c S-IC.B.4	Student does not calculate a margin of error but discusses possible sampling or nonsampling errors.	Student reports a margin of error, but it is not clear how that value is obtained, or uses an incorrect method (e.g., the number 2) or claims not enough information is provided.	Student eyeballs that most of the sample proportions are within 0.3 of the mean. OR Student uses $1SD$ as the margin of error or uses $SD = 0.022$ from the first graph.	Student calculates $2 \times SD = 0.032$ as the margin of error.
	d S-IC.B.4	Student does not make a claim about the change in the margin of error.	Student expects the margin of error to not change or to increase with the larger sample size.	Student expects the margin of error to be smaller with the larger sample size, but student's explanation reveals an inconsistency in logic.	Student expects the margin of error to be smaller with the larger sample size.
3	a S-ID.A.4	Student does not differentiate between the two distributions.	Student picks group 2 based on the mean being closer to the middle of the graph.	Student picks group 1 and gives a statistical argument such as $\text{mean} - 3SD$ is too close to zero for group 2.	Student picks group 2 based on the symmetric, bell-shaped nature of the distribution.

	b **S-IC.B.6**	Student does not describe a misleading aspect to the claim.	Student focuses on potential sampling or nonsampling errors in the data collection.	Student references the skewed nature of group 1 and indicates that means would not be an appropriate comparison. OR Student focuses on the discrepancy between 4.88 and 5.	Student recognizes that the statement does not reflect that these means are estimates. The statement also mixes *mean* and *individual* reaction times.
	c **S-IC.B.5**	Student does not describe a simulation approach.	Student focuses on random sampling rather than random assignment, and it is not clear on how a decision of significance will be made.	Student shows an understanding of how to use the simulation to evaluate 4.88, but details of the simulation are not clear.	Student includes random reassignment of the reaction times to two groups, recomputing of the difference in sample means, and use of the simulated distribution to judge how unlikely a result of 4.88 or more extreme is to occur. (How to use the simulation results could be described in part (d).)
	d **S-IC.B.5**	Student makes no use of the simulated distribution.	Student answers no because the center of the simulated distribution is close to 0.	Student attempts to integrate information from the graph and the observed statistic, but the response is inconsistent (e.g., "4.88 is not on the graph, so there is not a significant difference").	Student uses the observed statistic (4.88) and the simulated distribution to address the question of significance (e.g., "Yes, because 4.88 is not listed on the graph").

Name ______________________________ Date ______________

1. Suppose you wanted to determine whether students who close their eyes are better able to estimate when 30 seconds have passed than students who do not close their eyes. (You ask students to tell you when to stop a stopwatch after they think 30 seconds have passed.) You find the first 50 students arriving at school one day. For those 50, you flip a coin to decide whether or not they will close their eyes during the test. Then, you compare the amounts by which each group overestimated or underestimated.

   a. Did this study make use of *random sampling*? Explain your answer by describing what purpose random sampling serves in such a study.

      No. This study used a convenience sample of the first 50 students arriving at school. A random sample would give us a better representative sample by giving all students an equal chance to be selected for the sample rather than only the early risers.

   b. Did this study make use of *random assignment*? Explain your answer by describing what purpose random assignment serves in such a study.

      Yes. A coin was flipped to decide in which treatment condition each student would be placed. The goal is to create two groups that are "equal" with respect to all potential confounding variables. This will allow us to potentially draw cause-and-effect conclusions between the two treatments (i.e., eyes open or closed).

   c. Would the study described above be an *observational study* or an *experimental study*? Explain how you are deciding.

      This is an experiment because the explanatory variable (i.e., eyes open or closed) was randomly imposed by the researcher. An observational study would not control which method students use but would rather let the students choose for themselves (and would not even be told of different possible strategies).

2. A Gallup poll conducted July 10–14, 2013, asked a random sample of U.S. adults: "How much attention do you pay to the nutritional information that is printed on restaurant menus or posted in restaurants, including calories and sugar and fat content?" The sample results were that 43% of the respondents said they pay a "fair amount" or a "great deal" of attention. Suppose there had been 500 people in the study.

The following graph displays the results from 1,000 random samples (each with sample size 500) from a very large population where 50% of the respondents "pay some attention" and 50% "pay little or no attention."

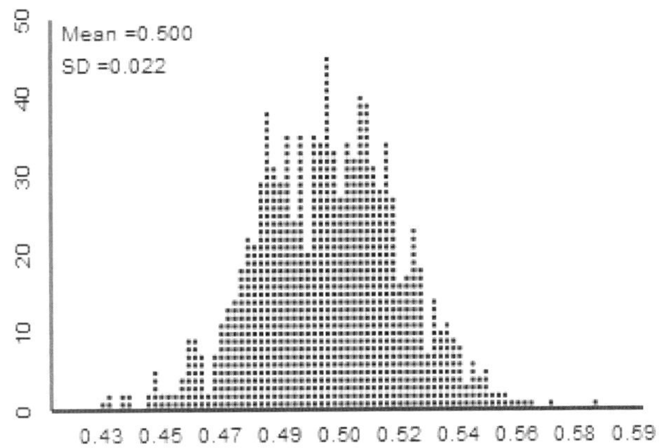

a. Based on the simulation results above, are the sample data (43% responding "pay some attention") consistent with the simulation? In other words, do these results cause you to question whether the population is 50/50 on this issue? Explain.

We see very few of the simulated sample proportions around 0.43. It falls in the tail of the distribution. So, this unusual outcome would lead us to question whether the actual sample proportion came from a population that was 50/50 on the issue. It would be surprising to get a sample proportion as extreme as 0.43 if the population proportion was indeed 0.50.

b. Do you believe it is reasonable to generalize the results from this study to all U.S. adults? Explain.

The Gallup poll claims to have collected a random sample, so I would feel comfortable generalizing these results to U.S. adults.

Suppose Gallup plans to conduct a new poll of a random sample of 1,000 U.S. adults on an issue where the population is evenly split between two responses. The following graph displays the results from 2,000 random samples (each with sample size of 1,000) from such a population.

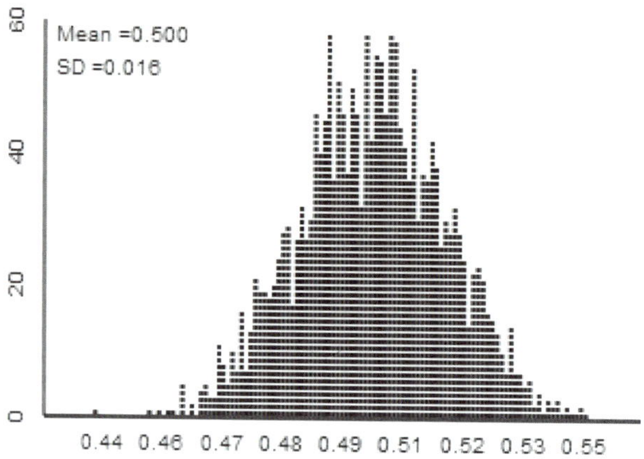

c. Based on these simulation results, estimate the expected margin of error for the Gallup poll. Explain how you developed your estimate.

The standard deviation of the simulated sample proportions is approximately 0.016. So, the approximate margin of error is two standard deviations or about 0.032.

d. Suppose the study used a sample size of 2,000 instead of 1,000. Would you expect the margin of error to be larger or smaller?

A larger sample size should lead to a smaller margin of error.

3. A randomized experiment compared the reaction time (in milliseconds) for subjects who had been sleep deprived (group 1) and subjects who had not (group 2).

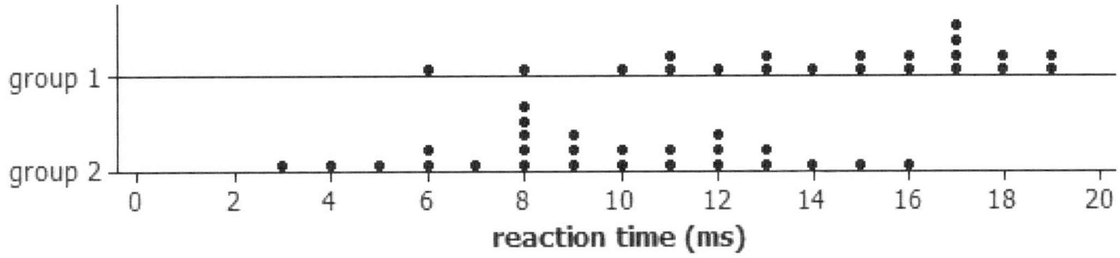

Variable	N	Mean	StDev
group 1	21	14.38	3.64
group 2	26	9.50	3.33

a. Based on the above output, for which group would it be more reasonable to use a normal curve to model the reaction time distribution? Justify your choice.

Group 2. The observations for group 2 appear to be more symmetric and mound shaped. Group 1 does not appear to be symmetric.

b. The difference in means is $14.38 - 9.50 = 4.88$. One of the researchers claims that the reaction time if you are sleep deprived is 5 ms greater than the reaction time if you are not sleep deprived. Explain one reason why this claim is potentially misleading.

The researcher is generalizing from the observed difference in means. It's possible that the actual difference in means is a bit smaller than 5. The researcher is also comparing reaction times rather than reflecting the conclusion as a comparison of means.

c. Describe how to carry out a simulation analysis to determine whether the mean reaction time for group 1 is significantly larger than the mean reaction time for group 2.

You could put all of the reaction times on index cards and then shuffle the cards and randomly deal them to two groups, one with 21 cards and one with 26 cards. You could also put the response values in a list and then flip a coin to put "group 1" or "group 2" by each value until you fill one of the groups.

Then, calculate the difference in the mean reaction time of the two groups.

Repeat this many times, and see how often the difference is as large as or larger than 4.88.

d. The graph below displays the results of 100 repetitions of a simulation to investigate the difference in sample means when there is no real difference in the treatment means. Use this graph to determine whether the observed mean reaction time for group 1 is significantly larger than the observed mean reaction time for group 2. Explain your reasoning.

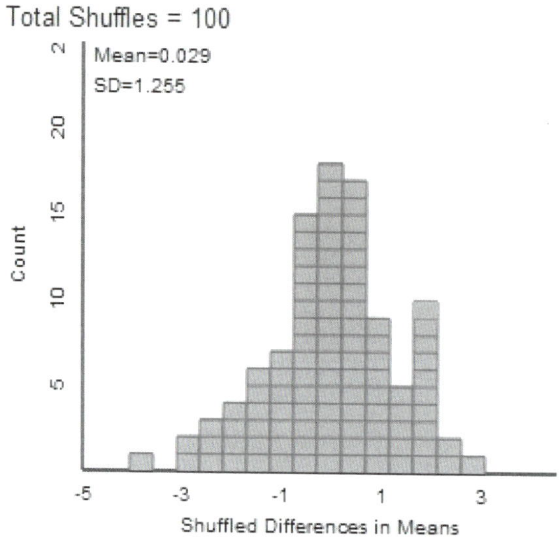

We see that 4.88 is in the tail of the distribution. We had no simulated differences in means that were as large as 4.88. This gives us reason to believe that our observed difference came from a process where there was a genuine difference between the treatments. Therefore, we have strong evidence that the average reaction time for those who are sleep deprived is larger than for those who are not sleep deprived (for subjects similar to those in this study).

This page intentionally left blank

This page intentionally left blank